MW01618060

Human Microbiome in Health, Disease, and Therapy

Pallaval Veera Bramhachari
Editor

Human Microbiome in Health, Disease, and Therapy

Editor
Pallaval Veera Bramhachari
Department of Biosciences & Biotechnology
Krishna University
Machilipatnam, Andhra Pradesh, India

ISBN 978-981-99-5113-0 ISBN 978-981-99-5114-7 (eBook)
https://doi.org/10.1007/978-981-99-5114-7

This Springer imprint is published by the registered company Springer Nature Singapore Pte Ltd.
The registered company address is: 152 Beach Road, #21-01/04 Gateway East, Singapore 189721, Singapore

Foreword

The microbial world is largely invisible to the human eye, but it is almost beyond imagination. There are hundreds, thousands of different kinds of bacteria (leaving aside other kinds of microbes: archaea, viruses, fungi, and protists), living in every possible environment including deep seabed, high in the clouds, in the boiling hot springs. Multicellular organisms created an entirely new set of habitats, in and on all those animals and plants.

Research data suggested that during the last two decades, extensive research has been carried out on endophytic fungi and several biologically active compounds have been isolated from endophytic fungi. This book makes all the readers generally conversant in the language of microbiomes and metagenomics. It also provides excellent examples of how microbial communities affect health and cure diseases and doles out typical practical examples of how medical interventions interact with the microbiome and change outcomes.

Human Microbiome Interactions: Targeted Therapeutic Interventions in Medicine. The volume published by Springer Nature is important, and I strongly believe that it will attract the readers working in the field. This volume has 16 chapters contributed by several competent academicians and scientists working on microbiome research throughout the world. I congratulate the editor of the book Dr. P. Veera Bramhachari, for bringing out this volume with excellent contributions from scientists working on the microbiome and their application in understanding the microbiome interactions in agriculture and the environment.

Krishna University
Machilipatnam, Andhra Pradesh, India

K. Rama Mohana Rao

Preface

Microbiome research has advanced rapidly over the past few decades and has now become a topic of great scientific and public interest. The study of the human microbiome and its links to disease has exploded in popularity in recent years. There is a rising interest in engineering microbiomes to alter the composition and function of the gut microbiome as a novel therapeutic approach as the significance of the association between human-associated microbial populations and disease development becomes clear. Many additional concerns regarding the interactions between human and microbial cells in the body delivered only partial answers to these questions. The researchers continue to make noteworthy and exhilarating contributions to our understanding of the basic biology of human health in the area of microbiome.

This unprecedented growth of data on the genetic makeup of bacteria has set the stage for groundbreaking new insights into microbial systems. We can now efficiently construct complex biological systems to execute desired activities, thanks to our increased understanding of the interconnecting networks of biological molecules, such as genes and proteins, at the systems level. With the advent of this technology and other essential enabling techniques, such as gene synthesis, the new interdisciplinary scientific subject of synthetic biology was born. This spectacular expansion of information regarding the genomic architecture of microbes has laid the foundation for truly revolutionary advances in our knowledge of microbial systems. We are now able to understand the interacting networks of biological molecules including genes and proteins at the systems level, and on the basis of this understanding, we can effectively engineer complex biological systems that perform desired functions.

Various initiatives are underway around the world to survey the human microbiota at several body sites, characterize them, understand their interactions with the human hosts, elucidate their role in diseases, and design possible therapeutic or dietary interventions. There is a new wave of studies mining the human microbiota for health-relevant bioactive compounds, for characterizing specific microbial strains and for cataloguing the human microbiota from all across the globe. This book will provide plentiful opportunities for researchers to learn about and to connect to important developments in studying the human microbiota.

Plethora of studies focused on the human microbiome across biological time and in nutrition and drugs, host–microbe interactions, integrative approaches in microbiome research and methodological advances in microbiome analysis. Even though we are still in the basal level of understanding the mechanisms involved in the cross-talk between the microbiota and the surrounding host environment, researchers are developing new therapeutic strategies to manipulate the gut microbiota has emerged as an evolving need in medicine, due to the important role of these microorganisms in the onset and the progression of many distant and local diseases. The functional meta-omics and synthetic biology approaches help resolve the limitations hindering the feasibility of engineered microbe therapy. However, experimental tools must still be specifically tuned for studying the microbiome, as synthetic biology emerged independently of microbiome studies. Mounting evidence suggests that the microbiota may be used for the development of novel therapeutic strategies.

Yet the practical translational applications of this fascinating and enthralling area of science are outstanding. The goal of this book is to provide a comprehensive understanding of advances in the microbiome–host relationship for human health and medicine. Additionally, this book aims to provide a nonexhaustive list of studies with a special focus on multiomics approaches and the cellular reprogramming of microbes to enable in-depth microbiome research on diagnostics, therapeutics, and bioinformatics. With these aims in mind, the material of this textbook has been structured from basic to more advanced topics in a sequential progression. Finally, this book also reviews advancement from fundamental research to relationships between immune microbiomes and human health microbiomes therapeutic applications.

We hope that your creativity is inspired by this book and wish you luck in your experiments. This book illustrates astonishingly the urgency with which numerous scientific brains are committed to the welfare of the scientific world. I am immensely grateful to the contributors for consistently paying attention to my request and expressing confidence in my skills. I will still be forever highly obliged to all the contributors forever. The worthlessness of their efforts cannot be explained by these terms.

Because of the heartfelt interest and painstaking effort of many other well-wishers whose names are not listed, but they are already in our hearts, we have effectively compiled our innovative and reflective research work. So, the reward for their sacrifices is definitely worth it. I want this book to be devoted to my mum, S. Jayaprada (late). From the bottom of our souls, I and the contributing authors hope this book will be a good guide and guidance for scientific studies to understand the host–microbiome relationships in human health.

Machilipatnam, Andhra Pradesh, India Pallaval Veera Bramhachari

Acknowledgments

My sincere thanks are extended to all the academicians and scientists who have contributed chapters and happily agreed to share their work on *Human Microbiome in Health, Disease, and Therapy* in this volume.

This book is a stunning reflection of the seriousness with which several scientific minds are dedicated to the research community. I am extremely thankful to the contributors for paying continuous attention to my requests and bestowing faith and confidence in my competencies and capabilities. I shall always remain highly grateful to all contributors forever. These words cannot justify the worthiness of their quintessential efforts. We appreciate the excellent work of the authors and co-authors who were invited to contribute chapters to this book. The credit for making this book a reality goes to them. The editor and the review team of the book especially appreciate sharing expertise with the contributors. Each chapter is pretty informative and written as stand-alone contributions from several research institutes so that the reader can begin reading anywhere in the book depending upon his/her interests and needs.

At the same time, I also express my deepest gratitude to my family members, especially my wife (Ramadevi Ramaswamy) and my kids (Ruthvik and Jayati) for their kind support which has prompted me to complete this assignment on time. I am also thankful to Krishna University administrative officials, and my colleagues in the Department of Biotechnology, Krishna University, for their intellectual support. I am equally thankful to the Springer Nature Publishing Group for their full cooperation during the peer review and production of the volume.

I am thankful to my beloved teachers and mentors, for their constant support and motivation at all stages of my progress.

Department of Biotechnology
Krishna University
Machilipatnam, Andhra Pradesh, India

Pallaval Veera Bramhachari

Contents

Editor and Contributors

About the Editor

Pallaval Veera Bramhachari is currently Associate Professor in the Department of Biosciences and Biotechnology at Krishna University, Machilipatnam, A.P State, India. He has earlier served as a postdoctoral research scientist at the Bacterial Pathogenesis Laboratory, Queensland Institute of Medical Research (QIMR-Berghofer Research Center), Brisbane, Queensland, Australia, and a DBT Postdoctoral Research Fellow at the Department of Microbiology and Cell Biology (MCBL), Indian Institute of Science (IISc), Bangalore. He completed his Ph.D. in Microbiology from Goa University, India. His major research interests are the investigations in the areas of applied research in microbiology and molecular cell biology, viz. exploring the biochemical mechanism of bacterial-biofilm formations, extracellular microbial polysaccharides (EPS), extracellular proteins (ECP), and the structure–function relationship of bacterial EPS, ECPs, biosurfactants (glycolipids and rhamnolipids), pathogenic and virulence characteristics of bacteria. He has published more than 150 research articles in Scopus indexed journals, authored and co-authored 9 international books and invited book chapters in prestigious peer-reviewed international and presented 50 abstracts at various national and international conferences. He has served as reviewer and editorial board for a number of International journals.

He is a member of numerous scientific societies and organizations, most importantly, the Indian Science Congress, Society of Biological Chemists, AMI and SBTI, India. He completed two major research projects from DST-SERB and DST-NRDMS, Govt. of India. He was awarded a Travel Scholarship from QIMR, Australia, for attending the 4th Indo-Australian Biotechnology Conference at Brisbane, Australia. He was awarded with a young scientist travel fellowship from the DST, Govt. of India, for attending XVII Lancefield International Symposium at Porto Heli, Greece. He was conferred with various prestigious awards, notably,

Science Education Research Board (SERB), Government of India-Young Scientist award (2011) with a research project and nominated as Associate Fellow of Andhra Pradesh Academy of Sciences (APAS)—2016, MASTER TRAINER: Andhra Pradesh English Communications Skills Project. British Council and APSCHE—2017, Andhra Pradesh State Best Scientist Award—2017, Dr. V. Ramalingaswamy Memorial Award for Biomedical Sciences—2019, KAPL Award (Industrial Biotechnology)-SBTI Annual awards—2022, and Dr. A.P.J. Abdul Kalam Teacher Excellence Award—2022. He also obtained two Indian patents. He has more than 18 years of teaching and research experience in Molecular Microbiology, Virology and Immunology, IPR, Bioprocess Engineering, Microbial and Environmental Biotechnology at the university level.

Contributors

P. Amudha Department of Biochemistry, School of Life Sciences, Vels Institute of Science, Technology & Advanced Studies (VISTAS), Chennai, Tamil Nadu, India

S. Anju Bhavan's Vivekananda College of Science, Humanities and Commerce, Sainikpuri, Secunderabad, Telangana, India

K. Anuradha Bhavan's Vivekananda College of Science, Humanities and Commerce, Sainikpuri, Secunderabad, Telangana, India

Krodi Anusha Department of Biosciences & Biotechnology, Krishna University, Machilipatnam, Andhra Pradesh, India

Yaser A. Arafath Department of Microbiology, School of Life Sciences, Pondicherry University, Puducherry, India

Chanda Vikrant Berde Marine Microbiology Program, School of Earth, Ocean and Atmospheric Sciences, Goa University, Taleigao, Goa, India

Pallaval Veera Bramhachari Department of Biosciences and Biotechnology, Krishna University, Machilipatnam, Andhra Pradesh, India

M. Subhosh Chandra Department of Microbiology, Yogi Vemana University, Kadapa, Andhra Pradcsh, India

Joseph G. Giduthuri Department of Epidemiology and Public Health, Swiss Tropical and Public Health Institute, University of Basel, Basel, Switzerland

Kishore Kumar Godisela Department of Biochemistry, SRR Govt. Arts & Science College, Karimnagar, Telangana, India

Sridevi Gopathy Department of Physiology, SRM Dental College, Chennai, Tamil Nadu, India

Kuraganti Gunaswetha Department of Microbiology, Kakatiya University, Warangal, Telangana, India

Vijay A. K. B. Gundi Department of Biotechnology, Vikrama Simhapuri University, Nellore, Andhra Pradesh, India

Saqib Hassan Department of Microbiology, School of Life Sciences, Pondicherry University, Puducherry, India

Division of Non-Communicable Diseases, Indian Council of Medical Research (ICMR), New Delhi, India

Department of Biotechnology, School of Bio and Chemical Engineering, Sathyabhama Institute of Science and Technology, Chennai, Tamil Nadu, India

Future Leaders Mentoring Fellow, American Society for Microbiology, Washington, DC, USA

M. Jayalakshmi Department of Biochemistry, Mahalashmi Women's College of Arts and Science, Chennai, Tamil Nadu, India

Satyanagalakshmi Karri Department of Biotechnology, Vikrama Simhapuri University, Nellore, Andhra Pradesh, India

George Seghal Kiran Department of Food Science and Technology, School of Life Sciences, Pondicherry University, Puducherry, India

Harish Babu Kolla Department of Biotechnology, Vignan's Foundation for Science, Technology and Research (Deemed to be University), Guntur, Andhra Pradesh, India

Anusha Konatala Department of Medical Biotechnology, University of Windsor, Windsor, ON, Canada

Langeswaran Kulanthaivel Department of Biomedical Sciences, Science Campus, Alagappa University, Karaikudi, Tamil Nadu, India

Hari Krishna Kumar Department of Microbiology, School of Life Sciences, Pondicherry University, Puducherry, India

B. Lalnundika Department of Microbiology, School of Life Sciences, Pondicherry University, Puducherry, India

Ishfaq Hassan Mir Department of Biochemistry and Molecular Biology, School of Life Sciences, Pondicherry University, Puducherry, India

Niharikha Mukala Department of Biotechnology, AUCST, Andhra University, Visakhapatnam, Andhra Pradesh, India

DhanaLakshmi Padi , Düsseldorf, North Rhine-Westphalia,, Germany

Surajit Pal Department of Microbiology, School of Life Sciences, Pondicherry University, Puducherry, India

Lakshmi Pethakamsetty Department of Microbiology, AUCST, Andhra University, Visakhapatnam, Andhra Pradesh, India

Padmaja Phani Department of Biotechnology, Vignan's Foundation for Science, Technology and Research (Deemed to be University), Guntur, Andhra Pradesh, India

Anushara Prabhakaran Department of Microbiology, School of Life Sciences, Pondicherry University, Puducherry, India

R. Prathiviraj Department of Microbiology, School of Life Sciences, Pondicherry University, Puducherry, India

A. M. V. N. Prathyusha Department of Biosciences & Biotechnology, Krishna University, Machilipatnam, Andhra Pradesh, India

Sabreena Qadri Department of Psychiatry, Asian Institute of Gastroenterology, Hyderabad, Telangana, India

M. Rajeswari Department of Food, Nutrition and Dietetics, AUCST, Andhra University, Visakhapatnam, Andhra Pradesh, India

Muthuvel Raju Department of Biochemistry, Karuna Medical College and Hospital, Chittur, Palakkad, Kerala, India

Ayushi Rambia Department of Microbiology, School of Life Sciences, Pondicherry University, Puducherry, India

Meenatchi Ramu Department of Biotechnology, SRM Institute of Science and Technology, Faculty of Science and Humanities, Kattankulathur, Chengalpattu, Chennai, Tamil Nadu, India

Neelapu Nageswara Rao Reddy Department of Biochemistry and Bioinformatics, GITAM Institute of Science, Gandhi Institute of Technology and Management (GITAM), Visakhapatnam, Andhra Pradesh, India

Prakash Narayana Reddy Department of Biotechnology, Vignan's Foundation for Science, Technology and Research (Deemed to be University), Guntur, Andhra Pradesh, India

Dr. V. S. Krishna Government Degree College (A), Maddilapalem, Visakhapatnam, Andhra Pradesh, India

Pavani Sanapala Department of Biotechnology, Andhra University, Visakhapatnam, Andhra Pradesh, India

J. Sarada Bhavan's Vivekananda College of Science, Humanities and Commerce, Sainikpuri, Secunderabad, Telangana, India

P. S. Seethalakshmi Department of Microbiology, School of Life Sciences, Pondicherry University, Puducherry, India

Joseph Selvin Department of Microbiology, School of Life Sciences, Pondicherry University, Puducherry, India

Srividya Seshadri Department of Biochemistry, Sathyabama Dental College and Hospital, Chennai, Tamil Nadu, India

Sriram Shankar Department of Microbiology, School of Life Sciences, Pondicherry University, Puducherry, India

D. S. L. Sravani Division of Nutritional Sciences, Cornell University, Ithaca, NY, USA

Gowtham Kumar Subbaraj Faculty of Allied Health Sciences, Chettinad Hospital and Research Institute, Chettinad Academy of Research and Education, Kelambakkam, Chennai, Tamil Nadu, India

Pola Sudhakar Department of Biotechnology, AUCST, Andhra University, Visakhapatnam, Andhra Pradesh, India

Edla Sujatha Department of Microbiology, Kakatiya University, Warangal, Telangana, India

Swarnakala Department of Marine Biotechnology, AMET Deemed to be University, Kanathur, Chennai, Tamil Nadu, India

Chinnasamy Thirunavukkarasu Department of Biochemistry and Molecular Biology, School of Life Sciences, Pondicherry University, Puducherry, India

Manohar Babu Vadela Department of Biotechnology, Vikrama Simhapuri University, Nellore, Andhra Pradesh, India

R. Vidya Department of Biochemistry, School of Life Sciences, Vels Institute of Science, Technology & Advanced Studies (VISTAS), Chennai, Tamil Nadu, India

Abbreviations

ABC	ATP-binding cassette
ACE2	Angiotensin-converting enzyme 2
ACTH	Adrenocorticotropic hormone
Acyl-HSLs	Acyl homoserine lactones
AD	Alzheimer's diseases
AD	Atopic dermatitis
AF	Aflatoxin
AF	Agave fructans
AHLs	N-acyl-homoserine lactones
AhR	Aryl hydrocarbon receptor
AI	Autoimmune diseases
AICR	American Institute for Cancer Research
AIEC	Adherent-invasive *E. coli*
AIs	Autoinducer
ALD	Alcoholic liver disease
ALs	Artilysins
AMPs	Antimicrobial peptides
AMR	Antimicrobial resistance
AN	Anorexia nervosa
ANS	Autonomic nervous system
ARB	Antibiotic-resistant bacteria
ARDS	Acute Respiratory Disease Syndrome
ARG	Antibiotic resistance genes
AS	Ankylosing spondylitis
ASC	Apoptosis-associated speck-like protein containing a CARD
ASDs	Autism spectrum disorders
AXOS	Arabinoxylan oligosaccharides
BAs	Bile acids
BBB	Blood-brain barrier
BCAAs	Branched-chain amino acids
BD	Bipolar disorder
BDNF	Brain-derived neurotrophic factor
BFT	*B. fragilis* toxin

B-GOS	Bimuno-galactooligosaccharides
BMI	Body mass index
BPs	Bacteriophages
cAMP	Cyclic AMP
CAR	Chimeric antigen receptor
CARD9	Caspase recruitment domain-containing protein 9
CCS	Circular consensus sequencing
CD	Crohn's disease
CDC	Centers for Disease Control and Prevention
CDF	Carrot dietary fiber
CDI	Clostridium difficile infection
CEACAMs	Human carcinoembryonic antigen-related cell adhesion molecule 6
CI	Clostridium infections
CLD	Chronic liver disease
CLRs	C-type lectin receptors
CN	Cellulose nanofiber
CNS	Central nervous system
COX-2	Cyclooxygenase-2
CRC	Colorectal cancer
CRD	Chronic respiratory disorder
CRF	Corticosterone-releasing factor
CRH	Hypothalamic corticotropin-releasing hormone
CRISPR	Clustered regularly interspaced short palindromic repeat
CSF	Colony-stimulating factor
CTLA-4	Cytotoxic T-lymphocyte antigen-4
CTLA-4	Cytotoxic T-lymphocyte-associated protein 4
CVD	Cardiovascular disease
DAMPs	Damage-associated molecular patterns
DCA	Deoxycholic acid
DCs	Dendritic cells
DGGE	Denaturing gradient gel electrophoresis
DMBA	1,3-dimethylbutylamin
E	Envelope
ECDC	European Centre for Disease Prevention and Control
ECs	Epithelial cells
EGFR	Epidermal growth factor receptor
EIF5A	Eukaryotic translation initiation factor 5A
ELs	Endolysins
EMA	The European Medicines Agency
ENS	Enteric nervous system
EPs	MDR-Efflux pumps
ERK	Extracellular regulated kinase
ESBL	Extended-spectrum beta-lactamase

EVs	Extracellular vesicles
FA	Food allergy
FBA	Flux balance analysis
FD	Ferredoxin
FDA	Food and Drug Administration
FFAR	Free fatty acid receptors 2/3
FGID	Functional gastrointestinal disorder
FISH	Fluorescent in situ hybridization
FIT	Fecal immunochemical test
FMT	Fecal microbiota transplantation
FMT	Microbiota transplantation
FOBT	Fecal occult blood test
FODMAPs	Fermentable oligosaccharides, disaccharides, monosaccharides, and polyols
FOS	Fructooligosaccharides
FP	Fusion peptide
Fqs	Fluoroquinolones
FXR	Farnesoid X-activated receptor
GALT	Gut-associated lymphoid tissue
GF	Germ-free
GIO	Glucocorticoid-induced osteoporosis
GIT	Gastrointestinal tract
GLP-1	Glucagon-like peptide
GM	Gut microbiota
GOS	Galactooligosaccharides
GPCR	G protein-coupled receptors
GPR43	G-protein-coupled receptor 43
GRAS	Generally recognized as safe
GWAS	Genome-wide association studies
Hb calyces	*Hibiscus sabdariffa*
HBM	Human breast milk
HCC	Hepatocellular carcinoma
HDACs	Histone deacetylase
HFD	High-fat diet
HGT	Horizontal gene transfer
HIF	Hypoxia-inducible factor
HM	Hematological malignancies
HMM	Human milk microbiome
HMP	Human microbiome project
HPA axis	Hypothalamic-pituitary-adrenal
HS	Hepatic steatosis
HS/HF	High sugar/high fat
HSCs	Hepatic stellate cells
HSP	Heat shock proteins
IBD	Inflammatory bowel diseases

IBS	Irritable bowel syndrome
ICIs	Immune checkpoint inhibitors
ICU	Intensive care unit
IDO	Indoleamine2,3-dioxygenase
IECs	Intestinal epithelial cells
IE-DAP	D-gamma-glutamyl-meso-DAP
IKK complex	I kappa B kinase
IKK	$I_{\kappa}B$ kinase
IL-10	Interleukin
IL-8	Anti-inflammatory cytokines
IL-8	Interleukin 8
IMG/ER	Integrated microbial genome-expert review
IPF	Idiopathic pulmonary fibrosis
IRAK4	IL-1 receptor-associated kinase 4
ISCs	Intestinal stem cells
ITS	Internal transcribed spacer
JCVI	J. Craig Venter Institute
JNK	c-Jun N-terminal Kinase
JP	Jujube polysaccharides
LABs	Lactic acid bacteria
LPS	Lipopolysaccharide
LR	*Lactobacillus reuteri*
LTA	Lipoteichoic acid
M	Membrane
MAFLD	Metabolic-associated fatty liver disease
MAGIC	Metagenomic alteration of gut microbiome by in situ conjugation
MAPK	Mitogen- activated protein kinase
MATE	Multidrug and toxic efflux
MD	Mediterranean diet
MDD	Major depressive disorder
MDP	Microbial muramyl dipeptide
MDP	Muramyl dipeptide
MDR	Multidrug-resistant
MF	Major facilitator
MGEs	Mobile resistance Genes
MIC	Minimum inhibitory concentrations
MKK	Mitogen-activated kinase kinase
MLNs	Mesenteric lymph nodes
MMPs	Matrix metalloproteinases
MR	Microbiome resilience
MRSA	Methicillin-resistant *Staphylococcus aureus*
MS	Multiple sclerosis
MSCs	Mesenchymal stem cells
MTT	Microbiome targeted therapies

MUFA	Monounsaturated fatty acids
MWAS	Microbiome/metagenome-wide association studies
N	Nucleocapsid
NAFLD	Non-alcoholic fatty liver disease
NASH	Non-alcoholic steatohepatitis
NCD	Non-communicable diseases
NCDs	Non-communicable diseases
NCDs	Several non-communicable diseases
NDCs	Non-digestible carbohydrates
NEC	Necrotizing enterocolitis
NEMO	NF- $_{K}$B essential modulator
NHGRI	National Human Genome Research Institute
NK	Natural killer
NLRs	NOD-like receptors
NOD	Nucleotide-binding oligomerization domain
NOS	Nitric oxide synthase
NPS	Neuropsychiatric disorders
NPY	Neuropeptide Y
OF	Oligofructose
OPs	Opportunistic pathogens
OSCs	Oligosaccharide carbohydrates
OTU	Operational taxonomic units
PABA	Para amino benzoic acid
PAMPs or MAMPs	Pathogen- or microbe-associated molecular patterns
PBPs	Penicillin-binding proteins
PD	Parkinson's disease
PD-1	Anti-programmed cell death protein 1
PD1	Programmed cell death 1 receptor
PD-1	Programmed cell death protein 1
PDGF	Platelet-derived growth factor
PD-L1	Programmed cell death ligand 1
PFC	Prefrontal cortex
PKB/Akt	Protein kinase B
PKC	Protein kinase C
PLPE	Phellinus linteus polysaccharide extract
PNS	Peripheral nervous system
PRRs	Pattern-recognition receptors
PSA	Polysaccharide A
PT	Phage therapy
PUFAs	Polyunsaturated fatty acids
QIIM	Quantitative Insights Into Microbial Ecology
QoL	Quality of life
QPS	Qualified Presumption of Safety
QQ	Quorum quenching
QS	Quorum-sensing

RA	Retinoic acid
RA	Rheumatoid arthritis
RAST	Rapid Annotation using Subsystems Technology
RBDF	Rice bran dietary fiber
RE-MS	Relapsing-remitting multiple sclerosis
RIs	Resistance islands
RLRs	RIG-I-like receptors
RNAi technology	RNA interference
RND	Resistance-nodulation-division
ROS	Reactive oxygen species
RS	Resistant starch
RSAI	Several Respiratory System-Associated Infections
S proteins	Spike
SARSCoV-2 (COVID-19)	Acute Respiratory Syndrome Coronavirus 2
SBS	Sequence-by-synthesis
SCFA	Short-chain fatty acids
SDS	Sodium dodecyl sulfate
SEC	Size exclusion chromatography
SFB	Segmented filamentous bacteria
SHDF	Soy hull dietary fiber
SIBO	Small intestine bacteria overgrowth
SIGA	Secretory IgA
SLE	Systemic lupus erythematosus
SlrP	Salmonella leucine-rich repeat protein
SMR	Small multidrug resistance
SNP	Single nucleotide polymorphism
SODs	Superoxide dismutase
SOLiD™ System	Solexa Genome Analyzer and the Applied Biosystems
SOPs	Standard operating procedures
SPDF	Sweet potato residue
SPF environment	Specific pathogen-free
SPF	Specific pathogen-free
STAT3	Signal transducer and activator of transcription
SUD	Substanc use disorder
T1D	Type 1 diabetes
T1R3	Type 1 taste receptor 3
T2DM	Type 2 diabetes mellitus
TAB levels	TET-assisted bisulfite sequencing
TAB1/2/3	TAK-binding proteins
TAK1	Ubiquitin-dependent kinase of MKK and IKK
TBK1	Serine/threonine-protein kinase 1
TCGA	The Cancer Genome Atlas
TGGE	Temperature gradient gel electrophoresis
TLR	Toll-like receptor
TMA	Trimethylamine

TMAO	Trimethylamine oxides
TMPRSS2	Transmembrane protease, serine 2
TNF	Tumor necrosis factor-α
TP	Triptolide
TRAF6	TNF receptor-associated factor 6
Tregs	T-regulatory cells
Ub	Tyrosine-protein kinase MET (c-MET) Ubiquitin
UC	Ulcerative colitis
VEGF	Vascular endothelial growth factor
VEGFR2	Vascular endothelial growth factor receptor 2
VIP	Vasoactive intestinal polypeptide
VN	Vagus nerve
VRE	Vancomycin-resistant *Enterococcus* spp.
VTA	Ventral tegmental area
WCRF	World Cancer Research Foundation
WHO	World Health Organization
ZnONPs	Zinc oxide nanoparticles

A Scoping Review of Research on the Unfolding Human Microbiome Landscape in the Metagenomics Era

1

Veera Bramhachari Pallaval

Abstract

Research on the microbiome is motivated by an urge to understand the underlying causes of disease and the microbiota's role in both causing and responding to illness. Microbiome and metagenomics studies have confirmed the significance of the microbiota for human health and disease. Human health is intricately intertwined with the vast microbial communities that call the human body home. Human-microbiota ecosystems have been shown to have an important role in maintaining health and developing various diseases, as has been demonstrated by human microbiome studies. Modern technological applications have yielded important new understandings of the intricate web of interactions between host and microbiome and the mechanisms behind the functions of microbiota and individual bacteria in influencing host health and disease. Mechanistic insights into the intricate relationship between the host and the microbiome have been gleaned from analyses of host-microbiome interactions. The microbiome is highlighted in this chapter as a potential therapeutic target. We have also emphasized how the microbiome's makeup and function can be altered using several different approaches for therapeutic purposes. More research is needed to refine therapeutic applications based on mechanistic insights into the microbiome's relationship to health.

Keywords

Microbiome · Metagenomics · Host-microbiota interactions · Therapeutics applications

V. B. Pallaval (✉)
Department of Biosciences & Biotechnology, Krishna University, Machilipatnam, Andhra Pradesh, India

P. Veera Bramhachari (ed.), *Human Microbiome in Health, Disease, and Therapy*, https://doi.org/10.1007/978-981-99-5114-7_1

1.1 Introduction

The term "human microbiome" refers to the collection of species living in and on humans, their combined genomes, their biological connections, and their hosts. Microorganisms such as bacteria, viruses, and fungi number in the trillions within the human body. The distribution of microbial cells surpasses the number of all human cells, including somatic and germ cells, throughout the human body (Clemente et al. 2012; Petersen and Round 2014). Furthermore, the makeup of the human microbiome is dynamic and can interact with its host and change over time to its conditions. However, the tendency of the microbiome to change over time varies from person to person. Thousands of microbes that have evolved with humans and are believed to impact human health and disease make up this microbiome, which has biomass comparable to the human brain in a healthy adult. Due to competitive ecological interactions, the microbiome is robust against perturbations and the introduction of new organisms (Ding and Schloss 2014; Al-Zyoud et al. 2020).

The human microbiota is dynamic and ever-evolving in response to several host factors like age, genetics, hormonal changes, diet, underlying disease, lifestyle, and the environment. The symbiotic relationship between the microbiota and host results in a complex "super-organism" that fundamentally improves the host. Relationships between hosts and their microbiota can be classified as commensal, symbiotic, or pathogenic based on the nature of the interactions between the host and its microbiota. The symbiotic relationship may be disrupted by environmental changes (infection, nutrition, or lifestyle) or by flaws in the host regulatory circuits that regulate bacterial sensing and homeostasis, which can promote disease (Schwabe and Jobin 2013; Petersen and Round 2014; Dietert & Dietert 2015).

Human health and disease are impacted by microbiota because they regulate crucial metabolic and immunomodulatory processes (Byrd et al. 2018; Al Bataineh et al. 2021). Disruptions in microbial hemostasis (dysbiosis) may contribute to life-threatening disorders (Shanahan et al. 2021), while balanced microbiota contributes significantly to healthy living (Ding and Schloss 2014). The HM significance for human biology is also underscored by the plethora of chronic disorders related to an unbalanced HM, ranging from inflammatory and metabolic conditions to neurological, cardiovascular, and respiratory ailments. To better understand the molecular mechanisms involved in regulating host cells and physiological processes, it will be necessary to characterize the bacterial taxa (both living bacteria and bacterial DNA) resident within tissues. This mechanism will pave the way for developing novel biomarkers and therapeutic approaches. The correlation between a dysbiotic HM and many diseases has elevated the microbiome to critical therapeutic targets, prompting the creation of a suite of microbiome-specific intervention techniques and an attempt to restore a healthful architecture. Recent biomedical literature has described the substantial role of the human microbiome in health and disease.

1.2 The Microbiome as a Precision Medicine Frontier

Since research into the microbiome is still relatively new, there is a good chance that more of the microbiome's crucial roles have yet to be uncovered. The same sequencing technology facilitating personal genomics is also driving these findings, and its price is dropping rapidly. Despite the microbiome's potential as a therapeutic tool, several challenges must be addressed before it can be used in precision medicine. Moreover, the full ramifications of microbiome therapy on a global scale remain unclear. While fecal transplants frequently increase with few reported adverse effects, their long-term repercussions on the Western population remain unknown.

Similarly, it is challenging to forecast how genes interact with the environment in genomic medicine. Because of this, studies need huge samples to be credible (Manuck and enMcCaffery 2014). Genomic medicine has a promising future, but progress is being impeded by several obstacles right now. Understanding the interplay between the microbiota and the human host can lead to important therapeutic advances. Studies that investigated the interplay between host biology and microbial gene regulation, secretion, and metabolite synthesis shall invariably yield useful results, last but not least, the creation of cutting-edge medicines may also alter host components in response to either commensal or pathogenic bacteria. Therefore, incorporating the microbiome into the field of precision medicine, which is currently dominated by genetic information, would significantly improve it.

1.3 Significance of the Human Microbiome Project

The composition of one's microbiome has been shown to affect one's vulnerability to infectious diseases and play a role in developing chronic gastrointestinal disorders. Certain microbial communities influence a person's response to a medication. The mother's microbiome may affect her offspring's well-being. The human microbiome has been mapped, thanks to multiple researchers identifying new species and genes. Researchers have shown associations between certain combinations of microbe species and human health issues based on genetic tests that examined the relative abundance of distinct species in the human microbiome. New treatments, such as those for common bacterial infections, could result from a better understanding of the microbial variety in the human microbiome. A better understanding of the microbiome's function in human health, nutrition, immunity, and disease can be gained using the human microbiome as a guide.

The human microbiome (HM) is crucial to human health because commensal microbes affect every aspect of human physiology. As a result of the HMP, venture investors are increasingly willing to put money into startups that study the human microbiome. Despite the obstacles, studies of the human microbiome are moving forward rapidly. This domain's data increases exponentially every year, thanks to improvements in whole-genome sequencing and the capacity to transplant and observe human microbial communities in mice. Due to several factors, the human microbiome therapeutics and diagnostics market is expected to expand rapidly in the coming years. These include increasing disease indications, the intriguing therapeutic potential of microbiome-based therapies, encouraging results from clinical trials, and the backing of investors. The microbial residents of the human microbiome have been methodically uncovered, cataloged, and analyzed over a decade with the support of various research institutes. To learn more about how the microbiota acts and interacts with its human hosts in health and disease, scientists are increasingly turning to "multi-omics" approaches, such as whole genome sequencing, proteomic, metabolomic, and transcriptome investigations. Disease prevention measures appear to benefit from combining metagenomics (diet's effect on microbiota), metaproteomics (microbes' gene expression), and metabolomics (microbes' metabolites). Their genes are believed to be more important to human survival than human genes. The human microbiome contains genes that play important roles in various biological processes, including but not limited to aging, digestion, immunity, central nervous system regulation, mood, and cognitive function. Because of this, the human microbiome can rightfully be called an "organ" of the body.

New sequencing technologies have advanced the study of microbial ecosystems and made it possible to examine microbial communities linked with the human body at a scale and resolution never before achieved (Eckburg et al. 2005). Inflammatory bowel disease, type II diabetes, and autoimmune diseases like rheumatoid arthritis and multiple sclerosis are all rising in Western societies, stoking interest and funding in microbiome research (Bäckhed et al. 2012).

Ongoing research topics include how these metabolites are made, which receptors they bind to, and their roles in the host. However, recently established large-scale metabolome screening techniques have increased our understanding of these signaling pathways by identifying novel interactions between microbially generated compounds and receptors (Chen et al. 2019).

1.3.1 Human Microbiome Research: Mounting Pains and Future Promises

The focus of human microbiome studies is shifting from elucidating interactions to determining how bioactive bacteria affect humans. Data-driven microbiome diagnostics and therapeutics are making headway despite obstacles and might usher in precision medical advances in the next decade. Genomics, transcriptomics, proteomics, and metabolomics allow researchers to assess the important but understudied nonbacterial commensal kingdoms and their complex interactome networks (Lin 2023) and uncover functional readouts beyond genomic sequencing. Exciting new research is being conducted on the roles that commensal and opportunistic viruses (including bacteriophages), fungi, and parasites may play in the commensal bacterial ecosystem and the human host. Research and analytical methods, including computational reference datasets, molecular exploitation techniques, and in vivo colonization models, need to be refined so that they may be used to understand these poorly understood commensal assemblages better. It is essential to comprehend host-microbiome interactions' enormous variety and modularity to move forward with novel microbiome research. When using the microbiome for diagnostic and therapeutic purposes in human disease, a greater awareness of molecular mechanisms and control, from macro-level descriptive community linkages to the microscale involvement of discrete bioactive therapeutic targets, is becoming increasingly relevant. Biochemical and structural elucidation of chemicals created, controlled, and removed by commensal bacteria, their human-binding analogs, and their downstream bioactive consequences on the human host would likely play a prominent role in such a study (Puschhof and Elinav 2023).

1.3.2 Human Microbiome Landscape in the Metagenomic Era

Recent advances in PCR and DNA hybridization have enabled the development of several culture-independent methods that can be used for both qualitative and quantitative identification. The study of microbial genomes inside varied environmental samples, known as metagenomics, has been made possible because of these straightforward techniques, which have radically changed our understanding of the human microbiome. Since its inception in 1998, metagenomics has transformed the study of the microbiota by producing a large library of sequences from microorganisms that live in different ecological niches within a host organism like humans. Shortly, treatment methods and/or vaccines will be developed, thanks to metagenomic sequencing's ability to characterize microbial populations comprehensively and objectively, including the virus spectrum (Finkbeiner et al. 2008). There is no longer a need to cultivate gut bacteria, thanks to recent developments in "omics" methodologies (i.e., genomes, transcriptomics, proteomics, and metabolomics) (Lamendella et al. 2012), which have led to the identification of certain intriguing approaches to prospective therapeutic and diagnostic applications.

In addition, although the human polyomavirus's presence on the skin had little impact on pathology, metagenomics techniques allowed for its discovery. Merkel cell carcinomas and normal skin carry this virus (Wieland et al. 2009; Schowalter et al. 2010). Characterizing the skin's viral microbiota can help identify microbiome patterns associated with specific skin disorders (Foulongne et al. 2012). Nonetheless, this is possible through high-throughput metagenomic sequencing (HTS), a highly comprehensive technology based on random sequencing of the whole DNA. Future research using this method (i.e., metagenomics) will be able to pinpoint health biomarkers associated with a diverse and stable gut microbiome. Furthermore, metagenomics can help us comprehend the significance of the interaction between us and our microbiome regarding our health.

1.3.3 The Way Forward

The microbiome study has progressed significantly over the past few decades and is now a major focus of scientific and popular attention. The study of microbiomes originates in environmental microbiome studies but has since expanded to include the concept of eukaryotes as integral members of the microbial community with which they coexist. Remember that you live in an environment with trillions of other little organisms. Therefore, microbiome interactions are crucial to human health, whether beneficial, harmful, or neutral. An estimated 50–100 times as the complex and diverse microbiome uses many genes as part of host genomes. These additional genes help regulate host physiology by encoding various enzymatic proteins that modify the host metabolism due to their effects on the metabolites generated. However, the definition of the microbiome has expanded to include not just the microbial population but also the full range of chemicals produced by the microbes, such as their structural components, metabolites, and compounds produced by the coexisting host. Microbiome metabolic pathways are strongly related to 34% of blood and 95% of fecal metabolites (Visconti et al. 2019), indicating a robust interplay of metabolites between the human microbiome and its host. Beyond metagenomics, additional multi-omics research is required to determine the precise mechanism by which microbial activity affects human health. To characterize the microbiome at the functional level in both healthy and pathological states, these include transcriptomics, proteomics, and metabolomics.

In this new era of microbiome therapeutics, we may aid in the clinical treatment of disease. Although our understanding of the microbiome is growing quickly, therapeutic approaches employing the microbiome are in their infancy. Understanding the underlying biological pathways of the microbiota will require future investigations employing animal models or epidemiological data generated from clinical trials to develop novel treatments. Yet current approaches need to combine new technologies to help microorganisms thrive in settings that replicate the human gastrointestinal tract. The notion of microbiome therapies has been proven effective by research efforts. Still, more work is needed to comprehend the microbiome and its relationships with the host before it can be advanced into clinical trials and used as a blueprint for efficient treatment. Therefore, the human microbiota holds great promise for ushering in a new era of biomarker research for diagnostic and therapeutic applications. Current difficulties in microbiome research stem from a lack of standardized research methodologies and knowledge regarding the connections between bacteria and human disease, which require the scientific community to work together to address.

1.4 Challenges and Future Directions

Novel approaches are desirable in microbiome research to standardize and mechanically validate the reported microbial gene clusters. Numerous diseases have been linked to bacteria and their metabolites. Appreciation must integrate multi-omics approaches that cataloged bacterial isolates, profiled new compounds, and measured host responses. Understanding the underlying processes through which microbes affect human health requires a revolutionary study of the links between the human microbiota and disease. This knowledge would push microbiome studies past the validation of biomarkers and into identifying therapeutic medication targets. In reality, bacterial metabolites may yield enormous mechanistic insights, which could hasten the development of new therapeutic options for a wide range of disorders, including managing impaired glucose metabolism in diabetes (Molinaro et al. 2020).

The previous two decades have seen uneven development in our understanding of microbiome research. Researchers from various fields have been able to study the metagenome to understand the effects of the microbiome better, thanks to the widespread availability of inexpensive second-generation sequencing technologies and readily available open-source bioinformatics software. Solutions to these and other issues may be found in further developing innovative and existing sequence-based technologies, which could lead to a deeper comprehension of the host-microbe and microbe-host interactions that are important to host health. Finally, we emphasized the mechanisms and novel therapeutic options related to this connection in the gut microbiome and explained the development of microbial interventions to increase therapeutic efficacy.

Acknowledgments The author gratefully acknowledge Krishna University, Machilipatnam, for the support extended.

Conflict of Interest The author declares that they have no competing interests.

References

Al Bataineh MT, Alzaatreh A, Hajjo R, Banimfreg BH, Dash NR (2021) Compositional changes in human gut microbiota reveal a putative role of intestinal mycobiota in metabolic and biological decline during aging. Nutr Healthy Aging 6:269–283

Al-Zyoud W, Hajjo R, Abu-Siniyeh A, Hajjaj S (2020) Salivary microbiome and cigarette smoking: a first of its kind investigation in Jordan. Int J Environ Res Public Health 17:256

Bäckhed F, Fraser CM, Ringel Y et al (2012) Defining a healthy human gut microbiome: current concepts, future directions, and clinical applications. Cell Host Microbe 12:611–622

Byrd AL, Belkaid Y, Segre JA (2018) The human skin microbiome. Nat Rev Microbiol 16:143–155

Chen H, Nwe PK, Yang Y et al (2019) A forward chemical genetic screen reveals gut microbiota metabolites that modulate host physiology. Cell 177(5):1217–1231.e18

Clemente JC, Ursell LK, Parfrey LW, Knight R (2012) The impact of the gut microbiota on human health: an integrative view. Cell 148:1258–1270

Dietert RR, Dietert JM (2015) The microbiome and sustainable healthcare. In Healthcare (Vol. 3, No. 1, pp. 100–129). MDPI

Ding T, Schloss PD (2014) Dynamics and associations of microbial community types across the human body. Nature 509:357–360

Eckburg PB, Bik EM, Bernstein CN et al (2005) Diversity of the human intestinal microbial flora. Science 308:1635–1638

Finkbeiner SR, Allred AF, Tarr PI, Klein EJ, Kirkwood CD, Wang D (2008) Metagenomic analysis of human diarrhea: viral detection and discovery. PLoS Pathog 4:e1000011

Foulongne V, Sauvage V, Hebert C, Dereure O, Cheval J, Gouilh MA, Pariente K, Segondy M, Burguière A, Manuguerra JC et al (2012) Human skin microbiota: high diversity of DNA viruses identified on the human skin by high throughput sequencing. PLoS One 7:e38499

Lamendella R, VerBerkmoes N, Jansson JK (2012) 'Omics' of the mammalian gut--new insights into function. Curr Opin Biotechnol 23:491–500

Lin T (2023) New techniques in microbiome research. Front Cell Infect Microbiol 13:161
Manuck SB, enMcCaffery JM (2014) Gene–environment interaction. Annu Rev Psychol 65:41–70
Molinaro A, Bel Lassen P, Henricsson M, Wu H, Adriouch S, Belda E, Chakaroun R, Nielsen T, Bergh P-O, Rouault C et al (2020) Imidazole propionate is increased in diabetes and associated with dietary patterns and altered microbial ecology. Nat Commun 11:5881
Petersen C, Round JL (2014) Defining dysbiosis and its influence on host immunity and disease. Cell Microbiol 16:1024–1033
Puschhof J, Elinav E (2023) Human microbiome research: growing pains and future promises. PLoS Biol 21(3):e3002053
Schowalter RM, Pastrana DV, Pumphrey KA, Moyer AL, Buck CB (2010) Merkel cell polyomavirus and two previously unknown polyomaviruses are chronically shed from human skin. Cell Host Microbe 7:509–515
Schwabe RF, Jobin C (2013) The microbiome and cancer. Nat Rev Cancer 13:800–812
Shanahan F, Ghosh TS, O'Toole PW (2021) The healthy microbiome-what is the definition of a healthy gut microbiome? Gastroenterology 160:483–494
Visconti A, Le Roy CI, Rosa F, Rossi N, Martin TC, Mohney RP, Li W, de Rinaldis E, Bell JT, Venter JC (2019) Interplay between the human gut microbiome and host metabolism. Nat Commun 10:4505
Wieland U, Mauch C, Kreuter A, Krieg T, Pfister H (2009) Merkel cell polyomavirus DNA in persons without merkel cell carcinoma. Emerg Infect Dis 15:1496–1498

Part I

Human Gut Microbiome Interactions

Elucidating the Role of Gut-Brain-Axis in Neuropsychiatric and Neurological Disorders

2

B. Lalnundika, Saqib Hassan, R. Prathiviraj, Hari Krishna Kumar, Sabreena Qadri, George Seghal Kiran, Pallaval Veera Bramhachari, and Joseph Selvin

Abstract

The gut-brain axis connects the enteric and central nervous systems in a bidirectional communication network. This involves an anatomical network and endocrine, humoral, metabolic, and immunological communication pathways. The gut and the brain are linked by the autonomic nervous system, the hypothalamic-pituitary-adrenal (HPA) axis, and nerves in the gastrointestinal tract, allowing the brain to influence intestinal activities, such as the activity of functional immune effector cells and the gut to influence mood, cognition, and mental health. Gut microorganisms may alter neurological development, modulate neurotransmission, and affect behavior, contributing to the pathogenesis and/or progression of

B. Lalnundika and Saqib Hassan contributed equally to this work.

B. Lalnundika · R. Prathiviraj · H. K. Kumar · J. Selvin (✉)
Department of Microbiology, School of Life Sciences, Pondicherry University, Puducherry, India

S. Hassan
Future Leaders Mentoring Fellow, American Society for Microbiology, Washington, DC, USA

Department of Biotechnology, School of Bio and Chemical Engineering, Sathyabama Institute of Science and Technology, Chennai, Tamilnadu, India

S. Qadri
Department of Psychiatry, Asian Institute of Gastroenterology, Hyderabad, Telangana, India

G. S. Kiran
Department of Food Science and Technology, School of Life Sciences, Pondicherry, India

P. V. Bramhachari
Department of Biosciences & Biotechnology, Krishna University, Machilipatnam, Andhra Pradesh, India

P. Veera Bramhachari (ed.), *Human Microbiome in Health, Disease, and Therapy*, https://doi.org/10.1007/978-981-99-5114-7_2

numerous neurodevelopmental, neuropsychiatric, and neurological illnesses. In this chapter, we provide an overview of recent data on the role of the microbiota-gut-brain axis in the pathophysiology of neuropsychiatric and neurological disorders, including depression, anxiety, schizophrenia, autism spectrum disorders (ASDs), Parkinson's disease, etc. We also discuss the role of prebiotics and probiotics in modulating the gut microbiome to benefit the host.

Keywords
Gut · Brain · Neurological disorders · Neuropsychiatric disorders · Microbiota

2.1 Introduction

The thirst for in-depth knowledge of the microbiome has existed for a long time among the scientific community and the commoners, as the idea of their relations with human health was proposed (Berg et al. 2020). The development of advanced culture-independent techniques like high-throughput and low-cost sequencing methods during the last 10 years has paved a great way to explore the symbiotic relationship between host and microbiota (Thursby and Juge 2017). The emergence of culture-independent techniques is a game-changer in the scientific community. According to Berg et al. (2020), a microbiome is a characteristic microbial community occupying a reasonably well-defined habitat with distinct physiochemical properties. The word microbiome here refers to the microorganisms involved and encompasses their theater of activity, resulting in the formation of specific "ecological niches." For instance, as far as a human is in concern, the collection of microorganisms naturally inhabiting human is called human microbiota, and the organs or sites of the body, like the gut, skin, etc., that harbor microbiota can be considered as their theater of activity (Berg et al. 2020).

Microorganisms can be found throughout the human body, and the natural microflora of humans are usually protozoa, archaea, eukaryotes, viruses, and predominantly bacteria. Therefore, most of the information on gut microbiota currently available is based on the analysis of gut bacteria (Morais et al. 2021). The human microbiota comprises $\sim 10^{13}$–10^{14} microbial cells, estimated to be around 1:1 microbial cells to human cells ratio. The gastrointestinal (GI) tract contains the largest number of microorganisms compared to other parts of the human body. It is mainly inhabitat by bacteria of three major phyla—*Firmicutes*, *Bacteroidetes*, and *Actinobacteria* (Kho and Lal 2018). Trillions of bacteria in the human body comprise about 4 million distinct bacterial genes, of which more than 95% belong to the large intestine. The complex microbiome in humans is a functional expansion of the host genomes. These genes encode enzymes and proteins not encoded by the host body, thus facilitating the functioning of the host and the metabolisms of the hosting body (Galland 2014).

As the taxonomic heterogeneity within the GI tract depends on numerous factors, including genetic, physiological, psychological, and environmental influences, the gut microbiome can vary in healthy individuals. Although each person's microbiota is distinct, researchers believe that humans have a core microbiome and that microbes colonize the GI tract similarly throughout their lives. The microbiota that lives in the gut can have a good or detrimental impact on human health. The gut microbiota has coevolved with the host and therefore plays a crucial role in the normal functioning of the host organism. Apart from digestion, the microbiota benefits the host in numerous ways, including gut health and epithelial morphology, harvesting of energy, defense, and host immunity modulation. Even if the adult microbiome is more stable than that of infants or older people, several factors can quickly alter its structure and composition. The use of antibiotics, stress, pathogenic infections, genetics, and food are examples of such influences. The alteration and disruption of normal gut microbiota composition can lead to a diseased state called dysbiosis (Mohajeri et al. 2018; Thursby and Juge 2017). Apart from the benefits they provide to the host, bacteria thriving in the gut, in return, get benefits from the nutritionally rich and safe environment of the human GI tract, thus, establishing a mutual relationship.

The presence of gut microbiota also impacts human mental health positively. Much available evidence and research data are derived from animal models (Mohajeri et al. 2018). Germ-free (GF) mice data revealed that brain development is abnormal without microbiota. Animal models fed with specific strains of bacteria showed changes in their behavior. It was then understood that the commensal gut microbiota could communicate with the brain, and altering their structures and composition can result in variations in mental health (Cryan et al. 2019). In the past decades, the interaction between gut microbiota and mental health has been bidirectional (Mohajeri et al. 2018), influencing each other. The gut microbiota can influence the shape and function of the brain, while the brain influences the gut microenvironment and microbial composition (Zhao et al. 2018). This bidirectional network of connection involving many biological systems that enable communication between gut microbiota and brain is termed as gut-microbiota-brain axis.

The gut-brain axis is important in maintaining homeostasis of the host's GI, central nervous, and microbial systems (Morais et al. 2021). Recent studies have suggested that efficient functional communication between gut microbiota and the brain is maintained via pathways like the hypothalamic–pituitary–adrenal (HPA) axis, the autonomic nervous system (ANS), the neuroendocrine system, the immune system, and metabolic pathways (Järbrink-Sehgal and Andreasson 2020).

Neuroactive compounds like neurotransmitters (like GABA), noradrenaline, dopamine and serotonin, amino acids, and microbial metabolites like SCFA are secreted by gut microbiota. These compounds interact with the host immune system, direct the host's metabolism, and affect the development and function of the enteric nervous system (ENS) and the vagus nerve (VN), which directly communicate with the brain (Morais et al. 2021). Alteration of neurotransmitters in GF mice is ascribed to a lack of microbial colonization, resulting in neuromuscular abnormality. This suggests that gut microbiota regulates the expression of enzymes involved in synthesizing and transporting neurotransmitters. Gut microbiota regulates the expression of brain-derived neurotrophic factor (BDNF), which involves the modulation of different brain activities and cognitive functions. Alteration of

BDNF expression can cause memory dysfunction, which was observed in GF animal studies. The signal generated from the brain to the gut can also result in the alteration of mucus and biofilm formation, gut motility, intestinal permeability, and immune functions, which ultimately affect the composition of microbiota in the gut (Carabotti et al. 2015). Changes in gut barrier integrity are found in neuropsychiatric conditions like anxiety, autism spectrum disorder, and depression (Morais et al. 2021).

Before the role of our commensal friends in the gut was determined, mental illnesses were thought to be mostly due to defects in the brain's functions. However, once the relationship between gut microbes and the brain is understood, this has led us to have a new perspective toward mental health research and lead us to new approaches to our research focusing on neuropsychiatric disorders associated with autism spectrum disorder (ASD) and schizophrenia which are known to cause development disorders; depression and anxiety which are known to influence mood disorders; Parkinson disease (PD); Alzheimer disease (AD); and multiple sclerosis (MS) associated with neurodegeneration. As the role of altered gut microflora in the causes of diseases has been revealed, many pieces of research have been carried out focusing on the role and capability of gut microbiota for ameliorating different diseases, including mental illnesses (Lee and Kim 2021).

2.2 Human Gut Microbiota

Human gut microbiota refers to the microorganisms naturally harbored by the GI system (Fan and Pedersen 2021). The presence of human gut microbiota in a GI niche creates a complex, diverse, and dynamic environment. Research has shown that the human gut microbiome remains unstable, and its composition constantly changes throughout life (Malan-Muller et al. 2018).

Previously, it was thought that the uterus was sterile; however, evidence has been emerging suggesting the colonization of the uterus by microorganisms. Therefore, humans are thought to be colonized by microbes during the very early stage of life (Rodríguez et al. 2015). The presence of bacterial species in the meconium of healthy neonates confirms microbial colonization. The microbiota composition of infants can also differ depending on the mode of delivery. Infants delivered via the vaginal birth canal receive their first major exposure from microbes inhabiting the birth canal and probably from fecal microbes, and *Bifidobacterium*, *Lactobacillus*, Enterobacteriaceae, and *Staphylococcus* mainly colonize the GI tract of such infants. Infants delivered via cesarean birth were first exposed to microbiota through their mothers' skin microbiota and the hospital environment. The gut microbiota composition of infants also varies between breastfed and formula fed. The microbiota of infants is unstable until they reach age 3 (Mohajeri et al. 2018; Borre et al. 2014; Dominguez-Bello et al. 2010; Biasucci et al. 2010).

The adolescent period is considered a susceptible stage for gut microbial composition. The gut microbiota composition in the adolescent period differs from that of infants and adults. However, the flexibility of gut microbiota composition during adolescence remains unclear; it is ascribed to hormonal changes, rapid physical body development, and exposure to new lifestyles. These factors are suspected of causing mental problems in adolescence. Studies analyzing the microbiota composition of adolescents have displayed that older adolescents are found to have microbial composition more similar to adults, and those at the early stage of their adolescent period have more gut microbial composition similar to infants. This finding has led to the idea that there is a transition and changes in the microbiota composition at different stages of life (Cryan et al. 2019).

In healthy adults, the gut microbial composition is found to be more stable, less diverse, and dominated by only four major phyla—*Firmicutes*, *Bacteroidetes*, *Proteobacteria*, and *Actinobacteria*, constituting roughly 64%, 23%, 8%, and 3% of the population, respectively, and lesser compositional changes are observed, if remain undisturbed. However, chances and several factors can always change and alter gut microbiota composition. Such factors are diet, antibiotic consumption, environment, infection, stress, and host genetics (Mohajeri et al. 2018).

The information generated by the Human Microbiome Project (HMP) showed interindividual differences in microbiota composition among healthy adults. Although all healthy guts have some degree of common microbial commensals to maintain a healthy and beneficial gut microbiome, each GI system can have slightly different microbial compositions, but not to the degree of altering the normal function of beneficial microflora. Such variations in composition can be due to racial differences, different environments (Malan-Muller et al. 2018), and intake of different foods (e.g., Western diet, which contains a high amount of sugar, salt, and fats, is found to be more associated with alteration of normal healthy microbiota compared to the Mediterranean diet which is rich in polyphenols) and physical fitness; it has been found out that moderate exercise is very effective in lowering stress levels and increasing immunity (Cryan et al. 2019).

2.3 Gut Microbiota and Brain Function

Long before, the relationship between gut microbiota and the brain was hypothesized without scientific evidence. Studies conducted in GF mice and mice in a specific pathogen-free (SPF) environment showed that GF mice displayed lesser anxiety-like behavior than their SPF mice counterpart. Adult GF mice, when moved to the SPF environment, did not increase the anxiety-like behavior of GF mice. Still, its offspring displayed an anxiety-like behavior similar to controls in the SPF environment. This finding suggested crucial time points in stages of life for the gut microbiota to influence the brain.

Several studies have revealed that gut microbiota communicates to the brain through the nervous, immune, metabolic, and endocrine systems. During the development of the brain, the influence of gut microbiota is limited only to a certain period; beyond the critical period, the action of microbiota does not have much influence on the mental development of the host (Wang and Wang 2016). Certain metabolites are synthesized by the influence of the gut microbiota, which can pass through portal circulation and interact with the neuroanatomical system, endocrine system, immunological system, and metabolic system to establish direct communication with the brain (Morais et al. 2021). However, the exact role of these metabolites in the brain is difficult to figure out due to the blood-brain barrier and other mechanisms that can interfere with the direct influence of these metabolites in the brain (Mohajeri et al. 2018).

In the neuronal pathway, the gut microbiota and the brain are physically linked (Morais et al. 2021). There are two neuroanatomical pathways in which the gut microbiota and brain communicate. The first is in the spinal cord by ANS and VN. Communication between the ENS in the gut and the ANS and VN in the spinal cord is the second pathway. The anatomical neural pathways are divided into four levels. ENS makes the first level, followed by prevertebral ganglia, which regulate peripheral visceral reflex responses; the third level is made up of the ANS in the spinal cord and brain stem nucleus tractus solitarius and dorsal motor nucleus of VN, which receive and give the origin of afferent and efferent fiber of VN, respectively, and finally in the fourth level is the higher brain centers. VN plays a major role in the direct communication between the brain and gut microbiota, in which bacteria stimulates the afferent neurons of ENS, and the vagal communication from the gut stimulates the anti-inflammatory response. According to studies in mice, the activation of the ENS pathway and regulation of gut motility is mediated by the gut microbial products like cell wall components, SCFAs, and other metabolites (Morais et al. 2021).

Studies have revealed that the gut microbiota regulates the normal development of the HPA axis. The HPA axis is considered to be the stress-efferent axis regulating the organism's adaptive responses under different kinds of stressors. HPA is part of the brain's limbic system that gets activated under stress. Studies in germ-free mice showed that activation of the HPA axis results in the secretion of stress hormones like cortisol and increased stress response. At the same time, there is decreased and improved anxiety-like behavior (Carabotti et al. 2015).

Gut microbiota also plays a crucial role in the maturation of the host immune system. Communication between the gut microbes and the host is facilitated by Toll-like receptors which can transport microbial metabolites into the nervous system (Wang and Wang 2016). Gut microbiota also regulates gut permeability, directing the gut-associated lymphoid tissue (GALT) to establish immunity against friendly commensals but develop defense barriers against pathogens (Skonieczna-Żydecka et al. 2018b). When comparing conventional mice and GF mice, microglia–conventional mice have more macrophage-like cells in the CNS than GF mice, suggesting that microbiota play an essential role in developing microglia-mediated immune systems. The expression of proteins like occludin and claudin five is found to be reduced in GF mice when compared to conventional control mice, which results in more permeability of the blood-brain barrier (BBB) and thus underscores the influence of gut microbiota in the development of immune systems (Morais et al. 2021). More permeability of BBB will result in more penetration of BBB by metabolites secreted by gut microbes and will have more impact on the brain and ultimately affect the CNS (Malan-Muller et al. 2018).

Gut microbiota can produce different kinds of neurotransmitters, neuropeptides, and their precursors, such as histamine, dopamine, acetylcholine, gamma amino acid, melatonin, serotonin, γ-aminobutyric acid, 5-HT, and butyric acid. These regulate the connection within the CNS and external connections with the endocrine and immune systems and play an important role in neural activation. Contrarily, neuropeptides like calcitonin gene-related peptide, substance P, somatostatin, neuropeptide Y (NPY), corticosterone-releasing factor (CRF), and vasoactive intestinal polypeptide (VIP) are known to regulate gut microbial activity and thus have an impact on the gut-brain axis. Bacterial products, SCFAs, are also important for the maturation of CNS and brain development. Studies in mice showed that SCFAs regulate genes that are part of microglia maturation, influence ENS activity, and regulate gut motility in rodents which induce morphology changes in mice (Zhao et al. 2018; Holzer 2016).

2.4 Factors Influencing the Microbiota-Gut-Brain Axis

Several factors (Fig. 2.1) have been proposed to influence the gut-brain axis (Cryan et al. 2019). Factors that are known to influence the microbiota gut brain axis are summarized below:

2.4.1 Host Genetics

This factor does not solely refer to the host genetics but also includes how the environment influences the host genes. Studies have shown that monozygotic twins' gut microbiota is more similar to dizygotic twins, suggesting that host genetics are important in building the microbiome (Kurilshikov et al. 2017). Gut microbiota is also known to be involved in the regulation of transcription, gene expression, and the synthesis of proteins. It has been observed that gut microbiota regulate miRNA expression in the amygdala and prefrontal cortex (PFC) of GF mice, and the expression is found to be declined. However, when the GF is conventionalized using gut microbiota, the expression becomes normal, similar to the conventional mice model (Hoban et al. 2017; Cryan et al. 2019).

2.4.2 Mode of Delivery at Birth

Delivery is considered to be the first major bacterial colonization of infants. Infants sliding through the vagina are exposed to vaginal microbiota. Studies have shown that the microbiota of babies delivered through the vaginal birth canal has a lot in common in their composition with the vaginal microbiota (Dominguez-Bello et al. 2010). When the babies do not pass through the vagina and delivery is done by C-section, the microbiota of such babies are found to be different from those delivered via the vagina. Babies delivered via C-section are considered to be colonized

Fig. 2.1 Factors affecting human gut microbiome

by microbes from the skin microbiota of the mother and hospital environments. The colonization is usually dominated by *Staphylococcus* spp. Unlike vaginal birth, C-sections are associated with decreased numbers of *Lactobacillus*, *Bifidobacterium*, and *Bacteroides* (Cryan et al. 2019). *Bifidobacterium* is considered to be important for the promotion of health. The gut of infants delivered via C-section is colonized more by *Clostridium* and *Lactobacillus* than babies born through the vaginal canal. Clostridia are considered to be harmful pathogens causing food poisoning and diarrhea. In addition, one study on 7 years old children has shown that microbial acquisition at the early stage of life greatly impacts further intestinal microbial development (Arboleya et al. 2018; Salminen et al. 2004).

It has been found that early colonization has a huge impact on later development. Some studies have concluded that C-section delivery is more associated with developing type 1 diabetes, obesity, and immune disorders such as allergies or asthma. Neuronal development is altered in babies delivered via C-section, and C-section is correlated with children's poor performance in schools. However, more studies are required to thoroughly understand the difference in the microbial composition between babies delivered through a vaginal birth and C-section and their impact on further development (Cryan et al. 2019).

2.4.3 Diet

In 1977, the association between diet and gut microbiota was first reported (Alfonsetti et al. 2022). It has been found that diet greatly influences the function and composition of the gut microbiome. Acute changes in dietary habits are found to have a larger impact on microbiota composition (Hansen and Sams 2018). Sudden changes in diet have been associated with quick changes in the gut microbial composition at species and family levels (Alfonsetti et al. 2022). Changes in dietary patterns have been reported to modulate the β-diversity of gut microbiota. Different diets have different influences on gut microbial composition. The structure and composition of gut microbiota can differ depending on the food type one consumes (Cryan et al. 2019).

The intake of standard Western diets rich in saturated and trans fats and low in mono- and polyunsaturated fats are associated with an increase in total anaerobic microflora and the relative quantity of *Bacteroides* and *Bilophila*. Studies on mice reported the richness of Actinobacteria, lactic acid bacteria, and Verrucomicrobia (Alfonsetti et al. 2022).

The Mediterranean diet has been found to reduce neurodegenerative disorders, psychiatric conditions, cancer, and cardiovascular disease (Cryan et al. 2019). The high content of polyphenols like flavonoids, anthocyanins, and phenolic acids contribute to increased *Bifidobacterium* and *Lactobacillus* genera in the gut. The antibacterial activities of these polyphenols also prevent the colonization of enteric pathogens like *Staphylococcus aureus* and *S. typhimurium* (Alfonsetti et al. 2022). Several studies have shown that the polyphenol content present in the Mediterranean diet reduces depression risk (Cryan et al. 2019).

A ketogenic diet is a type of food that is rich in fats (55–60%), moderate proteins (30–35%), and a very low carbohydrate diet (5–10%) (Batch et al. 2020). The ketogenic diet can modulate the expression of antioxidants and neurotransmitters and, thus, help in reducing the symptoms of neurological diseases like autism, depression, epilepsy, Alzheimer's, Parkinson's disease, and cancer. Consumption of a ketogenic diet is found to have some relation with the increasing abundance of *Akkermansia* and *Parabacteroides* which are found to promote ketogenic diet-mediated anti-seizure properties (Cryan et al. 2019).

Foods rich in glucose, fructose, and sucrose can significantly increase the abundance of *Bifidobacterium* with the decrease of *Bacteroides*. Indigestible carbohydrates are fermented by colon bacteria, which serve as a rich source of carbohydrates for the gut microbiota, thus, modulating the microbial composition in the gut (Cryan et al. 2019).

Protein consumption is positively correlated with overall microbial diversity. The consumption of proteins extracted from peas and whey increases the abundance of *Bifidobacterium* and *Lactobacillus*; moreover, they decrease the pathogenic *Bacteroides fragilis* and *Clostridium perfringens*. *Pisum sativum* (pea) proteins lead to an increase in intestinal SCFA levels. This illustrates the influence of proteins in the microbiota-gut-brain axis (Olson et al. 2018).

The consumption of healthy fats like polyunsaturated fatty acids (PUFAs) decreases the abundance of *Bacteroidetes* and increases *Firmicutes* and *Proteobacteria*. It has also been shown to reduce the possible onset of cardiovascular diseases and protects against depression, arthritis, cancers, and cognitive decline. On the other hand, it also supports cognitive, visual, social development, and motor in mice (Costantini et al. 2017; Cryan et al. 2019).

2.4.4 Physical Exercise

Physical exercise at moderate levels is known to benefit brain and mental health. Proper exercise in combination with healthy lifestyles is known to influence the composition of gut microbiota, improving the α-diversity of the gut microbiota. The sudden discontinuation of regular exercise harms human health and changes in plasma kynurenine and tryptophan metabolism levels, which are strongly related to depression (Cryan et al. 2019).

Rodents fed with high-fat content food were found to be prone to anxiety and cognitive problems. These problems can, however, be improved by exercise. Exercise during a juvenile period is found to have more impact on the gut microflora than in adults, with an increase in abundance of *Bacteroidetes* and a decrease of *Firmicutes*. Exercise also causes an improvement in sleep disorders (Monda et al. 2017).

2.4.5 Consumption of Medicines

Among different classes of medicines, antibiotics have the greatest ability to modulate and shape the gut microbiota. Some studies have shown that several nonantibiotic medicines like inflammatory bowel disease (IBD) medications, female hormones, benzodiazepines, osmotic laxatives, antihistamines, and antidepressants have antimicrobial properties. They influence the gut microbial composition and reduce microbial diversity (Cryan et al. 2019; Clarke et al. 2019; Falony et al. 2016).

Drugs like proton pump inhibitors (PPI) used to treat acid-related disorders are known to cause the translocation of oral microbiota to the gut. This may be due to gastric acid reduction from PPI drugs like pantoprazole and omeprazole. Therefore, these drugs can regulate gut microbiota composition to a certain level. In addition, pharmacomicrobiomics have been drawing the attention of researchers, as gut microbes can influence therapeutic efficacy and safety by enzymatically modifying drug structure and modifying drug bioavailability, bioactivity, or toxicity (Weersma et al. 2020).

2.4.6 Stress

The disturbance of the normal homeostasis of an organism can be referred to as stress. The HPA axis is recognized as a primary pathway of the stress response. Activation of HPA axis response to induced stress maintained homeostasis in the body. However, when exposed to chronic stress, this axis causes dysregulation of HPA and leads to diseased conditions. Stress causes the release of hypothalamic corticotropin-releasing hormone (CRH), and CRH promotes the release of adrenocorticotropic hormone (ACTH) from the pituitary gland. Subsequently, ACTH is secreted into the bloodstream to produce glucocorticoids (Suda and Matsuda 2022).

Experiments conducted on different animal models have revealed a connection between stress and the increasing abundance of *Lactobacilli* (Cryan et al. 2019). Chronic psychological stress has been found to have a connection with the abundance of *Helicobacter pylori* (Guo et al. 2009). Maternal stress during pregnancy also regulates infants' microbiota, which correlates with hyperreactivity of the HPA axis (Cryan et al. 2019; Hechler et al. 2019).

2.4.7 Environment

The environment can be considered one of the factors that have a significant impact on human development and health. According to studies, many gut microbiotas metabolize environmental chemicals, which modulate the gut microbial composition (Claus et al. 2016). Heavy metals, pollutants, and pesticides may be toxic to some microorganisms and lead to dysbiosis. The changes in composition and activity of the gut microbiota interfere with the normal intestinal epithelial-barrier function and increase the risk of causing mental health. Overuse of antibiotics in our environment results in the accumulation of antibiotics in rivers, lakes, agricultural land, etc., which can indirectly cause the alteration of gut microbial composition (Cryan et al. 2019).

2.4.8 Circadian Rhythms

With our modern lifestyle, there are many chances of disturbing our natural circadian rhythm, which is known to be associated with metabolic and psychiatric disorders (Arble et al. 2010). Studies have revealed that microbiota regulates the circadian clock, affecting peripheral and central clock changes. The disruption of the circadian clock also influences the composition of gut microbiota. Lifestyle change can alter the peripheral clock, resulting in dysregulation of the gut microbiome (Cryan et al. 2019; Voigt et al. 2014). The disruption of circadian rhythm is more associated with the population following the Western lifestyle, influenced by different factors like sleep schedule, work shifts, time of eating, exposure to light at night, and jet lags (Bishehsari et al. 2020).

2.4.9 Consumption of Alcohol

Consumption of alcohol is also found to be associated with the disturbance of the maintenance of gut microbial homeostasis. Once consumed, alcohol is converted into acetaldehyde which is harmful to gut microbiota. It decreases the abundance of SCFA-producing microbes, thus influencing the microbiota-gut-brain axis (Alfonsetti et al. 2022). Studies on rats identified the direct influence of alcohol on gut microbiota composition, which causes a decrease in α- and β-diversity, reduced abundance of *Lactobacilli* and increased *Bacteroidetes* (Lee and Lee 2021).

2.5 Meta-Omics and Gut Microbiota Analysis

The recent development in molecular biology and the emergence of meta-omics have paved the way for a better understanding of the role and function of commensals in the human gut microbiome—a complex environment (Wang and Wang 2016).

"Metagenomics is the study of genetic material retrieved directly from environmental samples including the gut, soil, and water etc.." It allows the characterization of the taxonomic composition and the functional metabolic potential of the microbiota and the reconstruction of microbial metabolic pathways, which were impossible through 16S rRNA (Malan-Muller et al. 2018). It is a culture-independent technique to study microbial colonies in the environment (Wang and Wang 2016). Metagenomics aims to catalog all the genes by randomly sequencing all DNA extracted from the environmental sample. In the case of gut microbiota, metagenomics helps us to analyze the genetic composition of the population in the target environment, understand the function and role of gut microbiota in metabolic pathways, estimate the diversity and abundance of the microbial population, discover and study novel genes with specialized functions, as well as explore the link between microbiota and the environment, the connection between the gut microbiota and the host. It can also be used to study individual medicine (Wang and Wang 2016).

According to MetaHIT and Human Microbiome Project (HMP), functional gene profiles are similar among different individuals. However, the taxonomic composition of the microbiota varies, which suggests that the functional core microbiome is more conserved when compared to the taxonomical core microbiome (Malan-Muller et al. 2018). Through metagenomics, we can now detect genes in the sample and describe the presence of microorganisms in the environment. However, it isn't easy to decipher their activity and expression using metagenomics. The development of other meta-omics approaches like metatranscriptomics, metaproteomics, and metabolomics can identify the functional activity of microbes present in the gut (Wang et al. 2015).

In metatranscriptomics, the expression of genes by the microbial population is analyzed. It provides information about the expression of genes at a specific point in time (Malan-Muller et al. 2018). By utilizing metatranscriptomics, energy production, synthesis of cellular components, and carbohydrate metabolism are found to be the main functional roles of the gut microbiota (Wang et al. 2015). However, the data acquired from metatranscriptomics are considered insignificant because the RNA transcript pool is very unstable and easily responds to environmental changes (Malan-Muller et al. 2018).

In metaproteomics, the analysis of proteins is done to have a deeper understanding of the functions of gut microbiota. The protein profiles are obtained and compared with different proteins at different physiological conditions (Malan-Muller et al. 2018). Metaproteomics also has some disadvantages related to the complexity of the protein matrix and the microorganisms expressing the proteins (Issa Isaac et al. 2019).

Metabolomics analyzes microbiota-derived metabolites in serum, urine, or fecal water, constructing a mass spectrometry-based library to enable either global metabolite analysis (untargeted approach) or the measure of a selected metabolite. Metabolomics is usually carried out using MS-based techniques like gas chromatography or liquid chromatography to discriminate metabolites based on their mass-to-charge (m/z) ratio and ^{1}H nuclear magnetic resonance (^{1}H NMR) spectroscopy (Malan-Muller et al. 2018).

2.6 The Emerging Role of the Gut-Brain Axis

Various studies based on animal models have shown that microbes residing in the gut bidirectionally communicate with the brain and play a crucial role in maintaining the proper function of the CNS. Studies in human and animal models have revealed that the lower abundance of *Bifidobacterium* in the gut is associated with obesity. Obesity and depression have been linked to low-grade inflammation. Therefore, these studies have suggested that the insufficient abundance of *Bifidobacterium* is strongly related to depression (Naseribafrouei et al. 2014). Desbonnet et al. (2014) also found that the gut microbiota is crucial for social development in mice, including social motivation and preference for social novelty. Such developments are affected in diseases like autism and schizophrenia which thus help us to understand better these neurodevelopmental disorders (Desbonnet et al. 2014).

It has been found that changes in the gut microbiome affect brain function and behavior. This results in the rise of the idea that the optimization of the gut microbiome can act as a therapeutic tool for the amelioration of certain mental diseases (Skonieczna-Żydecka et al. 2018a).

Currently, probiotics are commonly employed to optimize the composition of gut microbiota. Such probiotics are mainly equipped with *Lactobacillus* and *Bifidobacterium*. According to several studies, the consumption of probiotics is found to be related to the reduction of anxiety, reduction of emotional changes (Malan-Muller et al. 2018), fewer stress-induced symptoms like abdominal pain, nausea and vomiting, reduction of cognitive reaction to depression, and decreasing the severity of autism and also found to have a beneficial effect on mood disorders (Sivamaruthi et al. 2019). The beneficial aspects of probiotics on different disorders, including mental disorders, have been figured out. This has led to the emerging concept of developing next-generation probiotics by seeking broader candidates that could provide more benefits to humans (O'Toole et al. 2017).

2.7 Human Diseases and Gut-Brain-Axis

2.7.1 Gut-Brain Axis and Autism Spectrum Disorders (ASD)

ASD is a group of neurodevelopmental conditions characterized by stereotyped behaviors and activities and altered social communication and character (Xu et al. 2019). ASD is related to GI disorders like constipation, abdominal pain, gaseousness, diarrhea, and flatulence (Xu et al. 2019). Anxiety behavior has been found to have a strong link with ASD patients having GI disorders (Srikantha and Mohajeri 2019).

Patients having ASD show an alteration of gut microbiota (Xu et al. 2019). One study has shown that treating ASD children with Vancomycin improved behavioral symptoms (Cryan et al. 2019), suggesting that gut microbiota may be related to behavioral and GI symptoms correlated with the severity of ASD. The availability and diversity of nutrients and microbial metabolites can be affected by the changes in metabolic profiles due to changes in the microbiome (Sharon et al. 2019). Xu et al. (2019) showed that ASD patients have a lower abundance of *Bacteroides*, *Akkermansia*, *Bifidobacterium*, *Enterococcus*, and *E. coli*; a higher abundance of *Faecalibacterium* and *Lactobacillus*; and a slightly increased of *Ruminococcus* and *Clostridium*. The increase in abundance of *Faecalibacterium* has a strong relationship with the activation of type I interferon signaling and may be involved in immune dysfunction. The decrease in *Bifidobacterium* may result in the decrease of SCFAs involved in the development of ASD. The lower abundance of *Akkermansia* may also indicate an increase in gut permeability. In general, the alteration of gut microbial composition can also result in the production of neurotoxins which can worsen the symptoms of ASD (Xu et al. 2019).

2.7.2 Gut-Brain Axis and Depression

Major depressive disorder (MDD) is one of the major causes of disability, morbidity, and mortality worldwide (Liang et al. 2018). It is a serious mental health issue characterized by the symptoms like anhedonia, altered appetite, anxiety, depressed mood, fatigue, insomnia, irritability, and suicidal ideation (Suda and Matsuda 2022). MDD is not simply a mental problem but also a physiological disease having a clear biological foundation, as changes in the brain's normal functioning like abnormal neuronal circuitry, unbalanced neurotransmitters, hampered neurogenesis, and neuroplasticity decline are observed (Liang et al. 2018). It disrupts the normal program of neurotransmitters, neurogenesis, neural circuits (Chaudhury et al. 2015), and neuroplasticity (Liu et al. 2017). Experiments conducted on GF mice have shown that the absence of microbiota reduces depressive-like behavior (Cryan et al. 2019). MDD correlates with the HPA axis activation (Cryan et al. 2019). Abnormal stress is found to be correlated with MDD. The HPA-mediated stress response influences the development and progress of depressive symptoms (Suda and Matsuda 2022). Such hyperactivation of the HPA axis is found to decline when treated with probiotic strain microorganisms like *Lactobacillus*, which display the role and influence of microbiota in MDD. A clinical trial has shown that the administration of probiotics in combination with prebiotics decreases the Beck Depression Inventory (BDI) with a significant decrease and increase in kynurenine/tryptophan ratio and tryptophan/branch chain amino acids (BCAAs), respectively (Kazemi et al. 2019).

People having depression tend to have a lower abundance of *Faecalibacterium* and a reduction in the numbers of microorganisms (Cryan et al. 2019), but an increase in the abundance of *Actinobacteria*, *Eggerthella*, *Atopobium*, and *Bifidobacterium* (Knudsen et al. 2021). *Faecalibacteriumis* was reported to be the main producer of metabolites like butyrate in the gut. Butyrate regulates the level of BNDF and neurogenesis in the hippocampal. BNDF level is lower in patients with MDD, most probably due to the decrease in the abundance of *Faecalibacterium*. An increase in butyrate level can reduce depressive-like behavior; therefore, it is expected that the alteration of gut microbiota to increase the production of butyrate will improve the symptoms of MDD, especially depressive symptoms (Suda and Matsuda 2022).

2.7.3 Gut-Brain Axis and Schizophrenia

Schizophrenia is a debilitating psychiatric condition that causes many emotional, occupational, and cognitive problems (Szeligowski et al. 2020). It is characterized by complex, heterogeneous behavioral and cognitive syndrome with positive symptoms-delusions, hallucinations, the aberrant flow of thoughts, and negative symptoms—apathy, withdrawal, and slowness (Owen et al. 2016). Schizophrenia is found to be associated with the elevation of *Lactobacilli* (Szeligowski et al. 2020).

Some studies have revealed that altering gut microflora using probiotics can improve bowel problems. However, no complete successful treatment of the disease by altering gut microbiota composition has been reported (Cryan et al. 2019). Schizophrenia is associated with increased IL-6, IL-8, and TNF-α and reductions in the anti-inflammatory IL-10 (Miller et al. 2011). It has also been shown that *Roseburia*, *Coprococcus*, and *Blautia* were also reduced in schizophrenia patients (Shen et al. 2018), which are known to be involved in maintaining the intact intestinal barrier. Therefore, the increase in gut permeability may contribute to the inflammation associated with schizophrenia (Szeligowski et al. 2020).

Schizophrenia is also associated with the disturbance of the immune system by converting tryptophan to kynurenate. The conversion of tryptophan to kynurenate mediated by indoleamine2,3-dioxygenase (IDO) is known to weaken the immune system by reducing prepulse inhibition and increasing the firing rate burst-to-fire activity of ventral tegmental area (VTA) dopaminergic neuron. Studies in rats have shown that administering *Bifidobacterium* infants can increase the kynurenate. Therefore, the gut microflora may regulate the presence and abundance of tryptophan (Szeligowski et al. 2020).

2.7.4 Gut-Brain Axis and Bipolar Disorder (BD)

Bipolar disorder (BD) is another neuropsychiatric illness characterized by alternating recurrent manic and depressive (Järbrink-Sehgal and Andreasson 2020) or shifts in mood and energy throughout the disease. Episodes of mood shifts are often associated with low-grade peripheral inflammation (Flowers et al. 2020). BD patients with more depressive symptoms are reported to have less bacterial diversity (Bengesser et al. 2019). In addition, *Bacteroides*, *Clostridium*, *Bifidobacterium*, *Oscillibacter*, and *Streptococcus* are found to be more in abundance in a fecal sample of BD patients with more symptoms (Rong et al. 2019), which may improve inflammatory bowel disease, nonalcoholic steatohepatitis, and other psychiatric disorders like depression (Flowers et al. 2020).

2.7.5 Gut-Brain Axis and Addiction

Substance use disorder (SUD) is a mental condition affecting the brain (Russell et al. 2021) that alters circuitry involved in learning, memory, motivation, reward, and stress (Ren and Lotfipour 2020). SUDs represent one of the main public health challenges (Russell et al. 2021). The medications available for the treatment of SUDs are very limited, and the approved treatment of psychostimulants is not even available (Meckel and Kiraly 2019).

Substance addiction can affect many pathways that influence brain function. Out of all pathways, alteration of the dopaminergic system is common for all substance abuse. Therefore, substance abuse causes an increase in the level of dopamine. SUDs are associated with the disruption in the normal function of the mesolimbic pathway, sometimes also referred to as the reward system or pathway (Russell et al. 2021). SUDs are often related to increased intestinal permeability, which allows the transfer of location for the gut microflora (Leclercq et al. 2014). This translocation can thus lead to local and systemic inflammation. Tetrahydrocannabinol can also modify the composition of gut microbiota. These suggest the interconnection between addiction and gut microbes (Russell et al. 2021). The administration of methamphetamine in rodents was found to cause a decrease in the abundance of propionate-producing bacteria (Ning et al. 2017). Gut-brain communication through VN is critical in reward and motivation, and the gut microbes respond to rewards like drugs and foods. Natural rewards like food, sex, etc. are influenced by dopamine, GABA, VTA, etc.; however, the rewards driven by substances are unnatural and can cause neurophysiological changes that lead to addiction (Ren and Lotfipour 2020).

2.7.6 Gut-Brain Axis and Parkinson's Disease

Parkinson's disease (PD) is one of the most common neurodegenerative disorders. PD is considered a multisystemic disease affecting both the central nervous system (CNS) and peripheral nervous system (PNS), which results in non-motor symptoms like gastroparesis and constipation (Romano et al. 2021). The motor symptoms are usually preceded by GI illness. By the time the motor symptoms occur, the dopaminergic neurons in the substantia nigra have already been destroyed (Cheng et al. 2010).

Studies have shown that gut microbiota may be naturally related to the symptomatology and pathophysiology of PD. Fecal microbiota transplantation (FMT) from PD patients into GF mice showed overexpression of protein α-synuclein. This is followed by motor symptoms and neuroinflammation, which can be treated and improved by administering antibiotics. This finding highly suggests the influence of gut microbiota in PD. It has been further suggested that α-synuclein is transported to the brain via VN, which may be exerted by the influence of microbiota (Sampson et al. 2016).

PD patients have a strong relationship with increased gut permeability and neuroinflammation. This could be due to the decrease in the abundance of SCFAs-producing bacteria. Several studies have shown that the genus *Akkermansia*, *Bifidobacterium*, *Hungatella*, and *Lactobacillus* increased in PD patients. At the same time, the abundance of bacteria belonging to the Lachnospiraceae family and *Faecalibacterium* (Ruminococcaceae family) went down. However, all studies did not come to the same conclusion, and there are inconsistencies among those studies. In the normal gut environment, *Lactobacillus* strains are low-abundant members. PD patients with constipation symptoms have a higher abundance of *Akkermansia*. The increase in *Akkermansia* is due to the consequence of constipation. When the gut microbiota composition is unbalanced, it can also be suggested that *Akkermansia*, a mucin-degrading bacteria, might gain a chance to lead to the disruption of the intestinal mucus layer, decreased number of goblet cells, drier stools, and impaired intestinal barrier function (Romano et al. 2021).

2.7.7 Gut-Brain Axis and Anxiety

Anxiety is a debilitating psychiatric condition that is closely related to depressive disorders. The treatments available have been increasing quickly; however, the burden it causes remains the same. The gut-brain axis is a promising area of research to ameliorate anxiety disorders (Simpson et al. 2021).

The influence of gut microbiota on anxiety was shown by performing fecal transplantation from a high-anxiety mouse strain to low-anxiety mouse strain, which resulted in the development of anxiety-like behavior in a low-anxiety mouse strain (Malan-Muller et al. 2018; Bercik et al. 2011). The analysis of a fecal sample of patients with generalized anxiety disorders has shown fewer operational taxonomic units (OTU); lower bacterial α-diversity; lower abundance of *Firmicutes*, *Tenericutes*, and SCFAs producers; and excessive abundance of *Escherichia*, *Shigella*, *Fusobacterium*, and *Ruminococcus gnavus* (Järbrink-Sehgal and Andreasson 2020).

Anxiety disorders are strongly associated with the dysregulation of the HPA axis, which is casually known to be related to increased levels of cortisol and a pro-inflammatory state mediated by gut microbes. As the communication is bidirectional, the pro-inflammatory states can also negatively impact the gut, like increasing gut permeability which can result in the translocation of bacteria and their metabolites to the bloodstream causing CNS inflammation (Simpson et al. 2021; Foster et al. 2017).

2.7.8 Gut-Brain Axis and Anorexia Nervosa

Anorexia nervosa (AN) is a devastating eating disorder marked by a distorted body image, severe dietary restriction, significant weight loss, and mental comorbidities (Ghenciulescu et al. 2021). It has one of the highest mortality rates among mental illnesses (Cryan et al. 2019; Arcelus et al. 2011). AN patients are often associated with comorbid anxiety and depression. It is considered a multifactorial disease involving biological, physiological, and sociocultural factors (Schepici et al. 2019; Gorwood et al. 2016).

AN is strongly associated with a change in the habit of eating, which subsequently alters the gut microbial composition. The change in the gut microbial composition will lead to dysbiosis, which can then influence AN symptoms like greater weight loss, eating behavior, mood, and intestinal physiology. It has also been suggested that gut microbes may manipulate host appetite to gain benefits, as they completely depend on their hosts. The regulation of eating behavior by gut microbes may depend on the production of molecules that can modulate the neurohormones involved in mood and eating behavior or act directly as neurohormone-like molecules (Gorwood et al. 2016; Breton et al. 2019).

2.7.9 Gut-Brain Axis and Alzheimer's Disease (AD)

Alzheimer's disease (AD) is a neurodegenerative disorder and the leading cause of degenerative dementia, affecting about (Sharon et al. 2019) million people globally. AD is caused by aggregating polymerized forms of β-amyloid precursor protein (Ab) in the brain's soluble multimeric and/or insoluble amyloid deposits (Cryan et al. 2019). It is associated with impaired cognition and cerebral accumulation of amyloid-β peptides (Aβ) (Smith et al. 2013).

The stool microbial profile of AD patients showed a decrease in the abundance of *Firmicutes* and *Actinobacteria* and an increase in the abundance of Bacteroidetes. Among the *Firmicutes*, the families of *Ruminococcaceae*, *Turicibacteraceae*, and *Clostridiaceae* are found to be less abundant. Studies based on GF-APP/PS1 and conventional APP/PS1 transgenic mice have shown that the GF-APP/PS1 have lower levels of Aβ compared to conventional APP/PS1. Fecal transfer from conventional APP/PS1 to GF-APP/PS1 significantly increased GF-APP/PS1's cerebral Aβ pathology. It has also been found that co-housing and FMT can transfer the neuroinflammation and cognitive impairment from 5xFAD mice (diseased mice) to control mice. The decrease in the levels of SCFA produced is also found to be associated with AD. These findings support the connection between the gut-brain axis and the development of AD (Smith et al. 2013, Rutsch et al. 2020).

Studies in GF mice have shown a decrease in the expression of BNDF in the hippocampus, which is important for synaptic plasticity and cognitive function. BNDF is also found to be decreased in the brain and sera of AD patients. Antibiotic treatment of rats induced microbial dysbiosis, spatial memory impartments, increased anxiety-like behavior, and decreased N-methyl-D-aspartate (NMDA) receptor levels in the hippocampus, which can later be improved when the antibiotic treatment is disrupted with a change in diet with decreased BNDF levels. These findings also suggest the role of microorganisms in the pathogenesis of AD (Harach et al. 2017).

2.7.10 Gut-Brain Axis and Multiple Sclerosis(MS)

Multiple sclerosis (MS) is a chronic and inflammatory demyelinating CNS disease affecting more than 2 million people worldwide. The main pathological characteristic of MS is axonal loss, neuroinflammation, demyelination, and infiltration of lymphocytes into the CNS. Symptoms accompanying MS include ataxia, hyperreflexia, cognitive difficulties, loss of coordination, visual and sensory impairment, fatigue, and spasticity. Most patients experience relapsing-remitting multiple sclerosis (RE-MS) with a more severe neurological disorder (Smith et al. 2013). The factors involved in developing the disease can be environmental and genetic. Among the environmental factors, the intestinal microbiota is considered a potential pathogenic factor (Jiang et al. 2017).

Patients with active MS are known to have altered gut microbiota compared to patients with RE-MS. The microbial compositions of RE-MS are more related to healthy controls (Smith et al. 2013). Studies have found that transferring fecal matter from MS patients to GF mice caused the host GF mice to have MS hallmark symptoms, i.e., autoimmune encephalomyelitis (Cryan et al. 2019). FMT is also found to revert severe constipation and improve MS symptoms. These findings suggest the relationship between the gut-brain axis and MS disease development (Jiang et al. 2017).

Studies have shown that fecal samples of MS patients are strongly associated with a higher abundance of *A. muciniphila* and *Acinetobacter calcoaceticus*, and reduced levels of *Parabacteroides distasonis* are known to have anti-inflammatory properties (Cekanaviciute et al. 2017). Studies on murine models have shown that MS is associated with the lesser production of IL-10 regulatory cytokines than healthy controls, which suggests that gut microbiota may be responsible for disease severity and the modulation of the adaptive immune response during disease development. More research is required to understand better the mechanisms of different

Table 2.1 Neurological diseases and observed alteration of their gut microbial taxonomy

Disease	Microbes increased in diversity	Microbes decreased in diversity	References
Alzheimer's disease	Bacteroidetes	Firmicutes and Actinobacteria	Pistollato et al. (2016)
Depression	*Anaerofilum*, *Eggerthella*, *Holdemania*, *Gelria*, *Paraprevotella*, *Turicibacter*	*Dialister*and *Prevotella*	Kelly et al. (2016)
Parkinson's disease	*Blautia*, *Coprococcus*, *Proteobacteria*, and *Roseburia*	–	Keshavarzian et al. (2015)
Autism spectrum disorder (ASD)	*Clostridium* sp., *Bacteroidetes*, *Lactobacillus*, *Desulfovibrio*	*Bifidobacteria*	Adams et al. (2011), Song et al. (2004)
Anxiety	*Escherichia*, *Shigella*, *Fusobacterium*, and *Ruminococcusgnavus*	*Firmicutes*, *Tenericutes*	Järbrink-Sehgal and Andreasson (2020)
Bipolar disorder (BD)	*Bacteroides*, *Clostridium*, *Bifidobacterium*, *Oscillibacter*, and *Streptococcus*	–	Rong et al. (2019)
Schizophrenia	*Lactobacilli*	*Roseburia*, *Coprococcus*, and *Blautia*	Szeligowski et al. (2020), Shen et al. (2018)
Multiple sclerosis (MS)	*A. muciniphila* and *Acinetobacter calcoaceticus*	*Parabacteroides distasonis*	Cekanaviciute et al. (2017)

gut microbiota in different patients and develop effective therapeutics (Smith et al. 2013) (see Table 2.1).

2.8 Gut Microbiota–Inflammasome–Brain Axis

Inflammasomes are complexes of multi-protein whose main function involves the activation of caspase-1, which cleaves and activates inactive pro-IL-1β and pro-IL-18 (Chen 2017). The inflammasome is an innate immune signaling complex that gets activated and assembled by responding to pathogens or danger signals (Smith et al. 2013). Various pattern-recognition receptors (PRRs) in different families, including AIM2, NLRP1, NLRC3, NLRC4, NLRP6, and NLRP7, have been identified to play a role in inflammasome activation (Ma et al. 2019). Generally, inflammasome activation is initiated by two signals. The initial signal originates from outside the cell through pathogen- or danger-associated molecular patterns (PAMPS/DAMPS), which trigger the transcription of genes encoding inflammasome components and products. The second signal arises from internal danger signals such as adenosine triphosphate, uric acid, fatty substances that can cause lysosomal damage, or reactive oxygen species produced by nicotinamide adenine dinucleotide phosphate oxidase or mitochondria. Inflammasomes are formed and activated due to these processes (Smith et al. 2013).

Activation of inflammasome has been found to have a strong connection with neuroinflammatory conditions and an important role in the progression of neurological disorders, including AD, MS, PD, and neuropsychiatric disorders (NPS) (Smith et al. 2013). According to Wong et al. (2016), caspases-1-deficient mice show anxiety and depressive-like behaviors, while locomotor activity and skills are enhanced. Pharmacological caspase-1 antagonism with minocycline, which suppresses the inflammasome activation, ameliorated stress-induced depressive-like behavior in wild-type mice and altered the gut microbial composition. The microbiota composition alterations are similar to those in caspase-1–deficient mice. The relative abundance of *Akkermansia* spp. and *Blautia* spp. were observed in the gut microbiota, which is compatible with the beneficial effects of reduced inflammation, and rebalances, respectively, increase in Lachnospiraceae abundance was consistent with caspase-1 deficiency microbiota changes. These findings suggest that caspase-1 inhibition protects against depressive- and anxiety-like behavior by modulating the relationship between stress and gut microbiota composition. They also lay the groundwork for a gut microbiota–inflammasome–brain axis, in which the gut microbiota modulates brain function via inflammasome signaling. Inflammasome inhibition may also constitute a viable and direct therapeutic option in treating MDD and other neuropsychiatric illnesses with inflammatory components (Wong et al. 2016).

In the intestine, constant stimulation of inflammasome, which may have a distal effect on the brain, occurs due to gut microbes. It was found that producing IL-18 is vital for maintaining homeostasis in the gut (Macia et al. 2015). IL-1 and IL-18 are also important for physiological functioning in the CNS, as they are involved in cognitive, learning, and memory processes (Tsai 2017). The revelation that *Salmonella* leucine-rich repeat protein (SlrP) suppresses *Salmonella* virulence and the typical host anorexic response generated by infection was the first step in understanding intestinal inflammasome activation by gut microbiota and its effect on the brain. The *S. typhimurium* effector SlrP suppressed anorexia produced by IL-1β to the hypothalamus via the VN by inhibiting inflammasome activity. Pathogen-mediated anorexia inhibition boosted host survival rather than impairing host defenses (Rao et al. 2017).

2.9 Modulation of Gut Microbiome Using Probiotics and Prebiotics

Data from several studies have revealed that the gut-brain axis can be modulated by using probiotics and prebiotics or combining both probiotics and prebiotics, known as synbiotics which can give beneficial aspects to the brain and the host microbiome (Liu et al. 2015).

Probiotics are "live microorganisms that, when administered in adequate amounts, confer a health benefit on the host" (Hill et al. 2014). The probiotics available today for the commoners are mainly from *Lactobacillus* spp. and *Bifidobacterium* spp. It can also include *Saccharomyces*, *Bacillus* spp., *E. coli*, *Enterococci*, and

Weissella spp. (O'Toole et al. 2017). Microorganisms used as probiotics must be well-defined and cannot be extrapolated to other strains. It is necessary to identify the beneficial aspects the selected strains can provide (Sánchez et al. 2017). Several studies have shown that oral administration of commensal bacterial strains improves or reverses diseases. For instance, the administration of *Bacteroides fragilis* reversed the abnormalities in gut permeability and ASD-related behaviors. The beneficial effects of the administration of probiotic strains end up in the human gut and affect human brain activity. For example, the administration of *Lactobacillus casei* results in a significant increase of *Lactobacillus* and *Bifidobacterium* and a significant improvement in anxiety symptoms (Liu et al. 2015).

Treating *Citrobacter rodentium*-infected mice with the combination of *Lactobacillus rhamnosus* R0011 and *Lactobacillus helveticus* R0052 restored the expression of BNDF and hippocampal c-Fos. It improved the level of corticosterone and IFN-γ. Thus, this shows that probiotics can provide beneficial aspects through neural pathways. The gut microbiota can also produce a lot of important metabolites for the maintenance of two-way communication like GABA secreted by *Bifidobacterium* and *Lactobacillus*; norepinephrine secreted by *Bacillus*, *Escherichia*, and *Saccharomyces*; serotonin produced by *Candida*, *Enterococcus*, *Escherichia*, and Streptococcus; and dopamine from *Bacillus* and *Serratia* (Liu et al. 2015).

Studies have therefore shown that gut microbiome can be manipulated using probiotics. However, the communication and influence on the brain cannot be specified. It has been understood that gut-brain communication can occur via immune response, metabolite, and vagus nerve-mediated pathways (Liu et al. 2015).

According to Umu et al. (2017), "Prebiotics are a sub-group of dietary fibers with resistance to gastric acidity and the digestive enzymes of mammals, which confer various health benefits." The property of prebiotics showing resistance against gastric acids and digestive enzymes is considered advantageous over probiotics with survival limitations in the human GI tract (Liu et al. 2015). The main aim of prebiotics is to stimulate the growth and activities of commensal gut microbes, which can confer health benefits to the human host (Umu et al. 2017).

Dietary carbohydrates fermented by the gut microbiota take a large place in the human daily food intake. They are known to enhance the production of microbial metabolites, mainly acetate, butyrate, and propionate in the gut, which have a variety of health benefits (Umu et al. 2017). Plant polysaccharides like arabinoxylan influence the gut microbiome by increasing the abundance of butyrate-producing bacteria like *Roseburia intestinalis*, *Eubacterium rectale*, *Anaerostipes caccae*, etc. (Chen et al. 2019).

Prebiotics like Bimuno-galactooligosaccharides (B-GOS) lowers the cortisol awakening reactivity, which is found to be high in depressive persons and increases attentional vigilance. This suggests the modulation of HPA axis activity and can further state that B-GOS administration may have an anti-depression effect (Umu et al. 2017). Similar to probiotics, the administration of prebiotics is also known to increase the expression of BNDF and thus shows similar benefits to probiotics (Liu et al. 2015). Studies in pigs have shown that dietary fiber and algal polysaccharides such as alginates, agars, and carrageenans can increase the abundance of SCFA-producing bacteria like *Roseburia*, Ruminococcus, and *Lachnospira*. However, in humans, Bifidobacteria are the SCFA producer found to be increased (Umu et al. 2017).

Prebiotics can influence the gut-brain axis by modulating the composition of gut microbiota and their metabolites and influencing the secretion of neurochemicals (Liu et al. 2015). The construction of a biased microbiome that the beneficial microbiota colonized is targeted with prebiotics. However, more research and data would be required to understand the mechanisms by which the prebiotics influence the growth of beneficial microbes and identify the exact target to improve and maintain a healthy and beneficial gut microbiome (Shumin et al. 2020).

2.10 Conclusions

During the past decade, many studies have been conducted to understand how gut microorganisms affect communication between the gut and the brain. It is now well accepted that the gut microbiota plays a crucial role in brain function's proper growth and upkeep. Additionally, growing evidence links the microbiota to several mental, neurological, and neurodegenerative illnesses. This data comes from both animal and clinical investigations. Although much hypothesis exists, it is currently unknown whether alterations in the microbiota are crucial to the pathogenesis of at least certain mental and neurological illnesses. Targeting the microbiome has only been proven to enhance clinical outcomes in placebo-controlled trials for IBS, the only clinical condition.

Additionally, there are still many unanswered questions about psychobiotics, and much more research is needed to determine the best strain, dosage, and timing for therapeutic uses. Moving away from correlative research and toward prospective longitudinal investigations, causal and mechanistic analyses, and larger-scale trials of potential therapeutic techniques will be crucial for the field. Studies on a broad range of illnesses will undoubtedly be made available soon, which is a promising development for the potential of therapeutic uses for the microbiota. Identifying a healthy microbiome is one of the major problems with microbiota-based medicine. Targeting the microbiota with a "one size fits all" strategy is difficult because of the wide interindividual variations in microbiome makeup. However, it also presents prospects since, in the future, the microbiota might serve as a conduit for efficient, tailored medicinal techniques. Given the importance of diet in altering the microbiota, we may focus on a food-microbiota-gut-brain axis in regulating health and disease across the life span.

It is anticipated that a greater understanding of the close connection between the gut and the brain would significantly impact the field of psychiatry in particular. Given the growing body of evidence, future discussions of MH should consider immunology, microbiology, and GI pathophysiology. Their use will probably enhance both physical and mental health. We predict that newly found probiotics and other psychobiotic preparations will soon be regularly used in a psychiatrist's pharmacopeia. We regrettably do not yet have specific therapies aimed at the gut microbiota to suggest for treating particular mental illnesses, as is evident from our evaluation of the present literature. This should not stop researchers from investigating other ways to affect the intestinal microbiome in seeking mental symptom treatment, such as diet modification and psychobiotic supplementation.

References

Adams JB, Johansen LJ, Powell LD, Quig D, Rubin RA (2011) Gastrointestinal flora and gastrointestinal status in children with autism–comparisons to typical children and correlation with autism severity. BMC Gastroenterol 11:1–3

Alfonsetti M, Castelli V, d'Angelo M (2022) Are we what we eat? Impact of diet on the gut-brain axis in Parkinson's disease. Nutrients 14:380

Arble DM, Ramsey KM, Bass J, Turek FW (2010) Circadian disruption and metabolic disease: findings from animal models. Best Pract Res Clin Endocrinol Metab 24:785–800

Arboleya S, Suárez M, Fernández N, Mantecón L, Solís G, Gueimonde M et al (2018) C-section and the neonatal gut microbiome acquisition: consequences for future health. Ann Nutr Metab 73(Suppl 3):17–23

Arcelus J, Mitchell AJ, Wales J, Nielsen S (2011) Mortality rates in patients with anorexia nervosa and other eating disorders. A meta-analysis of 36 studies. Arch Gen Psychiatry 68:724–731

Batch JT, Lamsal SP, Adkins M, Sultan S, Ramirez MN (2020) Advantages and disadvantages of the ketogenic diet: a review article. Cureus 12:e9639

Bengesser SA, Mörkl S, Painold A, Dalkner N, Birner A, Fellendorf FT et al (2019) Epigenetics of the molecular clock and bacterial diversity in bipolar disorder. Psychoneuroendocrinology 101:160–166

Bercik P, Denou E, Collins J, Jackson W, Lu J, Jury J et al (2011) The intestinal microbiota affect central levels of brain-derived neurotropic factor and behavior in mice. Gastroenterology 141(599–609):609.e1–609.e3

Berg G, Rybakova D, Fischer D, Cernava T, Vergès MC, Charles T et al (2020) Microbiome definition re-visited: old concepts and new challenges. Microbiome. 8:103

Biasucci G, Rubini M, Riboni S, Morelli L, Bessi E, Retetangos C (2010) Mode of delivery affects the bacterial community in the newborn gut. Early Hum Dev 86:13–15

Bishehsari F, Voigt RM, Keshavarzian A (2020) Circadian rhythms and the gut microbiota: from the metabolic syndrome to cancer. Nat Rev Endocrinol 16:731–739

Borre YE, O'Keeffe GW, Clarke G, Stanton C, Dinan TG, Cryan JF (2014) Microbiota and neurodevelopmental windows: implications for brain disorders. Trends Mol Med 20(9) (Sharon et al., 2019):509–518

Breton J, Dechelotte P, Ribet D (2019) Intestinal microbiota and anorexia nervosa. Clin Nutr Exp 28:11–21

Carabotti M, Scirocco A, Maselli MA, Severi C (2015) The gut-brain axis: interactions between enteric microbiota, central and enteric nervous systems. Ann Gastroenterol 28:203–209

Cekanaviciute E, Yoo BB, Runia TF, Debelius JW, Singh S, Nelson CA et al (2017) Gut bacteria from multiple sclerosis patients modulate human T cells and exacerbate symptoms in mouse models. Proc Natl Acad Sci U S A 114:10713–10718

Chaudhury D, Liu H, Han MH (2015) Neuronal correlates of depression. Cell Mol Life Sci 72:4825–4848

Chen GY (2017) Regulation of the gut microbiome by inflammasomes. Free Radic Biol Med 105:35–40

Chen Z, Li S, Fu Y, Li C, Chen D, Chen H (2019) Arabinoxylan structural characteristics, interaction with gut microbiota and potential health functions. J Funct Foods 54:536–551

Cheng HC, Ulane CM, Burke RE (2010) Clinical progression in Parkinson disease and the neurobiology of axons. Ann Neurol 67:715–725

Clarke G, Sandhu KV, Griffin BT, Dinan TG, Cryan JF, Hyland NP (2019) Gut reactions: breaking down xenobiotic-microbiome interactions. Pharmacol Rev 71:198–224

Claus SP, Guillou H, Ellero-Simatos S (2016) The gut microbiota: a major player in the toxicity of environmental pollutants? NPJ Biofilms Microbiomes 2:16003

Costantini L, Molinari R, Farinon B, Merendino N (2017) Impact of Omega-3 fatty acids on the gut microbiota. Int J Mol Sci 18:2645

Cryan JF, O'Riordan KJ, Cowan CSM, Sandhu KV, Bastiaanssen TFS, Boehme M et al (2019) The microbiota-gut-brain axis. Physiol Rev 99(4) (Kelly et al., 2016):1877–2013

Desbonnet L, Clarke G, Shanahan F, Dinan TG, Cryan JF (2014) Microbiota is essential for social development in the mouse. Mol Psychiatry 19:146–148

Dominguez-Bello MG, Costello EK, Contreras M, Magris M, Hidalgo G, Fierer N, Knight R (2010) Delivery mode shapes the acquisition and structure of the initial microbiota across multiple body habitats in newborns. Proc Natl Acad Sci U S A 10:11971–11975

Falony G, Joossens M, Vieira-Silva S, Wang J, Darzi Y, Faust K et al (2016) Population-level analysis of gut microbiome variation. Science 352:560–564

Fan Y, Pedersen O (2021) Gut microbiota in human metabolic health and disease. Nat Rev Microbiol 19:55–71

Flowers SA, Ward KM, Clark CT (2020) The gut microbiome in bipolar disorder and pharmacotherapy management. Neuropsychobiology 79:43–49

Foster JA, Rinaman L, Cryan JF (2017) Stress & the gut-brain axis: regulation by the microbiome. Neurobiol Stress 7:124–136

Galland L (2014) The gut microbiome and the brain. J Med Food 17:1261–1272

Ghenciulescu A, Park RJ, Burnet PWJ (2021) The gut microbiome in anorexia nervosa: friend or foe? Front Psych 11:611677

Gorwood P, Blanchet-Collet C, Chartrel N, Duclos J, Dechelotte P, Hanachi M, Fetissov S, Godart N, Melchior JC, Ramoz N, Rovere-Jovene C, Tolle V, Viltart O, Epelbaum J (2016) New insights in anorexia nervosa. Front Neurosci 10:256

Guo G, Jia KR, Shi Y, Liu XF, Liu KY, Qi W, Guo Y, Zhang WJ, Wang T, Xiao B, Zou QM (2009) Psychological stress enhances the colonization of the stomach by helicobacter pylori in the BALB/c mouse. Stress 12:478–485

Hansen NW, Sams A (2018) The microbiotic highway to health-new perspective on food structure, gut microbiota, and host inflammation. Nutrients 10:1590

Harach T, Marungruang N, Duthilleul N, Cheatham V, Mc Coy KD, Frisoni G et al (2017) Reduction of Abeta amyloid pathology in APPPS1 transgenic mice in the absence of gut microbiota. Sci Rep 7:41802

Hechler C, Borewicz K, Beijers R, Saccenti E, Riksen-Walraven M, Smidt H, de Weerth C (2019) Association between psychosocial stress and fecal microbiota in pregnant women. Sci Rep 9:4463

Hill C, Guarner F, Reid G, Gibson GR, Merenstein DJ, Pot B, Morelli L, Canani RB, Flint HJ, Salminen S, Calder PC, Sanders ME (2014) Expert consensus document. The international scientific association for probiotics and prebiotics consensus statement on the scope and appropriate use of the term probiotic. Nat Rev Gastroenterol Hepatol 11(8) (Sharon et al., 2019):506–514

Hoban AE, Stilling RM, Moloney MG, Moloney RD, Shanahan F, Dinan TG, Cryan JF, Clarke G (2017) Microbial regulation of microRNA expression in the amygdala and prefrontal cortex. Microbiome 5:102

Holzer P (2016) Neuropeptides, microbiota, and behavior. Int Rev Neurobiol 131:67–89

Issa Isaac N, Philippe D, Nicholas A, Raoult D, Eric C (2019) Metaproteomics of the human gut microbiota: challenges and contributions to other OMICS. Clin Mass Spectrom 14(Pt A):18–30

Järbrink-Sehgal E, Andreasson A (2020) The gut microbiota and mental health in adults. Curr Opin Neurobiol 62:102–114

Jiang C, Li G, Huang P, Liu Z, Zhao B (2017) The gut microbiota and Alzheimer's disease. J Alzheimers Dis 58:1–15

Kazemi A, Noorbala AA, Azam K, Eskandari MH, Djafarian K (2019) Effect of probiotic and prebiotic vs placebo on psychological outcomes in patients with major depressive disorder: a randomized clinical trial. Clin Nutr 38:522–528

Kelly JR, Borre Y, O'Brien C, Patterson E, El Aidy S, Deane J, Kennedy PJ, Beers S, Scott K, Moloney G, Hoban AE (2016) Transferring the blues: depression-associated gut microbiota induces neurobehavioural changes in the rat. J Psychiatr Res 82:109–118

Keshavarzian A, Green SJ, Engen PA, Voigt RM, Naqib A, Forsyth CB, Mutlu E, Shannon KM (2015) Colonic bacterial composition in Parkinson's disease. Mov Disord 30:1351–1360

Kho ZY, Lal SK (2018) The human gut microbiome - a potential controller of wellness and disease. Front Microbiol 9:1835

Knudsen JK, Bundgaard-Nielsen C, Hjerrild S, Nielsen RE, Leutscher P, Sørensen S (2021) Gut microbiota variations in patients diagnosed with major depressive disorder-a systematic review. Brain Behav 11:e02177

Kurilshikov A, Wijmenga C, Fu J, Zhernakova A (2017) Host genetics and gut microbiome: challenges and perspectives. Trends Immunol 38:633–647

Leclercq S, Matamoros S, Cani PD, Neyrinck AM, Jamar F, Stärkel P et al (2014) Intestinal permeability, gut-bacterial dysbiosis, and behavioral markers of alcohol-dependence severity. Proc Natl Acad Sci U S A 111:E4485–E4493

Lee Y, Kim YK (2021) Understanding the connection between the gut-brain axis and stress/anxiety disorders. Curr Psychiatry Rep 23:22

Lee E, Lee J-E (2021) Impact of drinking alcohol on gut microbiota: recent perspectives on ethanol and alcoholic beverage. Curr Opin Food Sci 37:91–97

Liang S, Wu X, Hu X, Wang T, Jin F (2018) Recognizing depression from the microbiota⁻gut⁻brain axis. Int J Mol Sci 19:1592

Liu X, Cao S, Zhang X (2015) Modulation of gut microbiota-brain axis by probiotics, prebiotics, and diet. J Agric Food Chem 63:7885–7895

Liu B, Liu J, Wang M, Zhang Y, Li L (2017) From serotonin to neuroplasticity: evolvement of theories for major depressive disorder. Front Cell Neurosci 11:305

Ma Q, Xing C, Long W, Wang HY, Liu Q, Wang RF (2019) Impact of microbiota on central nervous system and neurological diseases: the gut-brain axis. J Neuroinflammation 16:53

Macia L, Tan J, Vieira AT, Leach K, Stanley D, Luong S et al (2015) Metabolite-sensing receptors GPR43 and GPR109A facilitate dietary fibre-induced gut homeostasis through regulation of the inflammasome. Nat Commun 6:6734

Malan-Muller S, Valles-Colomer M, Raes J, Lowry CA, Seedat S, Hemmings SMJ (2018) The gut microbiome and mental health: implications for anxiety- and trauma-related disorders. OMICS 22:90–107

Meckel KR, Kiraly DD (2019) A potential role for the gut microbiome in substance use disorders. Psychopharmacology 236:1513–1530

Miller BJ, Buckley P, Seabolt W, Mellor A, Kirkpatrick B (2011) Meta-analysis of cytokine alterations in schizophrenia: clinical status and antipsychotic effects. Biol Psychiatry 70:663–671

Mohajeri MH, La Fata G, Steinert RE, Weber P (2018) Relationship between the gut microbiome and brain function. Nutr Rev 76:481–496

Monda V, Villano I, Messina A, Valenzano A, Esposito T, Moscatelli F, Viggiano A, Cibelli G, Chieffi S, Monda M, Messina G (2017) Exercise modifies the gut microbiota with positive health effects. Oxidative Med Cell Longev 2017:3831972

Morais LH, Schreiber HL 4th, Mazmanian SK (2021) The gut microbiota-brain axis in behaviour and brain disorders. Nat Rev Microbiol 19:241–255

Naseribafrouei A, Hestad K, Avershina E, Sekelja M, Linløkken A, Wilson R, Rudi K (2014) Correlation between the human fecal microbiota and depression. Neurogastroenterol Motil 26:1155–1162

Ning T, Gong X, Xie L, Ma B (2017) Gut microbiota analysis in rats with methamphetamine-induced conditioned place preference. Front Microbiol 8:1620
O'Toole PW, Marchesi JR, Hill C (2017) Next-generation probiotics: the spectrum from probiotics to live biotherapeutics. Nat Microbiol 2:17057
Olson CA, Vuong HE, Yano JM, Liang QY, Nusbaum DJ, Hsiao EY (2018) The gut microbiota mediates the anti-seizure effects of the ketogenic diet. Cell 173:1728–1741.e13. (Foster, Rinaman&Cryan, 2017))
Owen MJ, Sawa A, Mortensen PB (2016) Schizophrenia. Lancet 388:86–97
Pistollato F, Sumalla Cano S, Elio I, Masias Vergara M, Giampieri F, Battino M (2016) Role of gut microbiota and nutrients in amyloid formation and pathogenesis of Alzheimer disease. Nutr Rev 74(10) (Foster, Rinaman&Cryan, 2017)):624–634
Rao S, Schieber AMP, O'Connor CP, Leblanc M, Michel D, Ayres JS (2017) Pathogen-mediated inhibition of anorexia promotes host survival and transmission. Cell 168(3) (Sharon et al., 2019):503–516.e12
Ren M, Lotfipour S (2020) The role of the gut microbiome in opioid use. Behav Pharmacol 31:113–121
Rodríguez JM, Murphy K, Stanton C, Ross RP, Kober OI, Juge N et al (2015) The composition of the gut microbiota throughout life, with an emphasis on early life. Microb Ecol Health Dis 26:260. (Sharon et al., 2019)
Romano S, Savva GM, Bedarf JR, Charles IG, Hildebrand F, Narbad A (2021) Meta-analysis of the Parkinson's disease gut microbiome suggests alterations linked to intestinal inflammation. NPJ Parkinsons Dis 7:27
Rong H, Xie XH, Zhao J, Lai WT, Wang MB, Xu D et al (2019) Similarly in depression, nuances of gut microbiota: evidences from a shotgun metagenomics sequencing study on major depressive disorder versus bipolar disorder with current major depressive episode patients. J Psychiatr Res 113:90–99
Russell JT, Zhou Y, Weinstock GM, Bubier JA (2021) The gut microbiome and substance use disorder. Front Neurosci 15:725500. (Sharon et al., 2019)
Rutsch A, Kantsjö JB, Ronchi F (2020) The gut-brain axis: how microbiota and host inflammasome influence brain physiology and pathology. Front Immunol 11:604179
Salminen S, Gibson GR, McCartney AL, Isolauri E (2004) Influence of mode of delivery on gut microbiota composition in seven year old children. Gut 53:1388–1389
Sampson TR, Debelius JW, Thron T, Janssen S, Shastri GG, Ilhan ZE et al (2016) Gut microbiota regulate motor deficits and neuroinflammation in a model of Parkinson's disease. Cell 167:1469–1480.e12
Sánchez B, Delgado S, Blanco-Míguez A, Lourenço A, Gueimonde M, Margolles A (2017) Probiotics, gut microbiota, and their influence on host health and disease. Mol Nutr Food Res 61:1600240
Schepici G, Silvestro S, Bramanti P, Mazzon E (2019) The gut microbiota in multiple sclerosis: an overview of clinical trials. Cell Transplant 28(12):1507–1527. (Sharon et al., 2019)
Sharon G, Cruz NJ, Kang DW, Gandal MJ, Wang B, Kim YM et al (2019) Human gut microbiota from autism spectrum disorder promote behavioral symptoms in mice. Cell 177:1600–1618.e17
Shen Y, Xu J, Li Z, Huang Y, Yuan Y, Wang J, Zhang M, Hu S, Liang Y (2018) Analysis of gut microbiota diversity and auxiliary diagnosis as a biomarker in patients with schizophrenia: a cross-sectional study. Schizophr Res 197:470–477
Shumin W, Yue X, Fengwei T, Jianxin Z, Hao Z, Qixiao Z, Wei C (2020) Rational use of prebiotics for gut microbiota alterations: specific bacterial phylotypes and related mechanisms. Funct Foods 66:103838
Simpson CA, Diaz-Arteche C, Eliby D, Schwartz OS, Simmons JG, Cowan CSM (2021) The gut microbiota in anxiety and depression - a systematic review. Clin Psychol Rev 83:101943
Sivamaruthi BS, Prasanth MI, Kesika P, Chaiyasu C (2019) Probiotics in human mental health and diseases - a mini-review. Trop J Pharm Res 18:890

Skonieczna-Żydecka K, Grochans E, Maciejewska D, Szkup M, Schneider-Matyka D, Jurczak A et al (2018a) Faecal short chain fatty acids profile is changed in polish depressive women. Nutrients 10:1939
Skonieczna-Żydecka K, Marlicz W, Misera A, Koulaouzidis A, Łoniewski I (2018b) Microbiome-the missing link in the gut-brain axis: focus on its role in gastrointestinal and mental health. J Clin Med 7:521
Smith MI, Yatsunenko T, Manary MJ, Trehan I, Mkakosya R, Cheng J, Kau AL et al (2013) Gut microbiomes of Malawian twin pairs discordant for kwashiorkor. Science 339:548–554
Song Y, Liu C, Finegold SM (2004) Real-time PCR quantitation of clostridia in feces of autistic children. Appl Environ Microbiol 70:6459–6465
Srikantha P, Mohajeri MH (2019) The possible role of the microbiota-gut-brain-axis in autism spectrum disorder. Int J Mol Sci 20:2115
Suda K, Matsuda K (2022) How microbes affect depression: underlying mechanisms via the gut-brain axis and the modulating role of probiotics. Int J Mol Sci 23:1172
Szeligowski T, Yun AL, Lennox BR, Burnet PWJ (2020) The gut microbiome and schizophrenia: the current state of the field and clinical applications. Front Psych 11:156
Thursby E, Juge N (2017) Introduction to the human gut microbiota. Biochem J 474(11) (Foster, Rinaman&Cryan, 2017):1823–1836
Tsai SJ (2017) Effects of interleukin-1beta polymorphisms on brain function and behavior in healthy and psychiatric disease conditions. Cytokine Growth Factor Rev 37:89–97
Umu ÖCO, Rudi K, Diep DB (2017) Modulation of the gut microbiota by prebiotic fibres and bacteriocins. Microb Ecol Health Dis 28:1348886
Voigt RM, Forsyth CB, Green SJ, Mutlu E, Engen P, Vitaterna MH, Turek FW, Keshavarzian A (2014) Circadian disorganization alters intestinal microbiota. PLoS One 9(5):e97500. (Sharon et al., 2019)
Wang HX, Wang YP (2016) Gut microbiota-brain axis. Chin Med J 129:2373–2380
Wang WL, Xu SY, Ren ZG, Tao L, Jiang JW, Zheng SS (2015) Application of metagenomics in the human gut microbiome. World J Gastroenterol 21:803–814
Weersma RK, Zhernakova A, Fu J (2020) Interaction between drugs and the gut microbiome. Gut 69:1510–1519
Wong ML, Inserra A, Lewis MD, Mastronardi CA, Leong L, Choo J et al (2016) Inflammasome signaling affects anxiety- and depressive-like behavior and gut microbiome composition. Mol Psychiatry 21:797–805
Xu M, Xu X, Li J, Li F (2019 Jul) Association between gut microbiota and autism spectrum disorder: a systematic review and meta-analysis. Front Psych 17(10):473
Zhao L, Xiong Q, Stary CM, Mahgoub OK, Ye Y, Gu L, Xiong X, Zhu S (2018) Bidirectional gut-brain-microbiota axis as a potential link between inflammatory bowel disease and ischemic stroke. J Neuroinflammation 15:339

Role of Gut Microbiome Composition in Shaping Host Immune System Development and Health

3

Padmaja Phani, Harish Babu Kolla,
Pallaval Veera Bramhachari, and Prakash Narayana Reddy

Abstract

The gut microbiome is a community of commensal microbes in the gastrointestinal tract that are ecologically, physiologically, and symbiotically associated with the host from the early days of life. Gut microbiota is analogous to endocrine glands. Microbial colonies in the gut produce certain microbial metabolites from nutrient metabolism. These gut-derived metabolites regulate the host's health and disease by influencing immunity and physiology. Gut microbiota protects the intestinal environment from invading non-native pathogens by immune modulation and direct competition with pathogens for nutrient access. The gut microbiome is essential in regulating the immune system through interaction with its microbial surface antigens and metabolites. Gut microbiota is coevolved with host development and varies among individuals. The proportion of gut microbiota is constant during health. This constancy is affected by factors such as diet, medications, environment, and mental status regulating the host's health. The dysbiotic microbiome is a risk signature of immune dysfunction and disorders in host physiology. The gut microbiome modulates the immune system locally and systematically; thus, its composition balances an individual's health

P. Phani · H. B. Kolla
Department of Biotechnology, Vignan's Foundation for Science, Technology and Research (Deemed to be University), Guntur, Andhra Pradesh, India

P. V. Bramhachari
Department of Biosciences & Biotechnology, Krishna University, Machilipatnam, Andhra Pradesh, India

P. N. Reddy (✉)
Department of Biotechnology, Vignan's Foundation for Science, Technology and Research (Deemed to be University), Guntur, Andhra Pradesh, India

Dr. V.S. Krishna Government Degree College (A), Maddilapalem, Visakhapatnam, Andhra Pradesh, India

P. Veera Bramhachari (ed.), *Human Microbiome in Health, Disease, and Therapy*, https://doi.org/10.1007/978-981-99-5114-7_3

and disease conditions. This chapter reviewed the link between microbiome composition and its outcome on host physiology, immune system development, metabolic syndromes, and cancer outcomes.

Keywords

Gut microbiome · Immune system · Metabolic disorders · Microbial metabolites · Innate immunity · Adaptive immunity

3.1 Introduction

The human body harbors trillions of microbial communities in the gastrointestinal (GI) tract called gut microbiome. Joshua Lederberg explained gut microbiota as "The community of microorganisms presents in the gastrointestinal tract of the host" (Bäckhed et al. 2005; Neish 2009). The gut microbiota is considered a completely evolved and established organ in the human body analogous to hormone-secreting endocrine glands. Gut microbiota regulates multifarious physiological and metabolic pathways via its derived metabolites as substrates and maintains immunohomeostatic comprehensive cellular functions through cell signaling and biochemical cascades (Cox and Blaser 2013). Gut microbiota is recorded as an extension to the host genome by 150 times more than the human genome. It is estimated to contain 3.3 million microbial genes that code for certain essential enzymes not included in the set of native human proteomes. The enzymes coded by genes in the microbiome catalyze several biochemical processes in nutrient absorption and metabolism (Rodríguez et al. 2015; Bäckhed et al. 2005). Recent studies from researchers relevant to metagenomics, molecular biology, and microbiology delineated the human body as a mutualistic superorganism of eukaryotic and prokaryotic microbial communities (Szablewski 2018). Hosts provide nutrition and shelter to the microbes in the gut; in turn, gut microbiota establishes its mutualistic and symbiotic nature by providing the host with better immunity and physiometabolic health. Trillions of microbes from hundreds of species constitute healthy microflora in the host gut. Of all the microbial communities, members of Bacteroidetes, Firmicutes, Actinobacteria, and Proteobacteria classes are the major contributors to gut microbiome composition (Senghor et al. 2018).

Balance among the proportion of gut microbiota and these members directs the fate of host health. Microbial colonization in the GI tract began before birth. Reports from the placental microbiome characterization showed similarities with the oral microbiome of healthy adults (Aagaard et al. 2014). In neonates, lactating has a defensive effect, deliberated by an intricate combination of lysozymes, sIgA, α-lactalbumin, free oligosaccharides, complex lipids, and other glycoconjugates (Gordon et al. 2012). Oligosaccharides such as fructans are prebiotic factors that help the growth of beneficial bacteria such as *Bifidobacterium* and *Lactobacillus*. Understanding the gut microbiome unravels microbial-mediated immune and metabolic regulation mechanisms in the host body. Studying the entire microbial communities in the host GI tract was a challenge to researchers and scientists during the initial days when gut microbiota composition was analyzed based on culture methods. These methods are inadequate to examine the total profile of the gut microbiome; as a result, only 10–50% of the intestinal microbes were probably cultured. In recent years, understanding of gut microbiota increased with advanced sequencing technologies adapting next-generation sequencing approaches, metagenomics, and advancements in bioinformatics tools to handle and analyze the downstream data from sequencers ensured in estimating several classes of microbes and their phylogenetic relationships. The qualitative analysis of gut microbiota is mostly delineated using techniques like DNA fingerprinting, terminal RFLP, 16 s ribosomal RNA amplicon sequencing, microarray technique, and whole genome sequencing, which provided enormous data about the total microbial population. High-throughput sequencing technologies like Roche/454, GS20, Illumina's Genome Analyzer IIx, Affymetrix microarray technique, and Qiagen's Gene Read are tremendously eminent.

Moreover, advanced bioinformatics tools have assisted in understanding and illustrating the downstream analysis of sequence data. The gut microbiota is highly reactive and adaptive to dietary alterations, medication choice, genetic factors, and the host's lifestyle. After weaning to solid foods, exorbitant modifications appear in the composition of their gut microbiota. Changes in the microbiota (Dysbiosis) could lead to numerous health disorders such as obesity, nonalcoholic fatty liver disease (NAFLD), diabetes, inflammatory bowel disease, ulcerative colitis, colorectal cancer, coronary heart disease, autoimmune diseases, and neurological disorders. The gut microbiome has become a major tool and a potential clinically important marker for diagnosing and treating many diseases in the body. Modulating or redirecting the gut microbiota to its native state (eubiosis) is an ideal and promising strategy for simulating host immunity. Reconstituting the gut microbial communities benefits the host with better health and immunity. Engineering gut environment with probiotic supplements, prebiotics, and functional foods effectively shapes host immunity. The importance of gut microbiota in immune system development and modulation, along with its fate in disease and health conditions, are discussed clearly in this chapter addressing the recent findings and outbreaks in gut microbiome research in correlation with host immunity and health.

3.2 Intestinal Microbiota and Host Immunity

The gut microbiota considerably regulates innate and adaptive immune system functions and development. The host's immune system has two protective mechanisms: innate immunity, specified as an immediate and nonspecific response against the pathogen. And another one is adaptive immunity which ensures both memory and specificity. In the innate immune response, secretory IgA (sIgA) plays a significant role and is a protective mechanism against infectious agents. The production of sIgA over various mucosal surfaces is through the entry of antigens and their subsequent capture through Peyer's patches, M cells, stimulation of T cells, dendritic cells (DCs), and changes in B cells to mesenteric lymph nodes (MLNs) recombination and lymphoid tissue connected to the gut. The group of cytokines such as IL-4, IL-5, IL-10 and including TGF-β increases the production of IgA. The sIgA binds to commensal bacteria and contributes to the gut barrier function and intestinal mucosal homeostasis (Chairatana and Nolan 2017).

The innate immune cells like DCs, natural killer (NK) cells, and macrophages express pattern recognition receptors (PRRs) that recognize specific molecular patterns on the bacterial surface, which are key mediators for communication between gut microbiota and the host (Pahari et al. 2018, 2019; Negi et al. 2019). These PRRs recognize pathogen- or microbe-associated molecular patterns (PAMPs or MAMPs) on the bacterial surface. PRRs majorly contain families of nucleotide-binding oligomerization domain (NOD)-like receptors (NLRs), toll-like receptors (TLRs), RIG-I-like receptors (RLRs), and C-type lectin receptors (CLRs) (Kumar et al. 2011). Identifying microbiota through PRRs promotes memory response on primary exposure (Mills 2011; Kleinnijenhuis et al. 2012). The TLRs exist on DCs, macrophages, intestinal epithelial cells (ECs), neutrophils, and other innate immune cells. The microbial products and metabolites alter the host immune system by stimulating different types of cells like intestinal epithelial cells (IECs), mononuclear phagocytes, innate lymphoid cells (ILCs), B cells, and T cells (Kabat et al. 2014).

The GI tract protects the host from habitat and interceding nutrient consumption. The IECs on the intestinal surface form a physical barrier that detaches the lumen from the lamina propria of commensal and intestinal microbes. Although IECs are not typical innate immune cells, these are essential in mucosal immunity. Despite this, IECs are armed with innate immune system receptors that could provide gut equilibrium by recognizing bacteria (Pott and Hornef 2012). ILCs are infrequent innate lymphocytes when correlated with adaptive lymphocytes, yet these are copious on the surface barrier of mucosal-connected tissues (Sonnenberg and Artis 2012). Several research studies demonstrated that specific microbiota metabolites could control ILCs (Lee et al. 2012) expressing IL-22 cytokine. The inadequacy of IL-22 is connected with various inflammations and metabolic diseases. IL-22 also elevates the antimicrobial peptides production (RegIIIγ and RegIIIβ) to reduce the SFB colonization, stimulate the surface proteins fucosylation to intensify the beneficial bacteria colonization, and increase the goblet cells proliferation for secretion of mucin (Goto et al. 2014). Based on T domain structures, the T cells can be further segregated into γδ T and αβ T cells. αβ TCR cells expressed by T cells are initially liable for antigen-specific cellular immunity, and γδ T cells are not MHC—restricted also engaged in initial immune responses (Pennington et al. 2005). Despite this, in the small intestine of murine, γδ TCR chains are expressed by a higher proportion of intraepithelial lymphocytes (IELs) (van Wijk and Cheroutre 2009). These γδ IELs specifically regulate IECs' continuous turnover and increase the growth of epithelial cells by keratinocyte secretion, an in vitro growth factor (Boismenu and Havran 1994). The γδ IELs also maintain the functions of the epithelial barrier by inhibiting pathogen reincarnation (Dalton et al. 2006). The association of innate and adaptive immune systems is involved in eliminating invasive pathogens and regulating symbiotic bacteria at mucosal sites. The antigen-presenting cells like naïve CD4+ αβ T cells ($CD4^+$ T cells) and DCs are further characterized into Th1, Th2, and Th17 or adaptive T-regulatory cells (Tregs). All these cells exist in intestinal lamina propria. Th17 is the group of $CD4^{+}T$ cells that secretes numerous cytokines (IL-22, IL-17F, and IL-17A) (Rossi and Bot 2013), including notable effects upon inflammation and immune homeostasis. These cells also retain various cytokine characteristic analyses and functions. Th1 and Th2 cells have a steady secretory analysis after differentiation. Tregs are essential mediators of immune tolerance, reducing an improper, immense inflammatory reaction, and their malfunctions lead to autoimmune disorders. Of interest, in the germfree mice administered with a lustrated dose of polysaccharide or by intestinal colonization with commensal bacteria, a non-toxigenic form of *B. fragilis* which expresses polysaccharide A (PSA), prevents the growth of experimental colitis by PSA-induced $Foxp3^{+}$-regulatory cells expressed by IL-10, through TLR2-dependent action (Round and Mazmanian 2010). In commensal microbiota, a few microbes have a higher effect on the responses of mucosal T cells. For instance, in Th17 cells of the small intestine, segmented filamentous bacteria (SFB) are effective stimulators; in germfree mice, it was observed by the lack of Th17 cells and their revival when SFB colonized in germfree mice (Ivanov et al. 2009a). In the gut, the abundant presence of retinoic acid (RA) activates lymphocyte gut-homing compounds and restricts the growth of Th17 cells (Mucida et al. 2007); still, the mechanism of regulation of Th17 gut tropism is unknown (Maynard and Weaver 2009). Therefore, intestinal lamina propria is an elemental site for the evolution of Th17, probably by the colonization of SFB and the expression of innate IL-23 in the intestinal microhabitat.

3.3 Gut Microbiota Metabolism

Intestinal microbiota regulates various host physiological mechanisms such as nutrient uptake, energy expenditure, and immune responses through metabolism. Gut microbiome-derived metabolites act as substrates for various cell signaling processes and host metabolic pathways and can alter the immune responses post-maturation and differentiation. Gut microbiota-derived metabolites are crucial in health and disease conditions in the host. According to recent investigations and reports, over 50% of the metabolome in stool and urine are derivatives of modulated gut microbiota. The microbiota metabolites are bioactive and intensely affect physiology and host immunity (Donia and Fischbach 2015). The following sections further discuss the role of gut microbial metabolites and their metabolic actions.

3.3.1 Retinoic Acid (RA) Metabolism

Retinoic acid (RA), a lipid metabolite of vitamin A, can regulate the equity among pro-inflammatory and anti-inflammatory immune reactions. RA inadequacy can run down the orchestration of gut microbiota and immune system activities. In constant conditions, RA is pivotal in maintaining intestinal immune homeostasis since it facilitates the regulatory T-cell progression by TGF-β and the formation of IgA through B cells (Mucida et al. 2007). It is perplexing that RA is also entangled in drawing out pro-inflammatory CD4+ T-cell reactions to diseases during inflammation—other vitamins like vitamin D extremely influence T-cell activation. Multifarious research analyses have associated vitamin D inadequacy with inflammatory bowel disease. The hook-up between the intestinal microbiota and vitamins is conspicuous in vitamins of B and K groups (Martens et al. 2002). The inadequacy of vitamin B12 leads to a reduced count of lymphocytes and induces NK cell functioning.

3.3.2 Tryptophan Metabolism

Inadequacy of innate immunity pathways results in malfunction of gut microbe. For instance, complex proteins and carbohydrates which are unable to degrade by the host can be digested by microbial colonies. Gut microbiota influences tissue-level immune development through the catabolism of tryptophan. The *Lactobacillus* utilizes tryptophan as a vitality source to form ligands of the aryl hydrocarbon receptors (AhR) like the metabolite indole-3-aldehyde (Nicholson et al. 2012).

3.3.3 Short Chain Fatty Acids (SCFAs) Metabolism

The microbiome provides mammalian enzymes to degrade dietary nondigestible carbohydrates (NDCs) adherent starch by fermentation into short-chain fatty acids (SCFAs) in the GI tract (Holscher 2017). SCFAs are known as carboxylic acids, including 1–6 aliphatic carbon tails such as acetate, propionate, and butyrate produced in a molar ratio of approximately 60:20:20, respectively (den Besten et al. 2013), and other end products consist of ethanol, succinate, formate, valerate, isobutyrate, and 2-methyl butyrate. SCFAs potentiate colonocytes, and inhabitant bacteria, decrease GI luminal pH to reduce pathogen growth, regulate anti-inflammatory and immunostimulatory properties, and endorse bile acid secretion, which aids in the digestion of dietary fats and increases mineral absorption (Schuijt et al. 2016). The SCFAs are proposed to engage certain G-protein-coupled receptors (GPR41, GPR43). GPR109a are stimulated through ionized SCFA, increasing the excretion of peptide YY, glucagon-like peptide (GLP-1), enhanced glucose usage, and reduced fatty acid metabolism (Koh et al. 2016). SCFAs are recorded to defend from diet-induced obesity, regulate gene expression, and induce anti-inflammatory reaction and apoptosis. In addition, SCFAs stimulate lipid metabolism by increasing lipogenesis and preventing fatty acid oxidation, as formerly recorded. SCFAs are crucial in colonic health, notably in protecting and differentiating epithelial cells. Some of the known well-characterized transporters and receptors of SCFAs are given in Table 3.1. SCFAs also regulate the expression of inflammatory cytokines like IL-6, IL-7, IL-8, IL-12, IL-1β, and TNF-α by colonic epithelial cells (Asarat et al. 2015), regulating blood pressure, leading to gut-barrier dysfunction. Butyrate is an energy

Table 3.1 Transporters and receptors of SCFAs

Transporters of SCFAs

Transporter molecule	Function	SCFAs	Model organism	Cell/tissue	References
MCT1	A H+-coupled transporter for SCFAs and related organic acids	Butyrate, pyruvate, lactate	Human	Distal colon> proximal colon>ileum>jejunum	Gill (2005)
			Mice, rat	Cecum>colon>stomach and small intestine	Kirat et al. (2009)
			Human	Monocytes, lymphocytes, and granulocytes	Murray et al. (2005)
SMCT1	A Na(+)-coupled transport of monocarboxylates and ketone bodies into various cell types	Butyrate > propionate > lactate >>acetate	Human, mice	Distal colon>proximal colon and ileum	Borthakur et al. (2010)

Receptors of SCFAs

Receptor	Function	SCFAs	Model organism	Cell/tissue	Reference
GPR109A	A receptor for C4 and niacin. cAMP regulation, suppression of adipocyte lipolysis, HDL metabolism, DC trafficking, antitumor activity and HDL metabolism	D-beta-hydroxybutyrate, nicotinic acid and butyrate	Human, mice	Adipose tissue	Tunaru et al. (2003)
			Human, mice	Colon	Thangaraju et al. (2009)
GPR43	A receptor for SCFAs. Secretion of PYY and GLP-1, adipocyte development, adipogenesis, suppression of lipolysis, epithelial innate immunity, antitumor activity, anti-inflammatory effect, and T-reg differentiation	Acetate=propionate= butyrate>pentanoate >hexanoate>formate	Human, mice	Colonic myeloid cells and Treg	Smith et al. (2013)
			Human	Intestinal epithelium	Agus et al. (2016)
GPR41	A receptor for SCFAs. Regulation of gut hormone, leptin production, and sympathetic activation, epithelial innate immunity	Propionate=pent-anoate=butyrate> acetate>formate	Human	Monocytes, monocyte-derived dendritic cells, and Nastasi neutrophils	Nastasi et al. (2015)

substrate that intensely affects the healthy colonic epithelial barrier and immunomodulatory effects. The gut microbiota synthesizes vitamins (like B vitamins, k, biotin, folates, riboflavin, and cobalamin) and amino acids and carries out bile transformation. The antimicrobial compounds produced from microbiota contend for nutrients and gut lining attachment, thereby inhibiting the growth of pathogens. As a result, it promotes to the reduction of the lipopolysaccharides and peptidoglycans synthesis that is pernicious to the host (Tlaskova-Hogenova et al. 2004).

3.3.4 Bile Acids Metabolism

Bile acids are steroid metabolites present in bile. They are produced in the liver from cholesterol. Bile acids ensure solubility and uptake of vitamins and fats. Bile acids directly synthesized from cholesterol in the liver are primary bile acids. Primary bile acids conjugate with glycine or taurine to form secondary bile acids. Gut resident microbes deconjugate secondary BAs to primary BAs and glycine or taurine again. BAs are signaling molecules to farnesoid X receptor (FXR) and GPCR TGR5 in controlling the uptake of fats and vitamins (Tolhurst et al. 2012; Velagapudi et al. 2010).

3.3.5 Choline Metabolism

Choline is a cell membrane component and also a cationic essential nutrient. Choline is found in meat and eggs. It is essential in lipid metabolism. Enzymatic degradation of choline in the liver yields TMA (trimethylamine). TMA further metabolizes into trimethylamine N-oxide (TMAO) (Spencer et al. 2011). TMA and TMA are toxic metabolites. Production of these metabolites is controlled and regulated by microbes in the gut microbiome. Disturbance in the gut ecosystem is associated with a rise in the levels of these metabolites, which further leads to immune dysfunction and cardiometabolic syndrome (Prentiss et al. 1961; Dumas et al. 2006). Hence, the gut microbiota is key in regulating host health and metabolism.

3.4 Gut Microbiota Dysbiosis and Disease

3.4.1 IBD

Inflammatory bowel disease (IBD) comprises Crohn's disease (CD). Ulcerative colitis (UC) is an incurable condition characterized by GI tract inflammation evoked by the consolidation of genetic, environmental, and microbial components typified by abdominal ache, diarrhea, and bloody feces (Wilson et al. 2016; Cosnes et al. 2011). The IBD is an exorbitant host immune system and gut flora stimulation in inherently affected patients (Wong and Ng 2013). The IL-23/Th17 deregulation is connected with numerous genetic sensitivity of single-nucleotide polymorphisms

(SNPs) in individuals affected with CD and UC due to deterioration of innate and adaptive immunity reactions (Yen et al. 2006). Remarkably, dysbiosis is linked with the exaggerated reproduction of the responsive oxygen category that consecutively results in alterations of intestinal microbiota composition, mucosal penetrability, and enhanced immune provocation. By way of illustration of how particular microbes produce intestinal inflammation and stimulate the pathogenesis of IBD exists in Bloom et al. (2011). In their examination, commensal Bacteroidetes strains have been secluded in IL-10r2 and Tgfbr2-insufficient mice (Bloom et al. 2011). There is a confirmation that the growth of IBD is a symbiotic impact of genetic and acquired components that results in the modulations in activities and arrangement of intestinal microbiota (Albenberg et al. 2012). Despite this, the metagenomic analysis explained that microbial ecosystem and intestinal flora were reduced in IBD-affected individuals compared with healthy adults (Hansen et al. 2010). Frequently, 25% lesser genes were discovered in the stool samples of IBD individuals than in healthy controls (Qin et al. 2010). In addition, humans with UC and CD have decreased the fewer microbes like Firmicutes (Sheehan et al. 2015; Frank et al. 2007; Walker et al. 2011) having anti-inflammatory and pro-inflammatory properties and enhanced the phyla of Bacteroidetes, Proteobacteria in mucosa-associated flora (Sokol et al. 2006). The CD-affected cases showed a lower abundance of *Faecalibacterium prausnitzii*, *Clostridium lavalense*, *Roseburia inulinivorans*, *Ruminococcus torques*, and *Blautia faecis* when correlated with healthy adults (Fujimoto et al. 2013; Takahashi et al. 2016).

Further, Sokol and co-workers demonstrated that mononuclear cells of human peripheral blood activated with *F. praunizii* to produce IL-10 and inhibit the formation of IL-12 and IFN-γ (Sokol et al. 2008). The other CD-associated *E. coli* AIEC (adherent invasive *E. coli*), which also contains pro-inflammatory features, a mucosa-associated *E. coli* with dynamic adhesive-invasive abilities, was initially isolated from CD-affected adults. Increased growth of AIEC has been observed in individuals of colonic CD, about 38% with effective ileal CD compared to healthy controls (Baumgart et al. 2007; Darfeuille-Michaud et al. 2004). As a result, the growth of pathogenic microorganisms that attach to gut epithelial cells influences intestinal penetrability, modulates the gut microbial configuration, and promotes inflammatory reactions by standardizing pro-inflammatory gene expression, eventually developing in colitis. In IBD patients, there is a reduction in the formation of SCFAs due to a decrease in the number of *F. praunizii*. *Clostridium* clusters IV, XIVa, and XVIII are butyrate-producing organisms in the gut, affecting the growth and differentiation of Tregs cells and an expansion of epithelial cells (Atarashi et al. 2013). Treg cells are CD4+ T cells that help to maintain gut homeostasis.

Furthermore, in IBD cases, there is a higher abundance of *Desulfovibrio*, which is sulfate-producing bacteria (Loubinoux et al. 2002; Zinkevich and Beech 2000). As a result, the formation of hydrogen sulfide harms the gut epithelial cells and stimulates mucosal inflammation (Loubinoux et al. 2002; Rowan et al. 2010). Accordingly, the above data indicate that gut microbial configuration changes are linked with IBD pathogenesis. The outcome of dysbiosis on IBD and the pathological changes are provided in Table 3.2.

Table 3.2 Dysbiosis in IBD and its pathological results

Dysbiosis in IBD	Outcomes of dysbiosis
↓ Firmicutes ↓ *F. praunizii* and *F. clostridium* cluster IV XIVa, XVIII ↑ *Desulfovibrio* bacteria ↑ Adherent/invasive *E. coli*	↓ Epithelial cells expansion and differentiation Change in Tregs cells differentiation ↑ Damage of epithelial cells Modulation in mucosal penetrability ↑ Bacterial invasion

↑ indicates increase, ↓ indicates decrease in level

3.4.2 Colorectal Cancer (CRC)

The World Cancer Research Foundation (WCRF), as well as the American Institute for Cancer Research (AICR), recognized that diet is one of the essential external components in CRC etiology (Dumas et al. 2016). The microbiota has been a prominent aspect of a few cancers such as breast, liver, biliary system, and CRC. Accommodating around 3×10^{13} microbes, the colorectum interplays with a huge population of microorganisms, and with that, the intestinal epithelium found a stable cross talk (Qin et al. 2010). The 16 s ribosomal RNA sequencing analyses were carried out to illustrate the CRC microbiota in stool and mucosal samples (Yu et al. 2017). Direct observation was that the microbiota of CRC individuals had undergone severe dysbiosis when correlated with the composition of healthy adults displaying numerous ecological microhabitats in individuals with CRC. In addition, certain microbes such as *Bacteroides fragilis*, *E. coli*, *Enterococcus faecalis*, and *Streptococcus gallolyticus* are independently associated with CRC in several combinations and systematic examinations. Metagenomic analysis revealed that gut microbiota associated with CRC, henceforth named CRC microbiota, consist of an abundance of species, a reduced plethora of *Roseburia*, and an enhanced myriad of procarcinogenic bacterial communities like Bacteroides, *Escherichia*, *Fusobacteria*, and *Porphyromonas* (Yu et al. 2017).

Recently, it was detected that intestinal bacteria could stimulate the development of CRC through chronic inflammation initiation, biosynthesis of genotoxin (meddle with the regulation of cell cycle), heterocyclic amine stimulation, or toxic metabolite synthesis of carcinogenic elements of pro-diet (Candela et al. 2014). Chronic inflammation is connected with the risk of evolving cancer and does through causing mutations, cell proliferation, and provoking angiogenesis or apoptosis inhibition (Medzhitov 2008; Grivennikov and Karin 2010). The microbiota dysbiosis benefits opportunistic pathogens that stimulate innate and adaptive immune system components, and bacterial shift, which results in chronic inflammation (Ivanov et al. 2009b). The commensal bacteria stimulate the innate immune system. As a result, dendritic cells, macrophages, and NK cells enhance the production of pro-inflammatory cytokines like TNF-α, IL-23, IL-12, and INFγ, with consequent stimulation of adaptive immune cells, including B cells, T cells, lymphocytes, and other mediators of inflammation (Keku et al. 2015). The inflammatory reaction to commensal bacteria is the stimulation of NF-κB transcription factor and (signal transducer and activator of transcription) STAT3 in epithelial cells (Greten et al. 2004; Guarner 2006; Hooper et al. 2014; Tian et al. 2003), the production of nitrogen and reactive oxygen species resulting in oxidative stress, damage of DNA, and the progression of CRC. In addition, colonic polyposis is linked with large microbial density compiled inside polyps that induce local inflammatory reactions. The development of polyps and the density of microorganisms may be inhibited through IL-10, a derivative of T cells and Tregs (Dennis et al. 2013). Therefore, it is concluded that the modulation of normal homeostasis among microbiota and host is important for inflammation and the subsequent alterations which cause colon carcinogenesis.

In the interaction between host and microbiota, metabolism is an essential factor. The microbial metagenome encrypts genes that digest more dietary components and host compounds like bile acids. The fecal bile acids increase through a high-fat diet, provoking their enterohepatic circulation and production. The 7α-dehydroxylating bacteria turn colonic initial bile acids into secondary bile acids that are cytotoxic to gut epithelial cells in animal models (Ridlon et al. 2006; Cheng and Raufman 2005). This conversion enhances these secondary bile acids' hydrophilicity (de Giorgio and Blandizzi 2010). Consuming animal protein and a high-fat diet increases the number of secondary bile acids like lithocholic acid, cholic acid, and deoxycholic acid, which causes a higher CRC risk. The deoxycholic acid damages the tract of the mucosa intestine, causes DNA damage, creates genomic instability, and assists the development of tumors. This process might influence bile acids' influence on the colon's carcinogenesis (Rubin et al. 2012).

In contrast, people with a low-fat diet are also affected by CRC, and the risk is through various factors like host health, genetic predisposition, and luminal interplay. Studies determined that CRC cases contained reduced butyrate-producing bacteria *F. prausnitzii*, *Eubacterium rectale*, and increased *Enterococcus faecalis*. Therefore, it is concluded that the colonic bacterial community is a factor that causes CRC.

Furthermore, *B. fragilis*, an enterotoxigenic strain, colonizes the mucosa of adults in an asymptomatic process. However, in a few cases, *these strains release B. fragilis toxin (BFT)*, which induces inflammatory diarrhea. The *B. fragilis* toxin stimulates NF-κB results in the expression of cytokines, which assist in mucosa inflammation (Sears 2009). Therefore, BFT is established as one of the major toxins in the progression of CRC; moreover, in CRC individuals, toxins are transcribed in tumors derived from *Shigella flexneri*, *E. coli*, and *Salmonella enterica*. The data indicate that enterobacterial toxins involve in tumorigenesis (Schwabe and Wang 2012).

The composition and activities of the intestinal microbiota are majorly affected by diet (Duncan et al. 2007). The colonic bacteria produce SCFAs like butyrate, which inhibits CRC progression, prevents histone deacetylases in colonocytes, and induces apoptosis in CRC cell lines (Leonel and Alvarez-Leite 2012; Zhang et al. 2010). Butyrate also stimulates the functions of the large intestine and prevents the growth of pathogens. In addition, butyrate and propionate were exhibited to alter colonic regulatory T cells and utilize an effective anti-inflammatory impact in

Table 3.3 Pathogenetic mechanism of bacteria linked with CRC in murine models

Bacteria	Pathogenetic mechanism	Association with murine model	Reference
Bacteroides fragilis	STAT3 activation; induction of Th-17 immune response IL-1 production; E-cadherin cleavage; stimulation of catenin signaling	Enterotoxigenic *B. fragilis* (ETBF) augments spontaneous colon cancer in multiple intestinal neoplasia (min) mice	Wu et al. (2009); Toprak et al. (2006)
Bacteroides vulgates	Stimulation of MyD88-dependent signaling NF-kB activation	Mono-association of AOM-IL10/ mice lead to mild colorectal tumorigenesis	Uronis et al. (2009)
Enterococcus faecalis	Production of ROS and DNA damage	Stimulates adenocarcinoma in IL-10 KO mice	Balamurugan et al. (2008)
Escherichia coli	Intracellular colonization	*E. coli* NC101 promotes invasive carcinoma in AOM-IL10/ mice; *E. coli* 11G5 enhances colonic polyps in multiple intestinal neoplasia (min) mice	Bonnet et al. (2014)

animal models (Chen et al. 2013). Research studies demonstrated that a fiber-containing diet affects the production of SCFAs (Tomasello et al. 2014). It is concluded that a high-fiber diet increases SCFAs production, with a subsequent decrease in intestinal pH that benefits fermentation in the colon, inhibits pathogen colonization, and reduces the absorption of carcinogen (Macfarlane and Macfarlane 2012) therefore decreasing the risk of CRC (Keku et al. 2015). Different pathogenic mechanisms linked with colorectal cancer are listed in Table 3.3.

3.4.3 Obesity

Obesity is a global condition that is likely accelerating its prevalence worldwide. It affected approximately 107.7 million young children and 603.7 million adults worldwide, and over 60% of fatality is associated with excess body mass index (BMI) (Afshin et al. 2017). Obesity is strongly related to numerous antagonistic comorbidities, such as cardiovascular disease, cancer, and type 2 diabetes mellitus. Obesity is recognized as a complex and multifactorial disorder primarily derivable to peril components of genetic history, lifestyle, and habitat (Hruby and Hu 2015). Over the last decades, the connection and induced role enacted through gut microbiota and obesity have been an astounding discovery. The gut microbiota of mice and humans is dominated by numerous bacterial microbiota containing Bacteroidetes, Firmicutes, and Actinobacteria. The initial observation exhibited distinctive gut microbial configuration in genetically obese (ob/ob) mice correlated to lean (ob/+) and wild (+/+) offsprings in a context of similar polysaccharide-enriched diet (Ley et al. 2005) through epitomizing the decreased plethora of Bacteroidetes and enhanced Firmicutes in obese patients. To characterize the impacts of gut microbiota from genetic alterations, Turnbaugh and co-workers relocated lean and obese microbiota to germfree mice; consequently, more enhancements in total body fat in recipients colonized by microbiota of obese were observed when correlated to lean microbiota (Turnbaugh et al. 2006). The malfunction of a gut ecosystem that leads to reduced microbiota certainty was linked with IBD and obesity (Qin et al. 2010; Turnbaugh et al. 2009). The initial observations on the association between the gut microbiota and obesity have revealed enhanced Firmicutes number, though a decrease in the number of Bacteroidetes in both humans and mice affected obesity when correlated with lean individuals (Furet et al. 2010). Fascinatingly, these changes can be reversed through weight loss through dietary habits. In the selection of microbiota, the immune system is also considered pivotal. The mice models with unusual TLR signaling or express bactericidal reactive oxygen microbes have exalted antibody serum titers to counteract one's commensal bacteria (Slack et al. 2009). The enriched serum titers are needed to retain the host's and gut microbiota's commensal association. The deficiency of TLRs in mutant mice showed a modified gut microbial composition. The deficiency of TLR-5 mice promotes obesity, metabolic disorders, and inflammation.

The gut microbiota of mice has an enhanced potential to harvest energy from the gut when correlated to their counterparts of germfree mice (Wostmann et al. 1983). Metagenomic gut microbiota studies in obese human and mouse models have identified enhanced carbohydrate fermentation ability (Turnbaugh et al. 2009). This transformation enhances the SCFAs production in the host to enhance the energy harvest. The SCFAs have been suggested to attach to certain GPC receptors such as GPR41, GPR43, FFAR2, and FFAR3, which could improve the nutrient consumption and/or progression of adipose tissue mass. The clinical analyses performed in mice with insufficient GPR41 proposed that the stimulation of GPR41 through SCFAs is responsible for the secretion of PYY gut hormone. Despite this, the mice with abundant expression of GPR43 are fed an obesogenic that enhances the propagation of adipocytes and prevents lipolysis in adipocytes. The mice with GPR43 deficiency are treated with enriched carbohydrates and an enriched fat diet containing a meager body mass and a myriad lean mass correlated with mice of wild type (Bjursell et al. 2011). In addition, the drastic modulations in the composition of gut microbiota, which appear aftermath of medication with antibiotics, can act defense against glucose sensitivity, obesity, and insulin resistance stimulated through enriched fat and a free carbohydrate diet (Cani et al. 2008). According to a recent hypothesis, the gut microbiota can retain the host's metabolic homeostasis. Metabolic disorders like T2DM and obesity are connected with low-level inflammation and modified microbial composition; a microbial strain might enact as a provoking factor in the progression of DM, obesity, as well as inflammation stimulated by a fat-enriched diet.

Low-level metabolic inflammation is considered a pivotal component of metabolic disorders. Numerous analyses illustrated that the metabolic system is unified with an enhanced pro-inflammatory cytokine-like TNF-α, common obesity-associated inflammation, and insulin resistance. Lipopolysaccharides (LPS) endotoxin, an important factor in Gram-negative bacterial cell walls such as Bacteroidetes, enhances the progression of adipose tissues, affecting insulin resistance and inflammation. LPS also acts as a stimulating factor of fat and enriches diet-activated metabolic disorders. However, metabolic endotoxemia provokes the production of TNF-α, IL-1, and IL-6. The research studies determined that metabolic endotoxemia exists because of the alterations in intestinal microbiota due to antibiotic medication that drastically decreased the native intestinal microbiota and reestablished the common plasma LPS values in the fat-enriched diet fed in mice models. Antibiotic medication suggests that bacteria in the gut affected by antibiotic consumption regulate intestinal penetrability; metabolic endotoxemia occurs. The deficiency of TLR4 (considered as LPS) is defensive against obesity from visceral and subcutaneous adipose tissue development, glucose resistance stimulated by a fat-enriched diet, and the endoplasmic reticulum stress is the major organ for digestion of lipids and glucose.

3.4.4 Diabetes

Genetic background, diet, and environmental conditions influence the gut microbial community. Any significant deviation of these factors influences the apparent habitat alterations. It is significantly stable in middle-aged humans. However, there is a high number of notable gut microbiota alterations that have been in interindividuals. The surfeit of biological reactions regulates with the assistance of modulated gut microbiota.

3.4.4.1 Type 1 Diabetes Mellitus (T1DM)

T1DM is a perennial autoimmune disorder diagnosed usually at a young age and distinguished by the demolition of immune-mediated responses of insulin formation from the pancreatic β-cells (Lamichhane et al. 2018). The ubiquity of T1DM is increasing globally because of a deficiency of suitable therapeutic procedures. The environmental factors associated with a genetic predisposition are eminent for the progression of T1DM (Battaglia and Atkinson 2015). The initial pathogenesis of T1DM is identified by insulitis, abundance expression of autoantibodies over β-cell antigens observed by decreased insulin production, and demise of β-cells (Battaglia and Atkinson 2015). The definite factors responsible for inducing T1DM pathogenesis are still unknown; moreover, the usual aspect that triggers T1DM is genetic history and habitat (Battaglia and Atkinson 2015). The triggering factors like viral infections, usage of antibiotics, consumption of cow milk proteins at early ages, deficiency in breastfeeding and vitamin D supplement, and disclosure to endocrine disrupting synthesis. The function of gut microbiota can affect intestinal mucosa, such as autoimmunity over β-cells. Clinical studies showed that in T1DM models, reduced Firmicutes and enhanced Bacteroides numbers are identified, exhibiting an association between T1DM and microbiota.

In contrast, the reduced number of Bacteroides and Firmicutes is linked with individuals correlated to lean individuals (Schwiertz et al. 2010). Modifications in microbial composition might occur because of differences in the glucose levels of host results due to diet and intestinal habitat. However, the definite mechanisms are yet unknown, though these alterations might be connected with the progression of T1DM, as reduced *Bifidobacterium* can impact the gut penetrability and mucosal immune reactions affecting autoimmune responses. The interplay between intestinal microbiota and the host immune system enacts an important role in the growth of T1DM. The immune system cells can perceive the metabolites and analytes of gut microbiota which can alter the role of immune cells, stimulating the development of T1DM pathogenesis.

Similarly, the pancreas also consists of its microbiota, and the modification of pancreatic microbiota is linked with the assistance of intrapancreatic immune reactions and initiation of diabetes in addition to pancreatic cancer and T1DM (Pushalkar et al. 2018). The dysbiosis and functions of gut metabolites can stimulate the immune system's GALT malfunctions, like unusual IgA excretion and reproduction of colonic regulatory T cells (Pabst and Mowat 2012). The microbiota-stimulated deterioration of the immune reactions in GALT can also affect systematic immune reactions.

As discussed above, the intestinal microbiota can be able to control the host immune reactions by certain mechanisms such as through stimulating the innate immune reactions by TLRs as well as through stimulating free fatty acid receptors 2/3 (FFAR) by gut metabolites like SCFAs (acetate, butyrate, and propionate) and lactic acid. Butyrate is recognized as linked with the distinction of endemic T cells into Tregs, although acetate and propionate are familiar to be fundamental for shifting Tregs to the intestine (Scott et al. 2018). Abundance stimulation of TLRs and usually less production of SCFAs, majorly butyrate, are noted to have reactive impacts on T1DM-related autoimmunity and might contribute essential therapeutic marks for the inhibition of T1DM. TLRs are imperative for identifying microbial compounds containing nucleic acids, proteins, and LPS. TLRs can also recognize the endogenous compounds produced from the injured tissues or cells by damage-associated molecular patterns (DAMPs) (Scott et al. 2018). In addition, a few analyses proposed that the TLRs (for instance, TLR 3, 7, and 9) are produced in the pancreas of individuals affected with T1DM. TLR mechanisms modify the transcription factor NF-kB and I kappa B kinase (IKK) complex (Xie et al. 2018). NF-kB also controls inflammatory intercessors like IL-Iβ, the usual stimulus of T1DM pathogenesis (Xie et al. 2018).

3.4.4.2 Type 2 Diabetes Mellitus

Globally, T2DM is a common chronic disease with an escalating predominance in several countries. However, the genetic history of individuals is pivotal, although the environmental aspects, lifestyle, and dietary habits are recognized as fundamental factors in T2DM individuals. An auspicious prospective path could use a symbiotic approach linking gut microbiota and diet to treat T2DM. T2DM is distinguished by the decrease in Firmicutes and an enrichment of Bacteroidetes and Firmicutes ratio due to variations in plasma glucose levels (Graessler et al. 2013; Larsen et al. 2010). In the patients affected with T2DM, obesity is proximately associated. Studies have shown that gut microbiota modifications are not similar between both category patients. The consumption of a high-fat diet enhances certain microbes in the gut, leading to increased levels of lipopolysaccharides and insulin tolerance. T2DM is a complex metabolic disorder characterized by insulin tolerance, hyperglycemia, and metabolic disruption of blood lipids. The gut microbiota is vital in preventing T2DM by modulating individuals' biological activities and metabolism. T2DM associated with the aberrant intestinal microbial composition initiates moderate inflammation. In T2DM individuals, the gut microbiota contains a decrease in the number of butyrate-expressing bacteria, specifically *Roseburia intestinalis* and *Faecalibacterium prausnitzii*; low-grade dysbiosis and pro-inflammatory habitat with enrichment in production of microbial genes responsible for oxidative stress; decreased genes expression entangled in the synthesis of vitamins; and enhanced serum LPS levels and increase in intestinal penetrability.

Furthermore, the major changes in the gut microbiota linked with T2DM contain a reduction in the levels of Firmicutes and an increase of Bacteroidetes and Proteobacteria (Roager et al. 2017). In T2DM patients, the microbiota contains high-grade levels of pathogens like *Clostridium clostridioformis*, *Bacteroides caccae*, *Clostridium ramosum*, *Clostridium hathewayi*, *E. coli*, *Clostridium symbiosum*, and *Eggerthella* spp. (Karlsson et al. 2013). The Gram-negative bacteria produce LPS, which can induce innate immunity by stimulating the TLRs and expressing inflammatory cytokines.

Moreover, LPS induces the expression of NF-kB and c-Jun-terminal kinase mechanisms; these two ways are associated with the progression of insulin resistance and the lack of insulin signaling in adipocytes, liver, and hypothalamus (Newsholme et al. 2016). The metabolites of gut microbiota, such as SCFAs like acetate, butyrate, and propionate, are responsible for the fermentation of dietary carbohydrates. Acetate and propionate are formed from the *Bacteroidetes* sp. The Firmicutes produce butyrate. Dysbiosis is associated with modifying SCFAs production, whereas butyrate progresses insulin resistance and secretion by activating the expression of GLP-1 and decreasing the adipocyte's inflammation (Ríos-Covián et al. 2016). More prominently, butyrate is pivotal for assisting T2DM symptoms.

3.4.5 Irritable Bowel Syndrome

Irritable bowel syndrome (IBS) is a functional gastrointestinal disorder associated with altered bowel discharges and severe abdomen pain. IBS is a downstream signature disease of gut microbiota dysbiosis. Existing evidence claims that gut dysbiosis is the main reason for IBS. The prevalence of IBS is up to 12% in the general population (Lovell and Ford 2012). Pathogenesis of IBS stems from a disturbance in the gut-brain axis due to psychological stress and visceral hypersensitivity. Microbial distribution in the GI tract of healthy individuals differs from IBS patients (Tap et al. 2017). Beneficial microbial communities reduce in the GI tract of IBS patients (Carroll et al. 2011). In 2017, Botschuijver et al. found a decline in the diversity of beneficial mycobiome (fungal communities) in IBS patients (Botschuijver et al. 2017). The overgrowth of pathogenic bacteria in the small intestine induces the pathogenesis of IBS. Small intestine bacteria overgrowth (SIBO) triggers clinical obligations such as visceral sensation and poor nutrition uptake (Coelho et al. 2000; Giannella et al. 1974). The correlation between SIBO and IBS is quite intuitive. The lack of diagnostic tools that could detect the markers of SIBO is a big problem in IBS clinical practice. But it is possible to screen and characterize the microbial communities during IBS pathogenesis with metagenomics and culture-independent tools. Understanding the gut microbiome dynamics of healthy and diseased individuals might be achieved by case-control studies with advanced genomic tools. Targeting gut microbes as markers enables the choice of therapy with pharmaceutical or non-pharmaceuticals and nutraceuticals.

3.4.6 Diarrhea

Diarrhea is one of the main clinical manifestations ranging from mild to severe gastrointestinal exacerbations. Diarrheal cases are reported in children under 5 years of age in poor and developing countries (MacGill 2017; Roman et al. 2017). Diarrhea is identified as a frequent discharge of loose bowels with high liquidity, nevertheless, poor hygienic practices, intake of contaminated consumables, and multiple factors (Liu et al. 2012). Pathogen invasion is the main reason for diarrheal infections. Certain pathogens, viz., *Salmonella*, *Campylobacter*, *Shigella*, and *Rotavirus*, are responsible for disturbing the gut ecosystem with their invasive mechanisms (Garthright et al. 1988). Non-typhoid *Salmonella enterica* serovar *Typhimurium* is associated with higher infectivity in dysentery cases. Next to *Salmonella*, *Campylobacter* is the pathogen that severely damages the balance in the gut ecosystem. *Shigella* is another food-borne pathogen that affects the small intestine and cause inflammatory diarrhea. *Shigella* can cause infection even at very low inoculums. *Shigella* invasion occurs through contaminated food and water intake, unhygienic sex practices, and poor sanitation. Besides, protozoans *Giardia* and *Entamoeba histolytica* (amoebic dysentery) and *Rotavirus* are responsible for diarrheal dysbiosis in the gut microbiome. Diarrheal infections can be inflammatory or non-inflammatory. Pathogens invade the GI epithelium and affect the intestine's colon and ileum. In non-inflammatory dysentery, the pathogen directly invades the small intestine and mediates its toxicity (Taylor et al. 2013). Antibiotics and oral rehydration solutions are used as therapeutic practices to treat dysentery damage in the gut. With the beneficial effects and nontoxic microbiome reconstructive properties of probiotics, they are currently being prescribed by doctors to combat these types of GI diseases (Kota et al. 2018).

3.5 Redirecting Gut Microbiome to Modulate Host Immunity and Health

Reconstituting the gut microbiome to its native state is called gut eubiosis. Therapeutic strategies with non-pharmaceutical active ingredients that target the host's gut microbiome to modulate the gut microbiome's composition offered promising and reliable outcomes in recent preclinical and clinical studies. Similar studies focusing on gut health and host immunity regulation with pre- and probiotics suggested strong and reliable observations toward exploring the gut microbiome as an operational tool for immunometabolic therapy. Prebiotics, probiotics, and some other functional foods which could favor nonpathogenic beneficial microbes in the GI tract to grow are mainly investigated and deeply studied as non-pharmaceutical factors to shape the gut environment by modulating the composition of the intestinal microbiome. Both pre- and probiotics are the better choices for immune and metabolic therapy because of their availability and accuracy. Probiotics and prebiotics used in human consumables are enlisted in Table 3.4. Probiotics are certain nonpathogenic microbes that may be residents of the GI tract. Probiotics alter the proportion of gut microbiome toward beneficial microbes and confer host with several benefits such as resistance to invading pathogenic groups and immunometabolic modulation. Prebiotics are the nutritional factors that help probiotic bacteria to grow. Dietary interventions with non-pharmaceutical factors such as probiotics and prebiotics offer an effective way of therapy to combat various gastrointestinal and non-gastrointestinal metabolic disorders by redirecting the gut microbiome to a native or eubiotic state.

Table 3.4 Mostly used prebiotic and probiotics in human consumables

Probiotics	Organism	Reference
Bacillus	*Bacillus subtilis*, *Bacillus coagulans*, *Bacillus laterosporus*	de Simone (2019)
Bifidobacterium	*Bifidobacterium animalis*, *Bifidobacterium bifidum*, *Bifidobacterium breve*, *Bifidobacterium catenulatum*, *Bifidobacterium longum*	
Enterococcus	*Enterococcus faecium*	
Lactobacillus	*Lactobacillus acidophilus*, *Lactobacillus bulgaricus*, *Lactobacillus casei*, *Lactobacillus crispatus*, *Lactobacillus gasseri*, *Lactobacillus plantarum*, *Lactobacillus paracasei*, *Lactobacillus rhamnosus*, *Lactobacillus reuteri*	
Saccharomyces	*Saccharomyces boulardii*	
Prebiotics		
Arabinoxylan		Chen and Karboune (2019)
Beta glucans		Velikonja et al. (2019)
Fructooligosaccharides (FOS)		dos Santos et al. (2019)
Galactooligosaccharides (GOS)		Fan et al. (2019)
Inulin		dos Santos et al. (2019)
Isomalto-oligosaccharides (IMO)		Wu et al. (2017)
Lactulose		Zeng et al. (2019)
Polydextrose		Ho et al. (2018)
Xylo-oligosaccharides (XOS)		Madhukumar and Muralikrishna (2012)
Xylo-polysaccharide (XPS)		Costa et al. (2019)

3.6 Conclusions and Future Perspectives

The human body harbors several microbial ecosystems called gut, oral, vaginal, skin, and respiratory microbiomes. Gut microbiota regulates the host immune system, and diversions from normal microbial development such as C-sections, formulated diet, antimicrobial usage, and sterile vaccine in neonates alter the progression of immune system outcomes and possibly predispose entities to several inflammatory disorders after that in life. According to immunological and clinical analysis, it is believed that intestinal microbiota dysbiosis might be a fundamental aspect of various inflammatory diseases. Intestinal dysbiosis decreases beneficial microbes resulting in the progression of several inflammatory responses and immune-interceded diseases. Hence targeting and engineering gut microbiota are an effective therapeutic strategy for immune and metabolic issues. Redirecting the gut microbiota from a dysbiotic to a eubiotic state is the main agenda and algorithm of gut engineering. Prebiotics and probiotics or their derived nutraceuticals alter the gut microbes toward the beneficial microbial communities and suppress the inflammatory responses and immune dysregulation, conferring host with boosted immunity and better health.

Acknowledgments Prakash Narayana Reddy thanks the INSPIRE division of the Department of Science and Technology, Government of India, for awarding the INSPIRE Faculty award and research grant (DST/INSPIRE/04/2017/000565). The authors thank the management and administration of Vignan's University and Krishna University for providing the necessary facilities during the preparation of this article.

Declaration of Competing Interests Authors declare no conflict of interests.

References

Aagaard K, Ma J, Antony KM, Ganu R, Petrosino J, Versalovic J (2014) The placenta harbors a unique microbiome. Sci Transl Med 6:237ra65. https://doi.org/10.1126/scitranslmed.3008599
Afshin A, Forouzanfar MH, Reitsma MB, Sur P, Estep K et al (2017) Health effects of overweight and obesity in 195 countries over 25 years. N Engl J Med 377:13e27. https://doi.org/10.1056/NEJMoa1614362
Agus A, Denizot J, Thévenot J, Martinez-Medina M, Massier S, Sauvanet P et al (2016) Western diet induces a shift in microbiota composition enhancing susceptibility to Adherent-Invasive E. coli infection and intestinal inflammation. Sci Rep 6:1–14. https://doi.org/10.1038/srep19032
Albenberg LG, Lewis JD, Wu GD (2012) Food and the gut microbiota in inflammatory bowel diseases: a critical connection. Curr Opin Gastroenterol 28:314–320. https://doi.org/10.1097/MOG.0b013e328354586f
Asarat M, Vasiljevic T, Apostolopoulos V, Donkor O (2015) Short-chain fatty acids regulate secretion of IL-8 from human intestinal epithelial cell lines in vitro. Immunol Investig 44:678–693. https://doi.org/10.3109/08820139.2015.1085389
Atarashi K, Tanoue T, Oshima K et al (2013) Treg induction by a rationally selected mixture of clostridia strains from the human microbiota. Nature 500(7461):232–236. https://doi.org/10.1038/nature12331
Bäckhed F, Ley RE, Sonnenburg JL, Peterson DA, Gordon JI (2005) Host-bacterial mutualism in the human intestine. Science 307(5717):1915–1920. https://doi.org/10.1126/science.1104816

Balamurugan R, Rajendiran E, George S et al (2008) Real-time polymerase chain reaction quantification of specific butyrate-producing bacteria, Desulfovibrio and Enterococcus faecalis in the feces of patients with colorectal cancer. J Gastroenterol Hepatol 23(8 Pt 1):1298–1303. https://doi.org/10.1111/j.1440-1746.2008.05490.x

Battaglia M, Atkinson MA (2015) The streetlight effect in type 1 diabetes. Diabetes 64:1081–1090. https://doi.org/10.2337/db14-1208

Baumgart M, Dogan B, Rishniw M, Weitzman G, Bosworth B, Yantiss R et al (2007) Culture independent analysis of ileal mucosa reveals a selective increase in invasive Escherichia coli of novel phylogeny relative to depletion of Clostridiales in Crohn's disease involving the ileum. ISME J 1:403–418. https://doi.org/10.1038/ismej.2007.52

Bjursell M et al (2011) Improved glucose control and reduced body fat mass in free fatty acid receptor 2-deficient mice fed a high-fat diet. Am J Physiol Endocrinol Metab 300:E211–E220. https://doi.org/10.1152/ajpendo.00229.2010

Bloom SM, Bijanki VN, Nava GM, Sun L, Malvin NP, Donermeyer DL et al (2011) Commensal Bacteroides species induce colitis in host-genotype-specific fashion in a mouse model of inflammatory bowel disease. Cell Host Microbe 9:390–403. https://doi.org/10.1016/j.chom.2011.04.009

Boismenu R, Havran WL (1994) Modulation of epithelial cell growth by intraepithelial gamma delta T cells. Science 266:1253–1255. https://doi.org/10.1126/science.7973709

Bonnet M, Buc E, Sauvanet P et al (2014) Colonization of the human gut by E. coli and colorectal cancer risk. Clin Cancer Res 20:859–867. https://doi.org/10.1158/1078-0432.CCR-13-1343

Borthakur A, Anbazhagan AN, Kumar A, Raheja G, Singh V, Ramaswamy K et al (2010) The probiotic Lactobacillus plantarum counteracts TNF-induced downregulation of SMCT1 expression and function. AJP Gastrointest Liver Physiol 299:G928–G934. https://doi.org/10.1152/ajpgi.00279.2010

Botschuijver S, Roeselers G, Levin E, Jonkers DM, Welting O, Heinsbroek S, de Weerd HH, Boekhout T, Fornai M, Masclee AA, Schuren F, de Jonge WJ, Seppen J, van den Wijngaard RM (2017) Intestinal fungal dysbiosis is associated with visceral hypersensitivity in patients with irritable bowel syndrome and rats. Gastroenterology 153(4):1026–1039. https://doi.org/10.1053/j.gastro.2017.06.004

Candela M, Turroni S, Biagi E, Carbonero F, Rampelli S, Fiorentini C et al (2014) Inflammation and colorectal cancer, when microbiota-host mutualism breaks. World J Gastroenterol 20:908–922. https://doi.org/10.3748/wjg.v20.i4.908

Cani PD et al (2008) Changes in gut microbiota control metabolic endotoxemia-induced inflammation in high-fat diet-induced obesity and diabetes in mice. Diabetes 57:1470–1481. https://doi.org/10.2337/db07-1403

Carroll IM, Ringel-Kulka T, Keku TO, Chang YH, Packey CD, Sartor RB, Ringel Y (2011) Molecular analysis of the luminal- and mucosal-associated intestinal microbiota in diarrhea-predominant irritable bowel syndrome. Am J Physiol Gastrointest Liver Physiol 301(5):G799–G807. https://doi.org/10.1152/ajpgi.00154.2011

Chairatana P, Nolan EM (2017) Defensins, lectins, mucins, and secretory immunoglobulin a: microbe-binding biomolecules that contribute to mucosal immunity in the human gut. Crit Rev Biochem Mol Biol 52(1):45–56. https://doi.org/10.1080/10409238.2016.1243654

Chen L, Karboune S (2019) Prebiotics in food and health: properties, functionalities, production, and overcoming limitations with second-generation levan-type fructooligosaccharides. Encyclopedia of Food Chem 3:271–279. https://doi.org/10.1016/B978-0-08-100596-5.21746-0

Chen HM et al (2013) Decreased dietary fiber intake and structural alteration of gut microbiota in patients with advanced colorectal adenoma. Am J Clin Nutr 97:1044–1052. https://doi.org/10.3945/ajcn.112.046607

Cheng K, Raufman JP (2005) Bile acid–induced proliferation of a human colon cancer cell line is mediated by transactivation of epidermal growth factor receptors. Biochem Pharmacol 70(7):1035–1047. https://doi.org/10.1016/j.bcp.2005.07.023

Coelho AM, Fioramonti J, Buéno L (2000) Systemic lipopolysaccharide influences rectal sensitivity in rats: role of mast cells, cytokines, and vagus nerve. Am J Physiol Gastrointest Liver Physiol 279(4):G781–G790. https://doi.org/10.1152/ajpgi.2000.279.4.G781

Cosnes J, Gower-Rousseau C, Seksik P, Cortot A (2011) Epidemiology and natural history of inflammatory bowel diseases. Gastroenterology 140:1785–1794. https://doi.org/10.1053/j.gastro.2011.01.055

Costa GM, Paula MM, Barão CE, Klososki SJ, Bonafe EG, Visentainer JV, Pimentel TC (2019) Yoghurt added with Lactobacillus casei and sweetened with natural sweeteners and/or prebiotics: implications on quality parameters and probiotic survival. Int Dairy J 97:139–148. https://doi.org/10.1016/j.idairyj.2019.05.007

Cox LM, Blaser MJ (2013) Pathways in microbe-induced obesity. Cell Metab 17:883–894

Dalton JE et al (2006) Intraepithelial gammadelta+ lymphocytes maintain the integrity of intestinal epithelial tight junctions in response to infection. Gastroenterology 131:818–829

Darfeuille-Michaud A, Boudeau J, Bulois P, Neut C, Glasser AL, Barnich N et al (2004) High prevalence of adherent-invasive Escherichia coli associated with ileal mucosa in Crohn's disease. Gastroenterology 127:412–421. https://doi.org/10.1053/j.gastro.2004.04.061

de Giorgio R, Blandizzi C (2010) Targeting enteric neuroplasticity: diet and bugs as new key factors. Gastroenterology 138:1663–1666. https://doi.org/10.1053/j.gastro.2010.03.022

de Simone C (2019) The unregulated probiotic market. Clin Gastroenterol Hepatol 17(5):809–817. https://doi.org/10.1016/j.cgh.2018.01.018

den Besten G, van Eunen K, Groen AK, Venema K, Reijngoud DJ, Bakker BM (2013) The roles of short-chain fatty acids in the interplay between diet, gut microbiota, and host energy metabolism. J Lipid Res 54:2325–2340. https://doi.org/10.1194/jlr.R036012

Dennis KL, Wang Y, Blatner NR, Wang S, Saadalla A, Trudeau E et al (2013) Adenomatous polyps are driven by microbe-instigated focal inflammation and are controlled by IL-10-producing T cells. Cancer Res 73:5905–5913. https://doi.org/10.1158/0008-5472.CAN-13-1511

Donia MS, Fischbach MA (2015) Small molecules from the human microbiota. Science 349(6246):1254766

dos Santos DX, Casazza AA, Aliakbarian B, Bedani R, Saad SMI, Perego P (2019) Improved probiotic survival to in vitro gastrointestinal stress in a mousse containing Lactobacillus acidophilus La-5 microencapsulated with inulin by spray drying. Lwt 99:404–410. https://doi.org/10.1016/j.lwt.2018.10.010

Dumas ME, Barton RH, Toye A, Cloarec O, Blancher C, Rothwell A, Mitchell SC (2006) Metabolic profiling reveals a contribution of gut microbiota to fatty liver phenotype in insulin-resistant mice. Proc Natl Acad Sci U S A 103(33):12511–12516. https://doi.org/10.1073/pnas.0601056103

Dumas JA, Bunn JY, Nickerson J, Crain KI, Ebenstein DB, Tarleton EK, Makarewicz J, Poynter ME, Kien CL (2016) Dietary saturated fat and monounsaturated fat have reversible effects on brain function and the secretion of pro-inflammatory cytokines in young women. Metabolism 65:1582–1588. https://doi.org/10.1016/j.metabol.2016.08.003

Duncan SH, Louis P, Microbiol FHJ (2007) Cultivable bacterial diversity from the human colon. Lett Appl 44:343–350. https://doi.org/10.1111/j.1472-765X.2007.02129.x

Fan J, Chen L, Mai G, Zhang H, Yang J, Deng D, Ma Y (2019) Dynamics of the gut microbiota in developmental stages of Litopenaeus vannamei reveal its association with body weight. Scientific Reports 9(1):734

Frank DN, St Amand AL, Feldman RA et al (2007) Molecular-phylogenetic characterization of microbial community imbalances in human inflammatory bowel diseases. Proc Natl Acad Sci U S A 104(34):13780–13785. https://doi.org/10.1073/pnas.0706625104

Fujimoto T, Imaeda H, Takahashi K et al (2013) Decreased abundance of Faecalibacterium prausnitzii in the gut microbiota of Crohn's disease. J Gastroenterol Hepatol 28(4):613–619. https://doi.org/10.1111/jgh.12073

Furet JP et al (2010) Differential adaptation of human gut microbiota to bariatric surgery-induced weight loss: links with metabolic and low-grade inflammation markers. Diabetes 59:3049–3057. https://doi.org/10.2337/db10-0253

Garthright WE, Archer DL, Kvenberg JE (1988) Estimates of incidence and costs of intestinal infectious diseases in the United States. Public Health Rep 103(2):107–115

Giannella RA, Rout WR, Toskes PP (1974) Jejunal brush border injury and impaired sugar and amino acid uptake in the blind loop syndrome. Gastroenterology 67(5):965–974

Gill RK (2005) Expression and membrane localization of MCT isoforms along the length of the human intestine. AJP Cell Physiol 289:C846–C852. https://doi.org/10.1152/ajpcell.00112.2005

Gordon JI, Dewey KG, Mills DA, Medzhitov RM (2012) The human gut microbiota and under nutrition. Sci Transl Med 6(4):137. https://doi.org/10.1126/scitranslmed.3004347

Goto Y, Obata T, Kunisawa J, Sato S, Ivanov II, Lamichhane A, Takeyama N, Kamioka M, Sakamoto M, Matsuki T (2014) Innate lymphoid cells regulate intestinal epithelial cell glycosylation. Science 345:1254009. https://doi.org/10.1126/science.1254009

Graessler J, Qin Y, Zhong H et al (2013) Metagenomic sequencing of the human gut microbiome before and after bariatric surgery in obese patients with type 2 diabetes: correlation with inflammatory and metabolic parameters. Pharmacogenomics J 13(6):514–522. https://doi.org/10.1038/tpj.2012.43

Greten FR, Eckmann L, Greten TF, Park JM, Li ZW, Egan LJ et al (2004) IKKbeta links inflammation and tumorigenesis in a mouse model of colitis-associated cancer. Cell 118:285–296. https://doi.org/10.1016/j.cell.2004.07.013

Grivennikov SI, Karin M (2010) Dangerous liaisons: STAT3 and NF-kappa B collaboration and crosstalk in cancer. Cytokine Growth Factor Rev 21:11–19. https://doi.org/10.1016/j.cytogfr.2009.11.005

Guarner F (2006) Enteric flora in health and disease. Digestion 73(Suppl 1):5–12. https://doi.org/10.1159/000089775

Hansen R, Thomson JM, El-Omar EM, Hold GL (2010) The role of infection in the aetiology of inflammatory bowel disease. J Gastroenterol 45:266–276. https://doi.org/10.1007/s00535-009-0191-y

Ho AL, Kosik O, Lovegrove A, Charalampopoulos D, Rastall RA (2018) In vitro fermentability of xylo-oligosaccharide and xylo-polysaccharide fractions with different molecular weights by human faecal bacteria. Carbohydr Polym 179:50–58. https://doi.org/10.1016/j.carbpol.2017.08.077

Holscher HD (2017) Dietary fiber and prebiotics and the gastrointestinal microbiota. Gut Microbes 8:172–184. https://doi.org/10.1080/19490976.2017.1290756

Hooper C, Jackson SS, Coughlin EE, Coon JJ, Miyamoto S (2014) Covalent modification of the NF-kappaB essential modulator (NEMO) by a chemical compound can regulate its ubiquitin binding properties in vitro. J Biol Chem 289:33161–33174. https://doi.org/10.1074/jbc.M114.582478

Hruby A, Hu FB (2015) The epidemiology of obesity: a big picture. PharmacoEconomics 33:673e89. https://doi.org/10.1007/s40273-014-0243-x

Ivanov II et al (2009a) Induction of intestinal Th17 cells by segmented filamentous bacteria. Cell 139:485–498. https://doi.org/10.1016/j.cell.2009.09.033

Ivanov K, Kolev N, Tonev A, Nikolova G, Krasnaliev I, Softova E et al (2009b) Comparative analysis of prognostic significance of molecular markers of apoptosis with clinical stage and tumor differentiation in patients with colorectal cancer: a single institute experience. Hepato-Gastroenterology 56:94–98

Kabat AM, Srinivasan N, Maloy KJ (2014) Modulation of immune development and function by intestinal microbiota. Trends Immunol 35(11):507–517. https://doi.org/10.1016/j.it.2014.07.010

Karlsson FH, Tremaroli V, Nookaew I, Bergström G, Behre CJ, Fagerberg B, Nielsen J, Bäckhed F (2013) Gut metagenome in European women with normal, impaired and diabetic glucose control. Nature 498(7452):99–103. https://doi.org/10.1038/nature12198

Keku TO, Dulal S, Deveaux A, Jovov B, Han X (2015) The gastrointestinal microbiota and colorectal cancer. Am J Physiol Gastrointest Liver Physiol 308:G351–G363. https://doi.org/10.1152/ajpgi.00360.2012

Kirat D, Kondo K, Shimada R, Physiol KS (2009) Dietary pectin up-regulates monocarboxylate transporter 1 in the rat gastrointestinal tract. Exp 94:422–433. https://doi.org/10.1113/expphysiol.2009.046797

Kleinnijenhuis J, Quintin J, Preijers F, Joosten LA, Ifrim DC, Saeed S et al (2012) Bacille calmette-guerin induces NOD2-dependent nonspecific protection from reinfection via epigenetic reprogramming of monocytes. Proc Natl Acad Sci U S A 109:17537–17542. https://doi.org/10.1073/pnas.1202870109

Koh A, De Vadder F, Kovatcheva-Datchary P, Bäckhed F (2016) From dietary fiber to host physiology: short-chain fatty acids as key bacterial metabolites. Cell 165:1332–1345. https://doi.org/10.1016/j.cell.2016.05.041

Kota RK, Ambati RR, Aswani Kumar YVV, Srirama K, Reddy PN (2018) Recent advances in probiotics as live biotherapeutics against gastrointestinal diseases. Curr Pharm Des 24(27):3162–3171. https://doi.org/10.2174/1381612824666180717105128

Kumar H, Kawai T, Akira S (2011) Pathogen recognition by the innate immune system. Int Rev Immunol 30:16–34. https://doi.org/10.3109/08830185.2010.529976

Lamichhane S, Ahonen L, Dyrlund TS, Siljander H, Hyoty H, Ilonen J, Toppari J, Veijola R, Hyotylainen T, Knip M et al (2018) A longitudinal plasma lipidomics dataset from children who developed islet autoimmunity and type 1 diabetes. Sci Data 5:180250. https://doi.org/10.1038/sdata.2018.250

Larsen N, Vogensen F, van den Berg F et al (2010) Gut microbiota in human adults with type 2 diabetes differs from non-diabetic adults. PLoS One 5(2):e9085. https://doi.org/10.1371/journal.pone.0009085

Lee JS, Cella M, Mcdonald KG, Garlanda C, Kennedy GD, Nukaya M, Mantovani A, Kopan R, Bradfield CA, Newberry RD (2012) AHR drives the development of gut ILC22 cells and postnatal lymphoid tissues via pathways dependent on and independent of notch. Nat Immunol 13:144–151. https://doi.org/10.1038/ni.2187

Leonel AJ, Alvarez-Leite JI (2012) Butyrate: implications for intestinal function. Curr Opin Clin Nutr Metab Care 15:474–479. https://doi.org/10.1097/MCO.0b013e32835665fa

Ley RE et al (2005) Obesity alters gut microbial ecology. Proc Natl Acad Sci U S A 102:11070–11075. https://doi.org/10.1073/pnas.0504978102

Liu L, Johnson HL, Cousens S, Perin J, Scott S, Lawn JE, Rudan I, Campbell H, Cibulskis R, Li M, Mathers C, Black RE, Child Health Epidemiology Reference Group of WHO and UNICEF (2012) Global, regional, and national causes of child mortality: an updated systematic analysis for 2010 with time trends since 2000. Lancet 379(9832):2151–2161. https://doi.org/10.1016/S0140-6736(12)60560-1

Loubinoux J, Bronowicki JP, Pereira IA et al (2002) Sulfate-reducing bacteria in human feces and their association with inflammatory bowel diseases. FEMS Microbiol Ecol 40(2):107–112. https://doi.org/10.1111/j.1574-6941.2002.tb00942.x

Lovell RM, Ford AC (2012) Global prevalence of and risk factors for irritable bowel syndrome: a meta-analysis. Clin Gastroenterol Hepatol 10(7):712–721.e4. https://doi.org/10.1016/j.cgh.2012.02.029

Macfarlane GT, Macfarlane S (2012) Bacteria, colonic fermentation, and gastrointestinal health. J AOAC Int 95:50–60. https://doi.org/10.5740/jaoacint.sge_macfarlane

MacGill M (2017) What you should know about diarrhea. https://www.medicalnewstoday.com/articles/158634.php

Madhukumar MS, Muralikrishna G (2012) Fermentation of xylo-oligosaccharides obtained from wheat bran and Bengal gram husk by lactic acid bacteria and bifidobacteria. J Food Sci Technol 49(6):745–752. https://doi.org/10.1007/s13197-010-0226-7

Martens JH, Barg H, Warren MJ, Jahn D (2002) Microbial production of vitamin B12. Appl Microbiol Biotechnol 58:275–285. https://doi.org/10.1007/s00253-001-0902-7

Maynard CL, Weaver CT (2009) Intestinal effector T cells in health and disease. Immunity 31:389–400. https://doi.org/10.1016/j.immuni.2009.08.012

Medzhitov R (2008) Origin and physiological roles of inflammation. Nature 454:428–435. https://doi.org/10.1038/nature07201

Mills KH (2011) TLR-dependent T cell activation in autoimmunity. Nat Rev Immunol 11:807–822. https://doi.org/10.1038/nri3095
Mucida D, Park Y, Kim G, Turovskaya O, Scott I, Kronenberg M, Cheroutre H (2007) Reciprocal TH17 and regulatory T cell differentiation mediated by retinoic acid. Science 317:256–260. https://doi.org/10.1126/science.1145697
Murray CM, Hutchinson R, Bantick JR, Belfield GP, Benjamin AD, Brazma D et al (2005) Monocarboxylate transporter Mctl is a target for immunosuppression. Nat Chem Biol 1:371–376. https://doi.org/10.1038/nchembio744
Nastasi C, Candela M, Bonefeld CM, Geisler C, Hansen M, Krejsgaard T et al (2015) The effect of short-chain fatty acids on human monocyte-derived dendritic cells. Sci Rep 5:1–10. https://doi.org/10.1038/srep16148
Negi S, Pahari S, Das DK, Khan N, Agrewala JN (2019) Curdlan limits mycobacterium tuberculosis survival through STAT-1 regulated nitric oxide production. Front Microbiol 10:1173
Neish AS (2009) Microbes in gastrointestinal health and disease. Gastroenterology 136:65–80. https://doi.org/10.1053/j.gastro.2008.10.080
Newsholme P, Cruzat VF, Keane KN, Carlessi BPI (2016) Molecular mechanisms of ROS production and oxidative stress in diabetes. Biochem J 473:4527–4550. https://doi.org/10.1042/BCJ20160503C
Nicholson JK, Holmes E, Kinross J, Burcelin R, Gibson G, Jia W, Pettersson S (2012) Host-gut microbiota metabolic interactions. Science 336:1262–1267. https://doi.org/10.1126/science.1223813
Pabst O, Mowat AM (2012) Oral tolerance to food protein. Mucosal Immunol 5:232–239. https://doi.org/10.1038/mi.2012.4
Pahari S, Kaur G, Negi S, Aqdas M, Das DK, Bashir H et al (2018) Reinforcing the functionality of mononuclear phagocyte system to control tuberculosis. Front Immunol 9:193. https://doi.org/10.3389/fimmu.2018.00193
Pahari S, Negi S, Aqdas M, Arnett E, Schlesinger LS, Agrewala JN (2019) Induction of autophagy through CLEC4E in combination with TLR4: an innovative strategy to restrict the survival of mycobacterium tuberculosis. Autophagy 8:1–23. https://doi.org/10.1080/15548627.2019.1658436
Pennington DJ, Silva-Santos B, Hayday AC (2005) Gammadelta T cell development—having the strength to get there. Curr Opin Immunol 17:108–115. https://doi.org/10.1016/j.coi.2005.01.009
Pott J, Hornef M (2012) Innate immune signalling at the intestinal epithelium in homeostasis and disease. EMBO Rep 13:684–698. https://doi.org/10.1038/embor.2012.96
Prentiss PG, Rosen H, Brown N, Horowitz RE, Malm OJ, Levenson SM (1961) The metabolism of choline by the germfree rat. Arch Biochem Biophys 94(3):424–429. https://doi.org/10.1016/0003-9861(61)90069-8
Pushalkar S, Hundeyin M, Daley D, Zambirinis CP, Kurz E, Mishra A, Mohan N, Aykut B, Usyk M, Torres LE et al (2018) The pancreatic cancer microbiome promotes oncogenesis by induction of innate and adaptive immune suppression. Cancer Discov 8:403–416. https://doi.org/10.1158/2159-8290.CD-17-1134
Qin J et al (2010) A human gut microbial gene catalogue established by metagenomic sequencing. Nature 464:59–65. https://doi.org/10.1038/nature08821
Ridlon JM, Kang DJ, Hylemon PB (2006) Bile salt biotransformations by human intestinal bacteria. J Lipid Res 47(2):241–259. https://doi.org/10.1194/jlr.R500013-JLR200
Ríos-Covián D, Ruas-Madiedo P, Margolles A, Gueimonde M, De Los Reyes-Gavilán CG, Salazar N (2016) Intestinal short chain fatty acids and their link with diet and human health. Front Microbiol 7:185. https://doi.org/10.3389/fmicb.2016.00185
Roager HM, Vogt JK, Kristensen M, LBS H, Ibrugger S, Maerkedahl RB, Bahl MI, Lind MV, Nielsen RL, Frøkiaer H, Gøbel RJ, Landberg R, Ross AB, Brix S, Holck J, Meyer AS, Sparholt MH, Christensen AF, Carvalho V, Hartmann B, Holst JJ, Rumessen JJ, Linneberg A, Sicheritz-Pontén T, Dalgaard MD, Blennow A, Frandsen HL, Villas-Bôas S, Kristiansen K, Vestergaard H, Hansen T, Ekstrøm CT, Ritz C, Nielsen HB, Pedersen OB, Gupta R, Lauritzen L, Licht

TR (2017) Whole grain-rich diet reduces body weight and systemic low-grade inflammation without inducing major changes of the gut microbiome: a randomised cross-over trial. Gut 68(1):83–93. https://doi.org/10.1136/gutjnl-2017-314786

Rodríguez JM, Murphy K, Stanton C et al (2015) The composition of the gut microbiota throughout life, with an emphasis on early life. Mi-crobEcol Health Dis 26(10):26050. https://doi.org/10.3402/mehd.v26.26050

Roman C, Solh T, Broadhurst M (2017) Infectious diarrhea. Physician Assist Clin 2(2):229–245

Rossi M, Bot A (2013) The Th17 cell population and the immune homeostasis of the gastrointestinal tract. Int Rev Immunol 32(5–6):471–474. https://doi.org/10.3109/08830185.2013.843983

Round JL, Mazmanian SK (2010) Inducible Foxp3+ regulatory T-cell development by a commensal bacterium of the intestinal microbiota. Proc Natl Acad Sci U S A 107:12204–12209. https://doi.org/10.1073/pnas.0909122107

Rowan F, Docherty NG, Murphy M et al (2010) Desulfovibrio bacterial species are increased in ulcerative colitis. Dis Colon Rectum 53(11):1530–1536. https://doi.org/10.1007/DCR.0b013e3181f1e620

Rubin DC, Shaker A, Levin MS (2012) Chronic intestinal inflammation: inflammatory bowel disease and colitis-associated colon cancer. Front Immunol 3:107. https://doi.org/10.3389/fimmu.2012.00107

Schuijt TJ, Lankelma JM, Scicluna BP et al (2016) The gut microbiota plays a protective role in the host defence against pneumococcal pneumonia. Gut 65(4):575–583. https://doi.org/10.1136/gutjnl-2015-309728

Schwabe RF, Wang TC (2012) Cancer-bacteria deliver a genotoxic hit. Science 338:52–53. https://doi.org/10.1126/science.1229905

Schwiertz A, Taras D, Schafer K, Beijer S, Bos NA, Donus C, Hardt PD (2010) Microbiota and SCFA in lean and overweight healthy subjects. Obesity (Silver Spring) 18:190–195. https://doi.org/10.1038/oby.2009.167

Scott NA, Andrusaite A, Andersen P, Lawson M, Alcon-Giner C, Leclaire C, Caim S, Le Gall G, Shaw T, Connolly JPR et al (2018) Antibiotics induce sustained dysregulation of intestinal T cell immunity by perturbing macrophage homeostasis. Sci Transl Med 10:eaao4755. https://doi.org/10.1126/scitranslmed.aao4755

Sears CL (2009) Enterotoxigenic Bacteroides fragilis: a rogue among symbiotes. Clin Microbiol Rev 22:349–369. https://doi.org/10.1128/CMR.00053-08

Senghor B, Sokhna C, Ruimy R, Lagier JC (2018) Gut microbiota diversity according to dietary habits and geographical provenance. Human Microbiome Journal 7:1–9. https://doi.org/10.1016/j.humic.2018.01.001

Sheehan D, Moran C, Shanahan F (2015) The microbiota in inflammatory bowel disease. J Gastroenterol 50(5):495–507. https://doi.org/10.1007/s00535-015-1064-1

Slack E et al (2009) Innate and adaptive immunity cooperate flexibly to maintain host-microbiota mutualism. Science 325:617–620. https://doi.org/10.1126/science.1172747

Smith PM, Howitt MR, Panikov N, Michaud M, Gallini CA, Bohlooly-Y M et al (2013) The microbial metabolites, short-chain fatty acids, regulate colonic Treg cell homeostasis. Science 341:569–573. https://doi.org/10.1126/science.1241165

Sokol H, Lepage P, Seksik P, Dore J, Marteau P (2006) Temperature gradient gel electrophoresis of fecal 16S rRNA reveals active Escherichia coli in the microbiota of patients with ulcerative colitis. J Clin Microbiol 44:3172–3177. https://doi.org/10.1128/JCM.02600-05

Sokol H, Pigneur B, Watterlot L et al (2008) Faecalibacterium prausnitzii is an anti-inflammatory commensal bacterium identified by gut microbiota analysis of Crohn's disease patients. Proc Natl Acad Sci U S A 105(43):16731–16736. https://doi.org/10.1073/pnas.0804812105

Sonnenberg GF, Artis D (2012) Innate lymphoid cell interactions with microbiota: implications for intestinal health and disease. Immunity 37:601–610. https://doi.org/10.1016/j.immuni.2012.10.003

Spencer MD, Hamp TJ, Reid RW, Fischer LM, Zeisel SH, Fodor AA (2011) Association between composition of the human gastrointestinal microbiome and development of fatty

liver with choline deficiency. Gastroenterology 140(3):976–986. https://doi.org/10.1053/j.gastro.2010.11.049
Szablewski L (2018) Human gut microbiota in health and Alzheimer's disease. J Alzheimers Dis 62(2):549–560. https://doi.org/10.3233/JAD-170908
Takahashi K, Nishida A, Fujimoto T et al (2016) Reduced abundance of butyrate-producing bacteria species in the fecal microbial community in Crohn's disease. Digestion 93(1):59–65. https://doi.org/10.1159/000441768
Tap J, Derrien M, Törnblom H, Brazeilles R, Cools-Portier S, Doré J, Störsrud S, Le Nevé B, Öhman L, Simrén M (2017) Identification of an intestinal microbiota signature associated with severity of irritable bowel syndrome. Gastroenterology 152(1):111–123.e8. https://doi.org/10.1053/j.gastro.2016.09.049
Taylor EV, Herman KM, Ailes EC, Fitzgerald C, Yoder JS, Mahon BE, Tauxe RV (2013) Common source outbreaks of campylobacter infection in the USA, 1997-2008. Epidemiol Infect 141(5):987–996. https://doi.org/10.1017/S0950268812001744
Thangaraju M, Cresci GA, Liu K, Ananth S, Gnanaprakasam JP, Browning DD et al (2009) GPR109A is a g-protein–coupled receptor for the bacterial fermentation product butyrate and functions as a tumor suppressor in colon. Cancer Res 69:2826–2832. https://doi.org/10.1158/0008-5472.CAN-08-4466
Tian J, Lin X, Zhou W, Xu J (2003) Hydroxyethyl starch inhibits NF-kappaB activation and prevents the expression of inflammatory mediators in endotoxic rats. Ann Clin Lab Sci 33:451–458
Tlaskova-Hogenova H, Stepankova R, Hudcovic T et al (2004) Commensal bacteria (normal microflora), mucosal immunity and chronic inflammatory and autoimmune diseases. Immunol Lett 93:97–108. https://doi.org/10.1016/j.imlet.2004.02.005
Tolhurst G, Heffron H, Lam YS, Parker HE, Habib AM, Diakogiannaki E, Cameron J, Grosse J, Reimann F, Gribble FM (2012) Short-chain fatty acids stimulate glucagon-like peptide-1 secretion via the G-protein-coupled receptor FFAR2. Diabetes 61(2):364–371. https://doi.org/10.2337/db11-1019
Tomasello G, Tralongo P, Damiani P, Sinagra E, Di Trapani B, Zeenny MN et al (2014) Dismicrobism in inflammatory bowel disease and colorectal cancer: changes in response of colocytes. World J Gastroenterol 20:18121–18130. https://doi.org/10.3748/wjg.v20.i48.18121
Toprak NU, Yagci A, Gulluoglu BM et al (2006) A possible role of Bacteroides fragilis enterotoxin in the aetiology of colorectal cancer. Clin Microbiol Infect 12:782–786. https://doi.org/10.1111/j.1469-0691.2006.01494.x
Tunaru S, Kero J, Schaub A, Wufka C, Blaukat A, Pfeffer K et al (2003) PUMAG and HM74 are receptors for nicotinic acid and mediate its anti-lipolytic effect. Nat Med 9:352–355. https://doi.org/10.1038/nm824
Turnbaugh PJ, Ley RE, Mahowald MA, Magrini V, Mardis ER, Gordon JI (2006) An obesity-associated gut microbiome with increased capacity for energy harvest. Nature 444:1027e31. https://doi.org/10.1038/nature05414
Turnbaugh PJ et al (2009) A core gut microbiome in obese and lean twins. Nature 457:480–484. https://doi.org/10.1038/nature07540
Uronis JM, Muhlbauer M, Herfarth HH et al (2009) Modulation of the intestinal microbiota alters colitis-associated colorectal cancer susceptibility. PLoS One 4:e6026. https://doi.org/10.1371/journal.pone.0006026
van Wijk F, Cheroutre H (2009) Intestinal T cells: facing the mucosal immune dilemma with synergy and diversity. Semin Immunol 21:130–138. https://doi.org/10.1016/j.smim.2009.03.003
Velagapudi VR, Hezaveh R, Reigstad CS, Gopalacharyulu P, Yetukuri L, Islam S, Felin J, Perkins R, Borén J, Oresic M, Bäckhed F (2010) The gut microbiota modulates host energy and lipid metabolism in mice. J Lipid Res 51(5):1101–1112. https://doi.org/10.1194/jlr.M002774
Velikonja A, Lipoglavšek L, Zorec M, Orel R, Avguštin G (2019) Alterations in gut microbiota composition and metabolic parameters after dietary intervention with barley beta glucans in patients with high risk for metabolic syndrome development. Anaerobe 55:67–77. https://doi.org/10.1016/j.anaerobe.2018.11.002

Walker AW, Sanderson JD, Churcher C et al (2011) High-throughput clone library analysis of the mucosa-associated microbiota reveals dysbiosis and differences between inflamed and noninflamed regions of the intestine in inflammatory bowel disease. BMC Microbiol 11:7. https://doi.org/10.1186/1471-2180-11-7

Wilson JC, Furlano RI, Jick SS, Meier CR (2016) Inflammatory bowel disease and the risk of autoimmune diseases. J Crohns Colitis 10:186–193. https://doi.org/10.1093/ecco-jcc/jjv193

Wong SH, Ng SC (2013) What can we learn from inflammatory bowel disease in developing countries? Curr Gastroenterol Rep 15:313. https://doi.org/10.1007/s11894-013-0313-9

Wostmann BS, Larkin C, Moriarty A, Bruckner-Kardoss E (1983) Dietary intake, energy metabolism, and excretory losses of adult male germfree Wistar rats. Lab Anim Sci 33:46–50. PMID: 6834773

Wu S, Rhee KJ, Albesiano E et al (2009) A human colonic commensal promotes colon tumorigenesis via activation of T helper type 17 T cell responses. Nat Med 15:1016–1022. https://doi.org/10.1038/nm.2015

Wu Y, Pan L, Shang QH, Ma XK, Long SF, Xu YT, Piao XS (2017) Effects of isomalto-oligosaccharides as potential prebiotics on performance, immune function and gut microbiota in weaned pigs. Anim Feed Sci Technol 230:126–135. https://doi.org/10.1016/j.anifeedsci.2017.05.013

Xie Z, Huang G, Wang Z, Luo S, Zheng P, Zhou Z (2018) Epigenetic regulation of toll-like receptors and its roles in type 1 diabetes. J Mol Med 96:741–751. https://doi.org/10.1007/s00109-018-1660-7

Yen D, Cheung J, Scheerens H, Poulet F, McClanahan T, McKenzie B et al (2006) IL-23 is essential for T cell-mediated colitis and promotes inflammation via IL-17 and IL-6. J Clin Invest 116:1310–1316. https://doi.org/10.1172/JCI21404

Yu J et al (2017) Metagenomic analysis of faecal microbiome as a tool towards targeted non-invasive biomarkers for colorectal cancer. Gut 66:70–78. https://doi.org/10.1136/gutjnl-2015-309800

Zeng J, Song M, Jia T, Gao H, Zhang R, Jiang J (2019) Immunomodulatory influences of sialylated lactuloses in mice. Biochem Biophys Res Commun 514(2):351–357. https://doi.org/10.1016/j.bbrc.2019.04.157

Zhang Y, Zhou L, Bao YL, Wu Y, Yu CL, Huang YX et al (2010) Butyrate induces cell apoptosis through activation of JNK MAP kinase pathway in human colon cancer RKO cells. Chem Biol Interact 185:174–181. https://doi.org/10.1016/j.cbi.2010.03.035

Zinkevich VV, Beech IB (2000) Screening of sulfate-reducing bacteria in colonoscopy samples from healthy and colitic human gut mucosa. FEMS Microbiol Ecol 34(2):147–155. https://doi.org/10.1111/j.1574-6941.2000.tb00764.x

Understanding the Probiotics and Mechanism of Immunomodulation Interactions with the Gut-Related Immune System

4

Kuraganti Gunaswetha, Edla Sujatha, Krodi Anusha, A. M. V. N. Prathyusha, M. Subhosh Chandra, Chanda Vikrant Berde, Neelapu Nageswara Rao Reddy, and Pallaval Veera Bramhachari

Abstract

Probiotics are live microorganisms that, when consumed in adequate amounts, give medical advantages to the host. The gastrointestinal lot is quite possibly the most microbiologically dynamic living habitats, and it is basic to the mucosal immune system's function (MIS). Because of their noteworthy capacity to contend with pathogenic microbiota for adhesion sites, alienate pathogens, or initiate, balance, and control the host's resistant reaction by enacting the actuation of specific genes in and outside the host intestinal tract, probiotics, prebiotics, and synbiotics have shown promising outcomes against different enteric microorganisms. Pattern recognition receptors, such as toll-like receptors and nucleotide-

K. Gunaswetha (✉) · E. Sujatha
Department of Microbology, Kakatiya University, Warangal, Telangana, India
e-mail: sujathaedla_1973@kakatiya.ac.in

K. Anusha · A. M. V. N. Prathyusha · P. V. Bramhachari (✉)
Department of Biosciences & Biotechnology, Krishna University, Machilipatnam, Andhra Pradesh, India

M. S. Chandra
Department of Microbiology, Yogi Vemana University, Kadapa, India

C. V. Berde
Marine Microbiology program School of Earth, Ocean and Atmospheric Sciences, Goa University,
Taleigao, Goa, India

N. N. R. Reddy
Department of Biochemistry and Bioinformatics GITAM Institute of Science, Gandhi Institute of Technology and Management (GITAM), Visakhapatnam, Andhra Pradesh, India

P. Veera Bramhachari (ed.), *Human Microbiome in Health, Disease, and Therapy*, https://doi.org/10.1007/978-981-99-5114-7_4

binding oligomerization domain-containing protein-like receptors, modulate key signaling pathways, such as nuclear factor-B and mitogen-activated protein kinase, to enhance or suppress activation and influence downstream pathways, as per growing evidence. A careful comprehension of these cycles will help in the choice of probiotic strains for specific applications and may even prompt the disclosure of new probiotic capacities. Thus, probiotics have shown remedial potential for an assortment of diseases, including allergy, migraines, viral disease, and potentiating vaccine reactions. The objective of this orderly survey that probiotics may give novel ways to deal with both disease counteraction and treatment and to investigate probiotic methods of activity zeroing in on how gut organisms impact the host.

Keywords

Probiotics · Microbiota · Human gut · Probiotic mechanism of action · Symbiotic relationship

4.1 Introduction

For a great many years, people have securely consumed microorganisms in the form of fermented foods. Immense quantities of these bacteria are seen as probiotics, which act through various systems to introduce a medical advantage to the host. It is broadly accepted that fermented products were likely found, or better to state, discovered impulsively. Probiotics and prebiotics are considered novel functional ingredients that can be applied to impact the host's microbiota, which thus assumes a significant part in the host's nourishment, advancement, health, and well-being. The World Health Organization (WHO) defined probiotics as live microorganisms, especially beneficial bacteria, when consumed as food products or supplements that react with host commensal microflora and gastrointestinal tract immune system (Johnson and Klaenhammer 2014). *The Prolongation of Life: Optimistic Studies*, written by Metchnikoff, promotes the probiotics idea by recommending that festering in the intestines corresponds with an abbreviated life expectancy (Call et al. 2015). In a scientific context, Metchnikoff quoted probiotics as altering microbial flora diversity in human bodies and replacing harmful microbes with useful ones (Metchnikoff 2004). Metchnikoff recommended that when lactic acid-producing microorganisms were eaten up may go about as against putrefactive agents in the gastrointestinal tract. Indeed, he assessed that the pathological manifestations might be ousted from old culture by changing the wild intestinal populace into a cultured populace. The length of man's existence may be broadly extended. Michel Cohendy, a colleague of Metchnikoff at the Pasteur Institute, supported his hypothesis by providing experimental data. *Lactobacillus delbrueckii* subsp. *Bulgaricus*, when feeding to human subjects, was recoverable from feces, diminished the pervasiveness of putrefactive poisons, also, upheld in the treatment of colitis following transplantation to the large intestine (Cohendy 1906a, b) (Fig. 4.1).

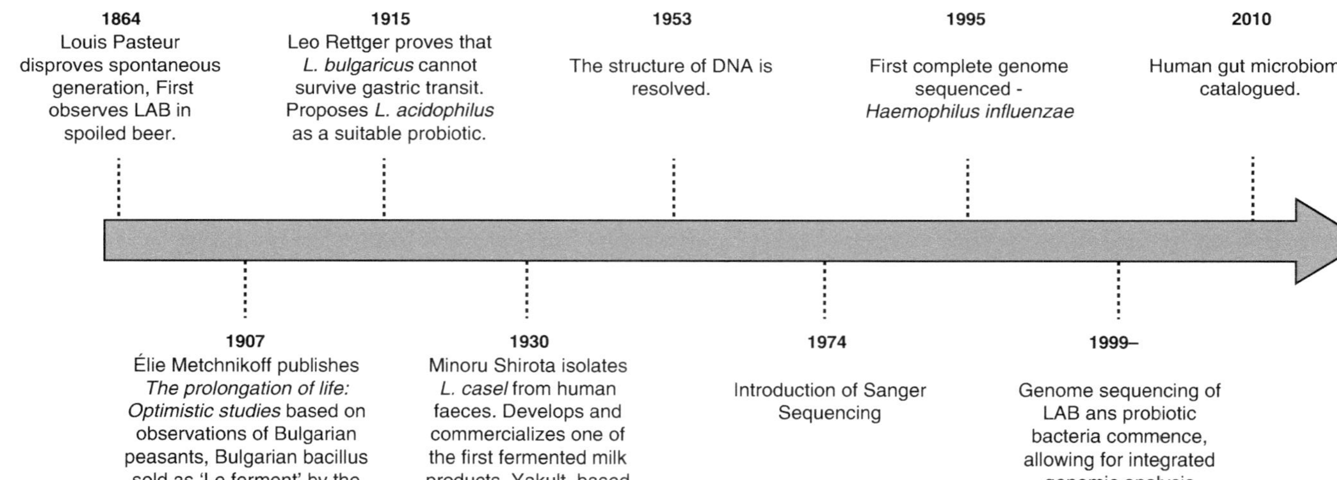

Fig. 4.1 Fundamental achievements adding to the utilitarian characterization of probiotic lactic acid bacteria

Studies on *Lactobacillus acidophilus* showed that it could endure gastric section and change the intestinal flora in constraining lactose and dextrin supplementation, making it a very prompt candidate for therapeutic applications; therefore, treatment with *L. acidophilus* originated (Walsh et al. 2017). Bacteria, including *Lactobacillus*, *Leuconostoc*, *Bifidobacterium*, *Enterococcus*, *Pediococcus*, and yeasts, are used as probiotics. Species of genera *Lactobacillus* and *Bifidobacterium* are safe and widely used in yogurts and other dairy products since they are also a part of the normal gastrointestinal microflora. Controlled gastrointestinal infections, improved lactose metabolism, reduction of cholesterol, anticarcinogenic and antimutagenic properties, improvement in inflammatory bowel disease, and immune system stimulation are some of the health benefits of probiotics (Aureli et al. 2011). Havenaar and Huisint Veld proposed a viable mono or mixture of the culture of bacteria as the high-level significance probiotic, when given to man, impacts the host advantageously by improving the properties of the indigenous flora (Kerry et al. 2018).

Heterogeneous and different groups of microorganisms live from a human microbiome superorganism since the human body acts as a reservoir. About 100–1000 microbial species resident in the human gut modulate the human internal environment, thereby playing a crucial role in host health. These symbiotic microorganisms are distinctive in defense function and impact brain-gut responses, eupepsia, catabolism, and anabolism (Kerry et al. 2018). By amplifying the number of gut microflora by probiotics, prebiotics and synbiotics initiate the activation of specific genes of the host intestinal tract that regulate and modulate host immune response and alienate pathogens by competing with enteric pathogenic microbiota for adhesion sites (Tripathi et al. 2019). There have been important advancements in the collection and characterization of particular probiotic cultures and significant health benefits from their use. Microbial species in the gastrointestinal tract (GIT) range from around 100 to 1000, with the total microbial population of the colony ranging from 10^{11} to 10^{12} cfu/g in each individual, which advances and changes over a long period, contingent upon the way of life of the host, antimicrobial use, genome, and on an unpredictable and dynamic exchange between the eating diet (Slavin 2013). By colonization resistance or a barrier effect against newly ingested microorganisms, including pathogens, gut microflora retains their existence and preserves their confer niche (Pérez-Cobas et al. 2015). Therefore, it's possible that controlling the gut microflora to expand the overall quantities of "beneficial bacteria" may improve immune capacity, digestion, processing, and brain-gut communication (Diop et al. 2016). Any shifts in their diversity can lead to various disorders and diseases.

Probiotics, prebiotics, and synbiotics are all microbes or communities of microorganisms that live in the gut and nourish the host body from within. Nonviable bacterial products or metabolic by-products from probiotic microorganisms are known as postbiotics. They also have biological activity in the host effects on signaling pathways and barrier function. Bacterial metabolic by-products like diacetyl, organic acids, bacteriocins, ethanol, hydrogen peroxide, and acetaldehydes, and certain heat-killed probiotics are exerted as postbiotics which they can also serve as an alternative to antibiotics because of their inhibitory property toward pathogenic bacteria since they are nonpathogenic and have resistance to hydrolysis by mammalian enzymes and nontoxic (Patel and Denning 2013; Ooi et al. 2015). For instance, postbiotics activate a2β1 integrin collagen receptors for improving angiogenesis in vitro and in vivo in epithelial cells against species like *Saccharomyces boulardii* by enhancing barrier function (Giorgetti et al. 2015). Prebiotics, which are obtained from probiotics that modify the gut microbial flora, are not effortlessly processed by people however assume a particular part in the incitement of helpful bacterial organisms in the gut. Prebiotics are a group of nutrients that include fructooligosaccharides (FOS) metabolized from sucrose, insulin with bifidogenic properties, oligofructose, and galactose- and xylose-containing oligosaccharides (Hukins et al. 2016). Prebiotics are found naturally in foods such as vegetables, fruits, and grains that we eat every other day. Prebiotics has numerous medical advantages besides expanding the rate and length of diarrhea, giving help from aggravation and different symptoms related to gastrointestinal bowel disorders, and applying defensive impacts toward colon malignant growth (Younis et al. 2015).

Synbiotics are a combination of probiotics and prebiotics items that work with the endurance and implantation of live microbial dietary enhancements in the gut because of advances in microbial examination. When both the probiotic and the prebiotic function together in the living environment, the synergistic effects are more effectively promoted (Westfall et al. 2018). Studies focused on determining new probiotic and prebiotic concoctions are critical to enhancing the nutritional and clinical health benefits of probiotics and prebiotics (Fig. 4.2).

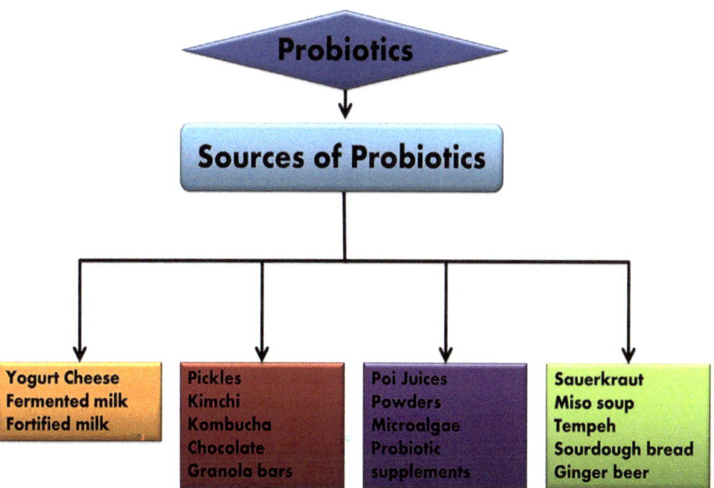

Fig. 4.2 Different sources of probiotics (Kerry et al. 2018)

4.2 Influence of Human Intestinal Microbiota on the Health of the Human Host

A robust microbiota is more ready to hold mucosal well-being, due to the great proximity of the GIT microbiota toward the gut lymphoid tissue and mucosa. At the same time, a variable composition seen in dysbiosis might expand the inescapability of diseases of the mucosal membrane just as inside the body, given solid interconnection with the gut immune system, the body's biggest immune organ (Vipperla and O'Keefe 2012). The body's first line of defense against pathogenic and toxic invasions from food is the intestinal mucosa. Orally administered antigens come into contact with the GALT after ingestion (gut-associated lymphoid tissue). The GALT's primary premise of defense is a humoral immune response intervened by secretory IgA (s-IgA), which keeps conceivably harmful antigens from entering the body while likewise interfering with mucosal pathogens without severe distress. Some probiotic strains have recently been reviewed for their potential to raise s-IgA and modulate the development of cytokines (mediators produced by immune cells) associated with activation, growth, immune cell regulation, and differentiation (Ashraf and Shah 2014). Probiotics advance the synthesis of bacteriocins and short-chain unsaturated fats, colonization site obstruction, invigorate mucosal barrier function and regulate the immune system, lower gut pH, colonize and fight for binding sites on gut epithelial cells, and complete accessible nutrients in the colon (Shah 2007). By facilitating phagocytosis, the innate and acquired immune responses are stimulated by probiotics, and secretory and systemic IgA are secreted, altering T-cell responses along with maintaining the homeostasis by enhancing Th1 responses and minimizing Th2 responses of Th1 and Th2 activities (Gourbeyre et al. 2011).

Components of Action of Probiotics: Increase in the epithelial barrier area, upgraded intestinal mucosa adhesion, and simultaneous restraint of microbe attachment are significant probiotic mechanisms of action, as are the competitive exclusion of pathogenic microorganisms, improvement of antimicrobial substances, and immune system regulation (Fig. 4.3).

4.2.1 Augmentation of the Epithelial Barrier

Amplifying the expression of genes engaged with close junction signaling may be one approach to improve intestinal barrier integrity. For instance, in a T84 cell barrier model, E-cadherin and β-catenin are the adherence junction proteins regulated and modulated by lactobacilli. Besides, when incubated with intestinal cells, lactobacilli influences the phosphorylation of adherence intersection proteins just as the protein kinase C (PKC) isoforms abundance, including PKCδ, decidedly adjusting epithelial barrier action (Hummel et al. 2012; Kelly et al. 2015). Avoiding cytokine-incited epithelial damage, typical in inflammatory bowel disease, can likewise assist with supporting mucosal obstruction when probiotics are utilized. Yan et al. (2007) and his colleagues demonstrated that p40 and p75 are the two proteins secreted by

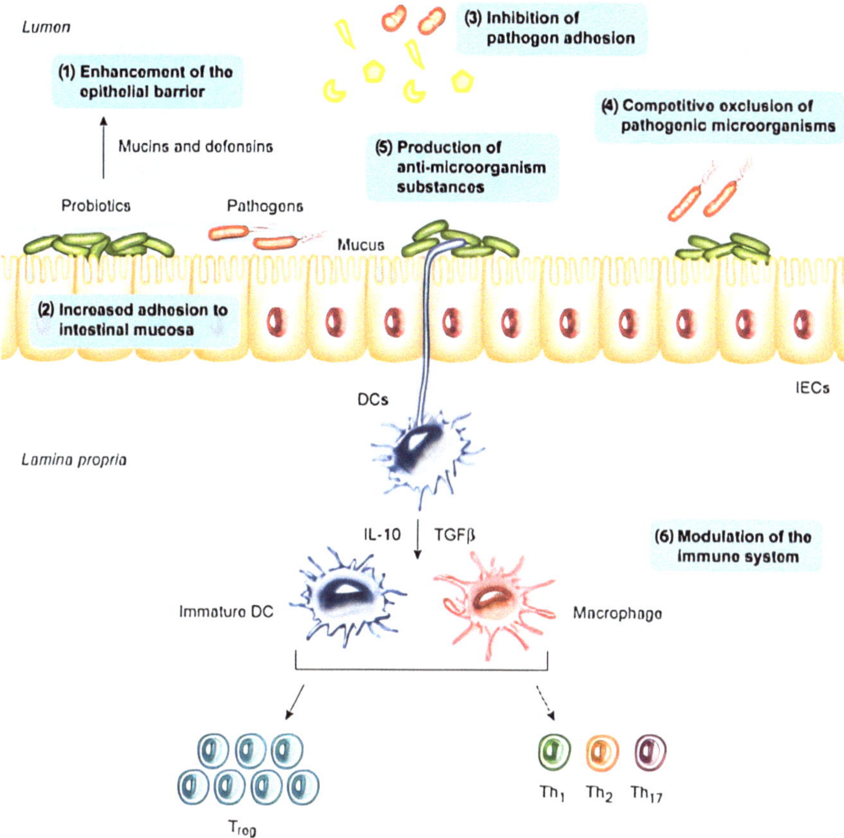

Fig. 4.3 Component of activity of probiotics (Bermudez-Brito et al. 2012)

Lactobacillus rhamnosus GG (LGG) in a phosphatidyl inositol-3`-a kinase-dependent pathway; these proteins forestall the cytokine-initiated cell apoptosis by enacting the antiapoptotic protein kinase B (PKB/Akt) and by restraining the favorable to apoptotic p38/mitogen-actuated protein kinase (MAPK). Probiotics can expand mucus secretion to enhance barrier function and pathogen removal since mucin glycoproteins (mucins) are the most critical macromolecular components of epithelial mucus. Moreover, they've been associated with both health and disease for a long time (Teichmann 2019).

4.2.2 Enhanced Adhesion to the Intestinal Mucosa

Among the best determination models for new probiotic strains has been adhesion. Adherence to the intestinal mucosa is considered essential for colonization and for probiotic strain interaction with the host, immune system modulation, and pathogen antagonism (Garcia-Gonzalez et al. 2018; Galdeano et al. 2019). Intestinal epithelial cells (IECs) discharge mucin, an unpredictable glycoprotein combination that is the vital part of mucous; accordingly, the pathogenic bacteria adhesion is prevented, and mucous gel is composed of lipids, immunoglobulins, free proteins, and salts (Morrin et al. 2019; Shang et al. 2020). Lactic acid bacteria (LABs) show different surface marker proteins associated with communication with mucus and intestinal epithelial cells (IECs). Surface adhesions displayed by bacteria intercede connection to the mucous layer, predominantly by proteins, even though lipoteichoic acids and saccharide moieties have likewise been ensnared (Sengupta et al. 2013). MUB (bodily fluid restricting protein) expressed by *Lactobacillus reuteri* is the most examined illustration of mucus-targeting bacterial adhesins. These proteins are surface-related proteins that are either harbored to the layer by a lipid moiety or installed in the cell wall. They may aid colonization of the human gut by facilitating close contact with the epithelium or by degrading the extracellular matrix of cells (Candela et al. 2007, 2009; Sanchez et al. 2010). Defensins are the protein molecules that stabilize the gut barrier function released by epithelial cells induced by probiotic strains (Wells et al. 2017). Antimicrobial proteins (AMPs) such as α- and β-defensins, cathelicidins, C-type lectins, and ribonucleases are increased in the host in response to pathogenic bacteria attacks, according to observations (Wang 2014).

4.2.3 Competitive Exclusion of Pathogenic Microorganisms

At the point when one type of microorganism contends all the more forcefully for receptor sites than another in the intestinal tract, this is known as "competitive exclusion." The components utilized by one bacterial species type to forestall or moderate the development of another are assorted. That includes an exclusion of available bacterial receptor sites, the establishment of a hostile microecology, competitive depletion of essential nutrients, and the synthesis of selective metabolites and antimicrobial substances secretion, among many others (García-Bayona and Comstock 2018). Explicit adhesiveness properties emerge from the association of surface proteins and mucins that repress pathogenic microorganisms' colonization and act as a source of antagonistic activity against gastrointestinal pathogen adhesion by some probiotic strains. Avoidance happens because of distinct variables and certain properties of probiotics that inhibit the adhesion of pathogens through substance production and IEC incitement (Oliveira and Reis 2017).

4.2.4 Development of Anti-microorganism Substances

The production of LMW compounds (<1000 Da), such as organic acids, and the synthesis of antibacterial substances regarded as bacteriocins (>1000 Da), are two of the proposed mechanisms engaged in the health incentives determined by probiotics. On the other hand, organic acids, notably acetic and lactic acids, show a substantial inhibitory effect on Gram-negative bacteria and have customarily been believed to be the key antimicrobial compounds liable for probiotics pathogen-inhibiting activity. The organic acid in its undissociated form dissociates within its cytoplasm when it enters the bacterial cell. The pathogen will perish if the intracellular pH declines too low or if the oxidized form of the organic acid absorbs too much through the cell (Bermudez-Brito et al. 2012; Reid 2016; Halloran and Underwood 2019). The disruption of target cells through pore formation and cell wall synthesis inhibition are two mechanisms of bacteriocin-mediated killing. Nisin, along with lipid II, forms a complex, which acts as an initial precursor, inhibiting the biosynthesis of the cell wall in spore-forming bacteria. The nisin and lipid II complex clumps together and integrates peptides to make a pore in the bacterial membrane (Bermudez-Brito et al. 2012; Cavera et al. 2015).

4.2.5 Probiotics and Immune Cells

The probiotic microorganisms can induce an immunomodulatory impact. These microorganisms interact with epithelial and dendritic cells (DCs), lymphocytes, monocytes, and macrophages. The immune system is partitioned into two sections: innate and adaptive (Gómez-Llorente et al. 2010). The adaptive immune response relies upon B and T lymphocytes, expressed for explicit antigens. On the other hand, the intrinsic, innate immune system responds to essential plans called pathogen-associated molecular patterns (PAMPs) shared by most microbes. Pattern recognition receptors (PPRs), which bind PAMPs, initiate the primary response to pathogens. The best thought about PPRs is toll-like receptors (TLRs). Besides, extracellular C-type lectin receptors (CLRs) and intracellular nucleotide-binding oligomerization domain-containing protein (NOD)-like receptors (NLRs) have been found to send signals when microbes are encountered (Lebeer et al. 2010). Via their PPRs, both IECs and DCs can interfere with and react to gut microorganisms. Recent studies on a probiotic mixture of *B. bifidum*, *L. acidophilus*, *L. reuteri*, *L. casei*, and *Streptococcus thermophilus* showed that the immune system is triggered via stimulating regulatory dendritic cells with high levels of IL-10, TGF-, COX-2, and indoleamine 2,3-dioxygenase, promoting the production of CD4 + Foxp3+ regulatory T cells (Tregs) from the CD4 + CD25 population and increasing the suppressor activity of naturally occurring CD4 + CD25 + Tregs (Yan and Polk 2011). Modulation and interaction of probiotics are summarized in Fig. 4.4.

4.2.6 Molecular Biological Methods for Studying Probiotics

With varying degrees of effectiveness, current techniques such as genetic fingerprinting, oligonucleotide probes, gene sequencing, and complex primer selection may distinguish closely related bacteria. Different molecular techniques being utilized to evaluate the constituents of complex microbiota in this field of exploration

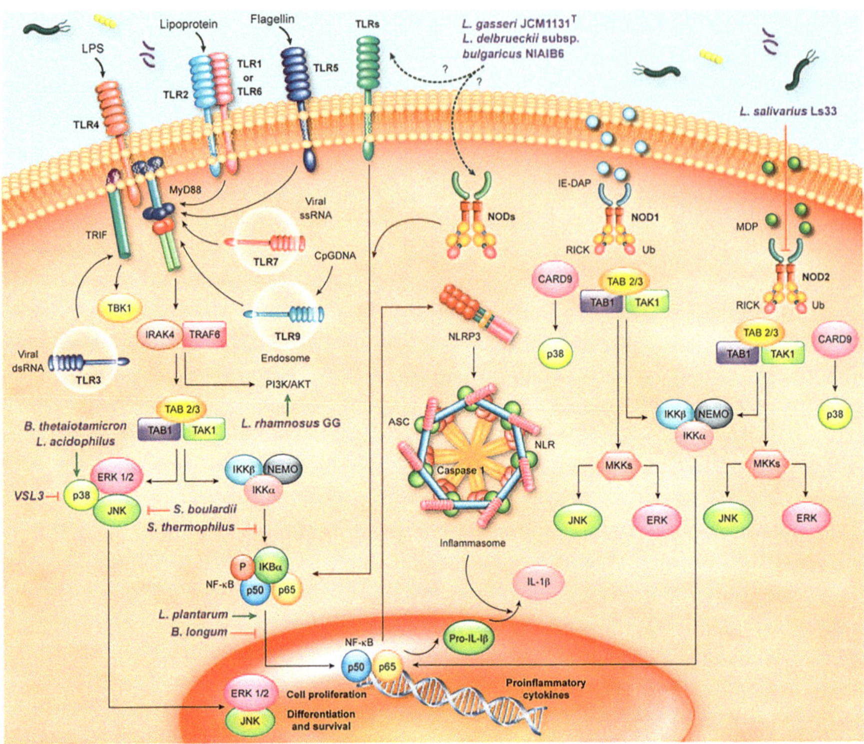

Fig. 4.4 Collaboration of probiotics with the gut-related immune system. ASC apoptosis-associated speck-like protein containing a CARD, CARD9 caspase recruitment domain-containing protein 9, ERK extracellular regulated kinase, IE-DAP D-gamma-glutamyl-meso-DAP, IKK $I_{K}B$ kinase, IRAK4 IL-1 receptor-associated kinase 4, JNK jun N-terminal kinase, MDP muramyl dipeptide, MKK mitogen-activated kinase kinase, NEMO NF- $_{K}B$ essential modulator, TAK1 ubiquitin-dependent kinase of MKK and IKK, TAB1/2/3 TAK binding proteins, TBK1 serine/threonine-protein kinase 1, TRAF6 TNF receptor-associated factor 6, Ub ubiquitin (Bermudez-Brito et al. 2012)

are populace study, denaturing gradient gel electrophoresis (DGGE)/temperature gradient gel electrophoresis (TGGE), fluorescent in situ hybridization (FISH), and probe grids (Franco-Duarte et al. 2019). In probiotics and human GI microflora studies, molecular techniques have been used for four main purposes: (1) enumeration of phylogenetically related groups of bacteria; (2) characterization of bacterial diversity within samples; (3) following or observing of specific organisms or populations, both quantitatively and subjectively; and (4) definitive identification of isolates, especially probiotics (Shekhar et al. 2020).

4.3 Conclusions and Future Directions

The intricate and dynamic collaborations between the intestinal epithelium and microscopic organisms on the luminal side, just as between the epithelium and the fundamental resistant framework on the basolateral side, must be accommodated in coculture tries different things with probiotics. The molecular elucidation of probiotic action in vivo can assist in identifying true probiotics and picking the most appropriate ones for the prevention and/or treatment of particular diseases. The different instruments are open for decisive assessment of the bacterial parts of probiotic products, strain integrity, and quality control. Future advancements in molecular biology ought to be pointed toward distinguishing and developing strategies to contemplate the situation of functional elements and biomarkers suitable for disentangling the action of the bacterial populace, explicit organisms, or potential qualities. The mission will be to exhibit the roles of the microflora and probiotic strains in vivo and not just to portray the populace or show the presence or nonappearance of particular organisms or groups. Future advancements in molecular biology should be pointed toward recognizing and progressing techniques to examine the situation with useful components and biomarkers appropriate for disentangling the action of the bacterial population, specific organisms, and/or genes. The mission will be to show the jobs of the microflora and probiotic strains in vivo and not just to describe the populace or exhibit the presence or nonappearance of particular organisms or groups.

Acknowledgments Kuraganti Gunaswetha, Edla Sujatha, Anusha, Prathyusha, and Pallaval Veera Bramhachari are grateful to Kaktiya University Warangal and Krishna University Machilipatnam for the support extended.

Conflict of Interest The authors declare that they have no competing interests.

References

Ashraf R, Shah NP (2014) Immune system stimulation by probiotic microorganisms. Crit Rev Food Sci Nutr 54(7):938–956

Aureli P, Capurso L, Castellazzi AM, Clerici M, Giovannini M, Morelli L et al (2011) Probiotics and health: an evidence-based review. Pharmacol Res 63(5):366–376

Bermudez-Brito M, Plaza-Díaz J, Muñoz-Quezada S, Gómez-Llorente C, Gil A (2012) Probiotic mechanisms of action. Ann Nutr Metab 61(2):160–174

Call EK, Goh YJ, Selle K, Klaenhammer TR, O'Flaherty S (2015) Sortase-deficient lactobacilli: effect on immunomodulation and gut retention. Microbiology 161(Pt 2):311

Candela M, Bergmann S, Vici M, Vitali B, Turroni S, Eikmanns BJ, Hammerschmidt S, Brigidi P (2007) Binding of human plasminogen to bifidobacterium. J Bacteriol 189(16):5929–5936

Candela M, Biagi E, Centanni M, Turroni S, Vici M, Musiani F, Vitali B, Bergmann S, Hammerschmidt S, Brigidi P (2009) Bifidobacterial enolase, a cell surface receptor for human plasminogen involved in the interaction with the host. Microbiology 155(10):3294–3303

Cavera VL, Arthur TD, Kashtanov D, Chikindas ML (2015) Bacteriocins and their position in the next wave of conventional antibiotics. Int J Antimicrob Agents 46(5):494–501

Cohendy M (1906a) Trial of treatment of acute mucomembranous enteritis by acclimatization of lactic acid bacteria in the large intestine. Reduced Acc Meet Mem Biol Soc 60:363

Cohendy M (1906b) Description du ferment lactique puissant capable de s' acclimater dans l'intestine de l'homme. CR Soc Biol 1906(60):558

Diop A, Khelaifia S, Armstrong N, Labas N, Fournier PE, Raoult D, Million M (2016) Microbial culturomics unravels the halophilic microbiota repertoire of table salt: description of Gracilibacillus massiliensis sp. nov. Microb Ecol Health Dis 27(1):32049

Franco-Duarte R, Černáková L, Kadam S, Kaushik S, Salehi B, Bevilacqua A, Corbo MR, Antolak H, Dybka-Stępień K, Leszczewicz M, Relison Tintino S (2019) Advances in chemical and biological methods to identify microorganisms—from past to present. Microorganisms 7(5):130

Galdeano CM, Cazorla SI, Dumit JM, Vélez E, Perdigón G (2019) Beneficial effects of probiotic consumption on the immune system. Ann Nutr Metab 74(2):115–124

García-Bayona L, Comstock LE (2018) Bacterial antagonism in host-associated microbial communities. Science 361(6408):eaat2456

Garcia-Gonzalez N, Prete R, Battista N, Corsetti A (2018) Adhesion properties of food-associated lactobacillus plantarum strains on human intestinal epithelial cells and modulation of IL-8 release. Front Microbiol 8(9):2392

Giorgetti G, Brandimarte G, Fabiocchi F, Ricci S, Flamini P, Sandri G, Trotta MC, Elisei W, Penna A, Lecca PG, Picchio M (2015) Interactions between innate immunity, microbiota, and probiotics. J Immunol Res 18:2015

Gómez-Llorente C, Munoz S, Gil A (2010) Role of toll-like receptors in the development of immunotolerance mediated by probiotics. Proc Nutr Soc 69(3):381–389

Gourbeyre P, Denery S, Bodinier M (2011) Probiotics, prebiotics, and synbiotics: impact on the gut immune system and allergic reactions. J Leukoc Biol 89(5):685–695

Halloran K, Underwood MA (2019) Probiotic mechanisms of action. Early Hum Dev 1(135):58–65

Hukins RW, Krumbeck JA, Bindels LB, Cani PD, Jr GY FG, Hamaker B, Martens EC, Mills DA, Rastal RA, Vaughan E, Sanders ME (2016) Prebiotics: why definition matter. Curr Opin Biotechnol 37:1–7

Hummel S, Veltman K, Cichon C, Sonnenborn U, Schmidt MA (2012) Differential targeting of the E-cadherin/β-catenin complex by gram-positive probiotic lactobacilli improves epithelial barrier function. Appl Environ Microbiol 78(4):1140–1147

Johnson BR, Klaenhammer TR (2014) Impact of genomics on the field of probiotic research: historical perspectives to modern paradigms. Antonie Van Leeuwenhoek 106(1):141–156

Kelly JR, Kennedy PJ, Cryan JF, Dinan TG, Clarke G, Hyland NP (2015) Breaking down the barriers: the gut microbiome, intestinal permeability and stress-related psychiatric disorders. Front Cell Neurosci 14(9):392

Kerry RG, Patra JK, Gouda S, Park Y, Shin HS, Das G (2018) Benefaction of probiotics for human health: a review. J Food Drug Anal 26(3):927–939

Lebeer S, Vanderleyden J, De Keersmaecker SC (2010) Host interactions of probiotic bacterial surface molecules: comparison with commensals and pathogens. Nat Rev Microbiol 8(3):171–184

Metchnikoff E (2004) The prolongation of life; optimistic studies. 1908. Trans. P. Chalmers Mitchell. In: Butler RN, Olshansky SJ (eds) Classics in longevity and aging. Springer Publishing Company, New York

Morrin ST, Lane JA, Marotta M, Bode L, Carrington SD, Irwin JA, Hickey RM (2019) Bovine colostrum-driven modulation of intestinal epithelial cells for increased commensal colonisation. Appl Microbiol Biotechnol 103(6):2745–2758

Oliveira JM, Reis RL. Advances in biomaterials for tissue regeneration: functionalization, processing and applications 2017

Ooi MF, Mazlan N, Foo HL, Loh TC, Mohamad R, Rahim RA, Ariff A (2015) Effects of carbon and nitrogen sources on bacteriocin-inhibitory activity of postbiotic metabolites produced by lactobacillus plantarum I-UL4. Malays J Microbiol 11(2):176–184

Patel RM, Denning PW (2013) Therapeutic use of prebiotics, probiotics, and postbiotics to prevent necrotizing enterocolitis: what is the current evidence? Clin Perinatol 40(1):11–25

Pérez-Cobas AE, Moya A, Gosalbes MJ, Latorre A (2015) Colonization resistance of the gut microbiota against clostridium difficile. Antibiotics 4(3):337–357
Reid G (2016) Probiotics: definition, scope and mechanisms of action. Best Pract Res Clin Gastroenterol 30(1):17–25
Sanchez B, Urdaci MC, Margolles A (2010) Extracellular proteins secreted by probiotic bacteria as mediators of effects that promote mucosa–bacteria interactions. Microbiology 156(11):3232–3242
Sengupta R, Altermann E, Anderson RC, McNabb WC, Moughan PJ, Roy NC (2013) The role of cell surface architecture of lactobacilli in host-microbe interactions in the gastrointestinal tract. Mediat Inflamm 2013:237921
Shah NP (2007) Functional cultures and health benefits. Int Dairy J 17(11):1262–1277
Shang J, Wan F, Zhao L, Meng X, Li B (2020) Potential immunomodulatory activity of a selected strain bifidobacterium bifidum H3-R2 as evidenced in vitro and in immunosuppressed mice. Front Microbiol 1(11):2089
Shekhar SK, Godheja J, Modi DR (2020) Molecular technologies for assessment of bioremediation and characterization of microbial communities at pollutant-contaminated sites. In: Bioremediation of Industrial Waste for Environmental Safety. Springer, Singapore, pp 437–474
Slavin J (2013) Fiber and prebiotics: mechanisms and health benefits. Nutrients 5(4):1417–1435
Teichmann J. The optimization of butyrate production by human gut microbiomes in the presence of resistant starches. 2019
Tripathi AD, Mishra R, Maurya KK, Singh RB, Wilson DW (2019) Estimates for world population and global food availability for global health. In: The role of functional food security in global health. Academic, pp 3–24
Vipperla K, O'Keefe SJ (2012) The microbiota and its metabolites in colonic mucosal health and cancer risk. Nutr Clin Pract 27(5):624–635
Walsh AM, Crispie F, Claesson MJ, Cotter PD (2017) Translating omics to food microbiology. Annu Rev Food Sci Technol 8:113–134
Wang G (2014) Human antimicrobial peptides and proteins. Pharmaceuticals 7(5):545–594
Wells JM, Brummer RJ, Derrien M, MacDonald TT, Troost F, Cani PD, Theodorou V, Dekker J, Méheust A, De Vos WM, Mercenier A (2017) Homeostasis of the gut barrier and potential biomarkers. American journal of physiology-gastrointestinal and liver. Physiology 312(3):G171–G193
Westfall S, Lomis N, Prakash S (2018) A novel polyphenolic prebiotic and probiotic formulation have synergistic effects on the gut microbiota influencing drosophila melanogaster physiology. Artif Cells Nanomed Biotechnol 46(sup2):441–455
Yan F, Polk DB (2011) Probiotics and immune health. Curr Opin Gastroenterol 27(6):496
Yan F, Cao H, Cover TL, Whitehead R, Washington MK, Polk DB (2007) Soluble proteins produced by probiotic bacteria regulate intestinal epithelial cell survival and growth. Gastroenterology 132(2):562–575
Younis K, Ahmad S, Jahan K (2015) Health benefits and application of prebiotics in foods. J Food Process Technol 6(4):1

Antimicrobial Agents Induced Microbiome Dysbiosis Its Impact on Immune System and Metabolic Health

5

K. Anuradha, J. Sarada, Y. Aparna, and S. Anju

Abstract

The role of antimicrobial agents like antibiotics and antiseptics in their clinical use in treating various diseases and in their non-medicinal uses in agricultural crops and animal farming is well known to all. Broad- and narrow-spectrum antibiotics when used against pathogenic bacteria not only show their action on pathogens but also show their indiscriminate action on commensal microbial flora. Over usage and misuse of antibiotics result in microbial resistance against antibiotics. Antibiotics, when administered to treat diseases, show their action on pathogenic bacteria, and simultaneously they show their mode of action on resident beneficial microbiota which in turn leads to alteration of microbiome composition. Excessive usage of antibiotics also negatively impacts human health and immunity. Microbiome, present in human beings, helps in various activities like nutrition, metabolism, etc. by producing amino acids, vitamins, and short chain fatty acids and also helps developing immunity against a wide range of pathogens. The constant exposure of human microbiome to various antibiotics and antiseptics results in disruption in its ecology and imbalance of microbial composition which is referred to as microbiome dysbiosis. Imbalance in the harmonic relationship between the host and the microbiome, caused due to various external and internal factors affecting the human body, results in disruption in homeostasis and causes various diseases. The diseases associated with microbiome dysbiosis include obesity, celiac disease, diabetes, inflammatory bowel disease, rheumatic arthritis, neurodegenerative disorder, depression, autism, malignancy, and cancer. Fecal microbiota transplantation and manipulation of microbiota using probiotics are trending tools in microbiome research and are

K. Anuradha (✉) · J. Sarada · Y. Aparna · S. Anju
Bhavan's Vivekananda College of Science, Humanities and Commerce, Sainikpuri, Secunderabad, Telangana, India

P. Veera Bramhachari (ed.), *Human Microbiome in Health, Disease, and Therapy*, https://doi.org/10.1007/978-981-99-5114-7_5

very useful applications in restoring microbial communities and correcting microbial dysbiosis.

Keywords
Microbiome · Antibiotics · Immunity · Metabolic health

5.1 Introduction

Various studies and research works are being carried out to understand the human microbiome, by using the technological advancements in molecular biology especially the methods like gene sequencing and metagenomics, to know the microbiome role and function (Human Microbiome Project Consortium 2012). The human microbiome constitutes with a wide variety of microorganisms like thousands of species bacteria mainly from the phylum Bacteroidetes and the phylum Firmicutes, fungi, etc., and all these microbes have more genetic complexity than the human genome (Utzschneider et al. 2016). The composition of microbiome is highly variable, and major changes come across in early age childhood especially in infancy (Palmer et al. 2007). The human body is home for hundred trillion microorganisms, and the majority of them lives in the intestine and plays critical role in homeostasis, physiological, metabolic, and immunological processes (Blumberg and Powrie 2012; Tremaroli and Bäckhed 2012; Greer et al. 2016; Sanz et al. 2015). In addition to this, the human microbiome synthesizes various amino acids, vitamins, etc. and helps host nutrition.

Various factors like by nutritious status of mother like obese or malnourished, gestation period, mode of childbirth, milk feeding types, diet, and use of antibiotics against pathogens in treating diseases would determine the gut microbiome of infants (Meropol and Edwards 2015). The origin and development of complex composition normal flora of infant have impact on the entire future life (Charbonneau et al. 2016).

Continuous therapeutic usage of antibiotics manipulates the gut microbiome as it changes the function of bacterial communities in the gut (Ferrer et al. 2017). Host gut microbe interactions and the effect of antibiotics on the gut microbial flora are better understood with antibiotic administration. Earlier studies focused to understand the role and use of different antibiotics like broad-spectrum antibiotics and antibiotics in combinations on bacterial pathogens of gut microbial function (Rodrigues et al. 2017; Strzępa et al. 2017). Not only antibiotic exposure, there are other factors like nutritional parameters-glucose tolerance, microbial diversity of gut microbiome, and body weight and bone growth influence the gut microbiome of human body (Rodrigues et al. 2017; Nobel et al. 2015; Ferrer et al. 2017; Mikkelsen et al. 2015). Normal flora of gut microbiome includes *Clostridium* sp., *E. coli*, *Bacteroides*, *Ruminococcus* sp., *Lactobacillus* sp., and *Akkermansia* sp., and any change in the gut flora were noticed to be associated with diabetes and obesity disorders in patients (Qin et al. 2012; Karlsson et al. 2013; Murri et al. 2013; Chakraborti 2015; Sanz et al. 2015). Excessive use of antibiotics also negatively impacts human health and immunity. Current studies on microbiome have established a clear concept on the perturbation of the gut microbiome by excessive use of antibiotic leading to imbalance of host-microbiome interactions, and such imbalance in turn changes immune reactions of the host and causes systemic spread of commensal bacteria and making host to become more susceptible to invasive pathogens. All these changes subsequently show impact on human health resulting in variety of noncommunicable health disorders (Zheng et al. 2020). Such dysbiosis, disruption, and change in composition flora of host result in various diseases, and disorders like dysbiosis include obesity, celiac disease, diabetes, inflammatory bowel disease, rheumatic arthritis, neurodegenerative disorder, depression, autism, malignancy, and cancer (Wang et al. 2017).

Human microbiota exists in harmony with their healthy host and plays important role in infection resistance, stimulation of immune system, and nutrient source stimulation of epithelial turnover (Fig. 5.1). Microbiota occupies binding sites on the surface of human body and secretes mucins and other metabolic products which may be noxious to pathogenic bacteria and contribute for infection resistance. Microbiota, specifically Gram-negative bacteria, acts as immunogens and stimulates immune system and produces antibodies in response to commensals and confers protection against pathogenic microorganisms. Microbiota also provides vitamin K, a cofactor essential for clotting factors (Virella 1997).

The emergence of antibiotic resistance due excessive and prolonged usage of antibiotics is a serious health concern on which the whole world is working to control the drug resistance. Physiological and metabolic functions of gut microbiome and antibiotic use in treatment of bacterial infections are the two key elements which need to be understood in depth especially the important genetic factors like mobile genetic elements (MGE) and gene responsible for antibiotic resistance (ARGs) are to be clearly studied to understand the effect of antibiotic exposures on microbiome (Xu et al. 2020).

5.2 Antibiotic-Induced Microbiome Alterations

Aggregated complex microbial populations inhabited on the human host exhibit symbiotic relationship as host provides physical, environmental, and nutritional support, and intestinal microbiome provides infection resistance and easy food absorption (Hooper et al. 2002; Vollaard and Clasener 1994). Several research findings on human microbiome confirm that administration of antimicrobial agents causes disturbance in composition and function of microbial flora resulting in microbiome dysbiosis (Fig. 5.2).

Usage of broad-spectrum antibiotics lowers one third of the existence of the bacterial flora of gut and significantly changes the microbial diversity (Dethlefsen et al. 2008; Dethlefsen and Relman 2011). In flexibility presented by microbiota of the human gut after for about 5 days of amoxicillin antibiotic treatment, fecal microbiota profile tends to reappear to its normal original composition within 2 months (De La Cochetière et al. 2005). Indeed, antibiotic-induced microbiome variations persist

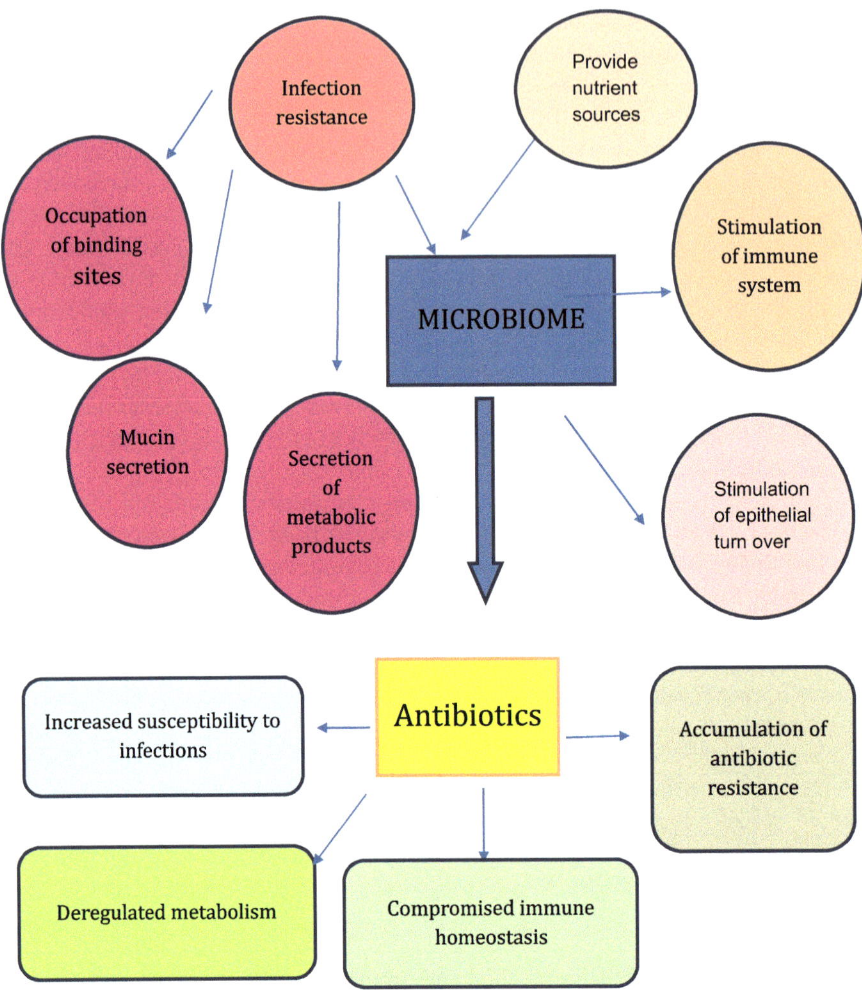

Fig. 5.1 Antibiotic-induced microbiome alterations on immune system and human health

after prolonged exposures for an extended period of time (Jernberg et al. 2007). At a very early stage of life, i.e., in the first year and second year of childhood life, drastic and dramatic changes take place in the gut of infants. The use of antibiotics critically affects the origin and evolution of indigenous human neonate microbiota (Tanaka et al. 2009). Combined antibiotic therapy and combinational use of gentamycin and ampicillin in childhood that too at very early stage of life affect the origin and distribution of gut normal microbiota and also show its impact on health implications and problems in the long run. Comparatively very few findings suggested retrieval of normal infant intestinal flora after therapeutic use of antibiotics (Fouhy et al. 2014a). Similar studies carried on microbiome of infants with antibiotic administrations clearly indicated that there was reduced population of

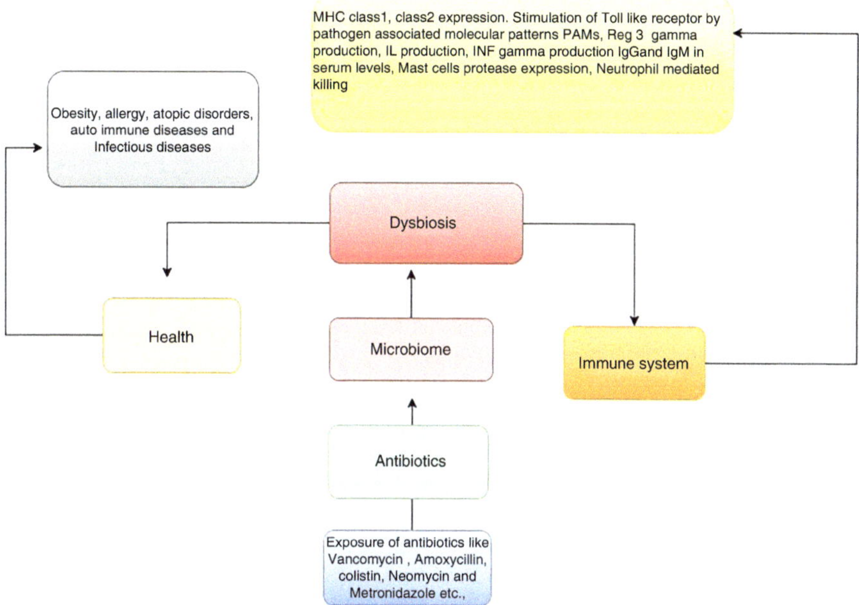

Fig. 5.2 Overview of antibiotic-induced dysbiosis of microbiome

Bifidobacterium and increased population count of Proteobacteria when the results were compared with microbial flora of infant gut population which was untreated with antibiotics to that of their mothers whose microbiome was exposed to antibiotics before the childbirth had shown same effect as that of infant's gut which was exposed to antibiotics (Tanaka et al. 2009; Franzosa et al. 2015; Fouhy et al. 2014b).

DNA sequencing strategies and combining many forms of molecular data in an integrated framework obtained from omic studies like metabolomics, proteomics, transcriptomics, and metagenomics are used in classifying and understanding the microbial variations of functional activity and sequential dynamics at community level and at strain level. These types of research indicated that antibiotics exposure on gut microbiome brings changes in the genetic expression and further influences the microbial metabolism. Such changes can occur rapidly when compared to natural microbial community replacements in minimizing microbial populations (Pérez-Cobas et al. 2013). Genetic changes induced in the genome of human microbiome influence the expression of virulence factors which are capable of causing disease and thereby leading to disease conditions driving toward pathogenicity of the microbiota; otherwise, this microbiota was not pathogenic. In this frame of reference, the exposure of β-lactam antibiotics on microbiota has influenced the expression of carbohydrate-degrading enzymes, and enzymatic activity of such enzymes resulted in deregulation of carbohydrate metabolism, and similar observation was noticed with gut flora of individuals who were obese (Hernández et al. 2013).

The bacterial taxa of large intestine and small intestine differ in their microbiome composition. The bacterial flora that constitutes gut microbiome of the large intestine is more diversified and denser, whereas the small intestinal flora constitutes limited flora (Garner et al. 2009; Ubeda et al. 2010). Within the same anatomical site, i.e., in the intestine, the microbial populations of lumen and mucus layer of the intestine indicated different bacterial flora representing the diversity (Eckburg et al. 2005). Based on genome sequencing techniques, major population of resident bacterial flora intestine belongs to phylum Firmicutes and phylum Bacteroidetes and remaining minor populations of the phylum Actinobacteria, phylum Proteobacteria, phylum Fusobacteria, and phylum Verrucomicrobiota (Eckburg et al. 2005). Antibiotic administration against microbial pathogens not only shows cidal and static activity on pathogens but also perturbs the intestinal microbial flora and influences the immune reactions (Hill et al. 2010) and further promotes disease conditions like, diabetes, inflammatory bowel disease, atopy, and arthritis (Brandl et al. 2008).

Experimental investigations established that antimicrobial agents like antibiotics quickly influence metabolic and physiology of the microorganisms present in the microbiota. Antibiotic-treated fecal suspension samples the microbial populations exhibited damaged membranes and also expressed genes that conferred stress response to maintain nutritional imbalance, phage induction as virulence factor, and antibiotic resistance to inactivate enzymes (Maurice et al. 2013). Not only such genetic changes increased the expression of genes for antibiotic resistance against various antibiotics like macrolides and tetracycline antibiotics. Conditions like anesthesia, surgery, and treatment of bacterial infections with an antibiotic in isolation or in combination like cocktail of antibiotics when used caused resulted not only lowered microbial population but also reduced diversity of microorganism. Significant microbiota diversity changes were noted in the microbiota of host gut (Zhang et al. 2019; Wang et al. 2020).

5.3 Antimicrobial Agents-Induced Changes in the Microbiome Impact on Immune and Metabolic Health

Administration of broad-spectrum antimicrobial agents can disturb the equilibrium of the microbiome and changes the composition of microbiome. Both nonresistant microorganisms and potential pathogens express and present different microbial-associated molecular patterns to the receptors located in epithelial cells as well as cell of immune system of the human body. This leads to stimulation of receptors such as Toll-like receptors (TLRs) and NOD1 receptors that mediate cascade of variety of immune processes including lymphoid follicle stimulation, differentiation of T-lymphocyte, activation of neutrophils and cytokine release, and production of antibacterial compounds (Fig. 5.3), thereby alter the effectiveness of innate and adaptive immunity (Ubeda et al. 2012). Table 5.1 summarizes the antibiotic-induced microbiome alterations on immune system.

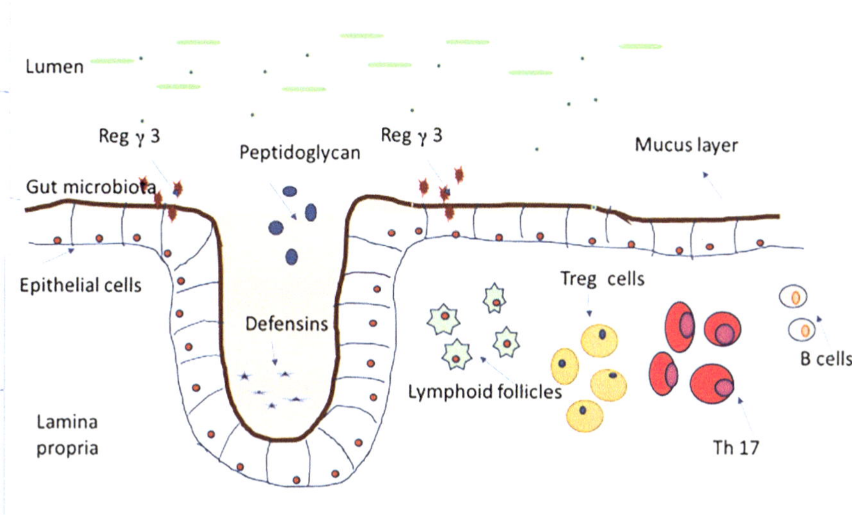

Fig. 5.3 Antibiotic-induced gut microbiome alterations impact on immune system

Colonization of bacteria in the gut layers of infants at an early life period makes the gut lining as an efficient barrier for luminal antigens. Administration of antibiotics which are broad spectrum in nature can vary the composition of bacterial colonization of the gut and diminish commensal bacterial flora and damage the gut barrier function development. Suckling Sprague-Dawley rats gut microbiome treated with Clamoxyl—an amoxicillin-based antibiotic—was tested intestinal microbiota and gene expression for products of immune response. *Lactobacillus* sp. was more susceptible to drug action and got eliminated completely and reduced the colonic count off both aerobic and anaerobic bacteria which included particular species of *Enterobacteriaceae* and *Enterococcus*. The gene expression profiles of immune products were studied and clearly understood that there was downregulation of gene expression coding from artilysin, phospholipase A2, and defensins. It was also observed that there was downregulation of gene expression of class I and II genes of major histocompatibility complex (MHC), but mast cell protease expression was upregulated (Schumann et al. 2005; Brandl et al. 2008) under similar condition of antibiotic exposures, and significant downregulation of RegIIIc was observed which regulates C-type lectin acts on Gram-positive bacterial cell wall and shows cidal activity even on vancomycin-resistant *Enterococcus* sp. Oral administration of lipopolysaccharides stimulates intestinal Toll-like receptor 4 which reactivates and reinduces RegIIIc and boosts innate immunity in vancomycin-treated mice or antibiotic-treated mice. Such receptor stimulation in the intestine becomes an effective therapeutic tool to reduce pathogen colonization and infection of drug-resistant bacteria (Brandl et al. 2008).

Mice treated with streptomycin and metronidazole perturbs the intestinal microbial flora and disturbs intestinal immune response and homeostasis prior to

Table 5.1 Antibiotic-induced microbiome alterations on immune system

Antibiotics	Effect on microbiome	Immune response	Reference
Enrofloxacin, cephalexin, paromomycin, and clindamycin	Fermentative bacteria *Enterobacteriaceae*	Increased neutrophils and Th17 cells increased soluble CD14 in plasma	Manuzak et al. (2020)
Enrofloxacin, vancomycin polymyxin	Gut microbiota	Cytokines production	Sun et al. (2019)
Amoxicillin metronidazole, vancomycin, and neomycin gentamycin	Microbiota of large intestine	Basophil production interferon and IL production Ig E levels in serum	Hill et al. (2010, 2012)
Metronidazole	Bacteroidales, *Clostridium* sp., and *Lactobacillus* sp.	IL production increase macrophages NK cells in colon Reg γ 3 expression	Wlodarska et al. (2011)
Amoxicillin metronidazole, vancomycin and neomycin	Microbiota depletion	Neutrophil mediated killing interferon and IL production Reg γ 3 expression	Clarke et al. (2010)
Metronidazole, vancomycin, neomycin	*Enterobacteriaceae* members and *Bacteroidetes*	Reg γ 3 expression small intestine	Ubeda et al. (2010) and Brandl et al. (2008)
Colistin	Gram negative bacteria	IL production	Bouskra et al. (2008)
Vancomycin	*Enterobacteriaceae* Gram-positive bacteria	Treg cells of colon Th 17 of small intestine Il production	Bouskra et al. (2008) and Ivanov et al. (2009)
Amoxicillin	*Lactobacillus* sp. aerobic and anaerobic bacteria of intestinal flora	MHC class1 and class2 in expression in small and large intestine, mast cells protease expression	Schumann et al. (2005)
Amoxicillin clavulanate	Gram-negative bacteria	IgG in serum levels	Dufour et al. (2005)

Citrobacter rodentium infection. Both antibiotic exposures altered microbial populations. Treatment with metronidazole increased the expression of Reg3gamma and IL-25 in the colon of murine. It resulted in more numbers of NK cells, and macrophages in the large intestine lowered the Muc2 expression and regulate the key content of the intestine mucin layer. The thinner the mucin layer, the more the adherence, and contact of epithelial cells and the bacterial flora triggers immunity and results in inflammation tone of intestine. Antibacterial effect of metronidazole was shown on *Clostridium coccoides* and *Bacteroidales* without lowering the bacterial density and increasing *the* count of aerobic *Lactobacillus* sp. (Wlodarska et al. 2011).

Lymphoid structure that represents gut immunity includes Payer's patches, cryptopatches, mesenteric lymph nodes, and isolated lymphoid follicles (ILFs). Nod1 receptor stimulation by peptidoglycan of Gram-negative bacterial cell wall plays a major role in gut homeostasis and immunity of the gut. Antibiotic exposure of microbiota affects lymphoid structures of intestine especially ILFs. Mice exposed with broad-spectrum antibiotics like vancomycin and colistin killed both Gram-positive and Gram-negative bacteria which show a smaller number of ILFs as there was no stimulation of Nod 1 receptor. Experimental results reveal that the antibiotic colistin had shown greater effect (Bouskra et al. 2008).

Peptidoglycan content of bacterial cell wall systematically triggers the innate immune responses and plays critical role in killing the pathogens. Receptors like Nod1, Nod2, and Toll-like receptors specifically bind to different sites of peptidoglycan layer of bacteria. Nucleotide-binding oligomerization domain-containing protein-1 Nod1 binds to the meso-diaminopimelic acid of side chain pentapeptide of Gram-negative cell wall; Nod 2 recognizes both Gram-positive and Gram-negative bacteria cell walls, and Toll-like receptor 4 recognizes lipopolysaccharide (LPS) content of Gram-negative bacterial cell wall. Exposure of antibiotics like neomycin, metronidazole, ampicillin, and vancomycin depletes the microbiome and thereby diminishes peptidoglycan levels; otherwise, such circulating peptidoglycan induces the killing and elimination of pathogens like *Streptococcus aureus* and *Staphylococcus pneumoniae* by neutrophils driven by bone marrow (Clarke et al. 2010).

Administration of amoxicillin/clavulanate changes the levels of different immunoglobulins in the serum. It lowered immunoglobulin G without showing any effect on affecting immunoglobulin A or immunoglobulin M (Dufour et al. 2005). Commensal resident flora and probiotics bacteria manipulate immune pathways to enforce the gut tolerance as these bacteria and their products are able to induce T regulatory cell induction (Kline 2007; Baba et al. 2008). Kanamycin sulfate-administered mice inoculated with *Enterococcus faecalis*, *Lactobacillus acidophilus*, or *Bacteroides vulgatus* an efficient probiotic bacterium after antibiotic treatment improved, balanced the intestinal flora, and prevented the Th 2-shifted immunity (Sudo et al. 2002). Administration of broad-spectrum antibiotics 15 days old mice showed lowered the expression of TLRs and cytokines profiles and promotes Th 2 response (Dimmitt et al. 2010).

Antimicrobial agents' exposures alter the gut microbiome and induce more pro-inflammatory molecules like Gram-negative bacterial lipopolysaccharides play a major role causing metabolic disorders like obesity and diabetes. Excessive adiposity attributed by the host contributes to the inflammation and may also progresses toward metabolic deviation of health from obesity to metabolic disorders. All this resulted because of high-fat diet, increasing more bacterial population with lipopolysaccharide cell walls, as these bacterial lipopolysaccharides act like proinflammatory molecules and trigger inflammation and other metabolic health-associated disorders.

The above disorders resulted due to the changed microbial gut environment caused by high-fat diets that enhance bacteria containing lipopolysaccharides, finally resulting in high concentrations of proinflammatory molecules in the serum (Cani et al. 2007).

Antibiotic-induced dysbiosis impacts on various immune and metabolic pathways, affects the intestinal environment, increases inflammatory tone, and lowers the intestinal immunoglobulin IgA; however, a pathogen elimination is carried out by noninflammatory antibody (Cerutti and Rescigno 2008: Rautava et al. 2004). Gut microbiota contact and establishment depend on its anchoring onto epithelial lining of the intestine. The thinner the mucin layer, the more would be the contact of bacterial flora with mucin layer. Administration of metronidazole causes lower expression of Muc2 as it is important in maintenance of mucin content (Wlodarska et al. 2011). Decreased expression leads to thinning of the mucin layer; as a result, direct contact of gut microbiome with that of the epithelial layer takes place triggering the innate immunity and manifests inflammation.

Experiments conducted on mice indicated that antibiotics exposures trigger inflammation by bacterial translocation through passage of viable resident bacteria from the gastrointestinal tract through epithelial mucosa into the splanchnic and systemic circulation. This results the translocation bacterial flora from its site of establishment to the other site and results in week microbial signaling to cells of immune systems like colonic goblet cells and dendritic cells (Knoop et al. 2016). A deficiency in Toll-like receptor 5 in microbiome results in obesity, dyslipidemia, insulin resistance, and other metabolic disorders. Such disorders caused by the microbial dysbiosis antibiotic exposures were demonstrated by transplantation of microbiota of Toll-like receptor 5-deficient mice in germ-free mice (Vijay-Kumar et al. 2010).

Microbial dysbiosis noted in wild-type mice indicated high levels of pro-inflammatory molecules IL-1β and tumor-necrotizing factors TNFα in setting up the inflammation followed by intestinal metabolic disorders. Another mechanism which altered microbiome shows impact on metabolic health through fatty acid metabolism. Gut microbial flora utilizes nondigestible carbon sources to produce short chain fatty acids like ethanoate, propanoate, and butanoate used by colonocytes of intestinal epithelium and finally transported into the blood, as these fatty acids are anti-inflammatory and antitumorigenic which play vital role in metabolic activities of human gut and build physiological homeostasis and also help in triggering immunity (Bindels et al. 2012; Tan et al. 2014).

Gut microbiome secretes SCTAs, and these short chain fatty acids recognize cell surface receptors like G-protein-coupled receptors to regulate lipid metabolism (Samuel et al. 2008). They also involve in secretion of insulin by controlling the concentration levels. Glucagon-like peptide 1 hormone (Tolhurst et al. 2012). The defined and distributed bacterial flora that constitutes part gut microbiome changes the concentration levels and also types of SCFA like acetate, propionate, and butyrate that can be formed during various physiological processes of their metabolism (Macfarlane and Macfarlane 2011). The gut bacterial flora influence liver activities like conversion of the primary bile acids into secondary bile acids; it also promotes glucose homeostasis by G-protein-coupled receptors. Upon antibiotic administration, the disturbed microbiome alters physiological activities particularly insulin sensitivity, and bile acid metabolism is demonstrated (Thomas et al. 2009; Vrieze et al. 2014).

Antibiotic-induced host-microbiome interactions result in spread and establishment of commensal bacterial flora and pathogenic dissemination, and abnormal microbiome host immune responses are resulted and known to exhibit different gastrointestinal diseases like obesity, celiac disease (Valitutti et al. 2019), diabetes (Wen et al. 2008), metabolic syndrome (Belizário et al. 2018), inflammatory bowel disease (Zhang et al. 2017), rheumatic arthritis (Maeda and Takeda 2019), neurodegenerative disorder (Main and Minter 2017), malignancy and cancer (Gopalakrishnan et al. 2018), depression, and autism (Fung et al. 2017). Patients who are critically ill treated with antimicrobial agents are likely to show disruption of gut microbiota and microbial dysbiosis in their intestine and which makes the patients suffer from bacterial infections. *Clostridium difficile* is an opportunistic pathogen and is the best example that causes the disease colitis which can be treated by introducing the microbiota (Britton and Young 2014). The use of broad-spectrum antibiotics can disturb the equilibrium of normal microbiota and creates conditions favorable for the overgrowth of pathogens and also favor antibiotic-resistant bacterial strains. Colitis caused by *Clostridium difficile* and mucosal candidiasis caused by *Candida* sp. are two such examples that get established after the antibiotic administrations and disturbance of microbiome. Gut microbiota by enlargement influences many aspects like physiology of human host like nutrient synthesis like vitamins and amino acids, metabolic activities, infection resistance, and immunity.

5.4 Conclusion

Wide usage of antibiotics in treating infectious diseases leads to so many threats like emergence of bacterial resistance against antibiotics which posed a big challenge to the scientific world (Francino 2016). Excessive usage of antibiotic not only shows cidal and static activity on pathogen but also on microbiota of the human body. Microbial imbalance caused by antibiotics impacts on human health and immunity. Antibiotic-induced microbiota dysbiosis affects systemic immunity and causes many health disorders. Fecal microbiota transplantation and manipulation of microbiota using probiotics are trending tools in microbiome research and a very useful application in restoring microbial communities and correcting microbial dysbiosis (Mekonnen et al. 2020). This review on antibiotic-mediated microbiome dysbiosis and its impact on immune system and health will provide a better understanding on antibiotic-driven microbiome-host interaction that impacts on immunity and human health and also through challenges on therapeutic tools that would correct dysbiosis of microbiota.

Acknowledgment Authors thank principal and management of Bhavan's Vivekananda college of Science, Humanities and Commerce for the encouragement and support.

Declaration of Competing Interest The authors declare that there is no conflict of interest.

References

Baba N, Samson S, Bourdet-Sicard R, Rubio M, Sarfati M (2008) Commensal bacteria trigger a full dendritic cell maturation program that promotes the expansion of non-Tr1 suppressor T cells. J Leukoc Biol 84(2):468–476

Belizário JE, Faintuch J, Garay-Malpartida M (2018) Gut microbiome dysbiosis and immunometabolism: new frontiers for treatment of metabolic diseases. Mediat Inflamm 2018:2037838

Bindels LB, Porporato P, Dewulf EM, Verrax J, Neyrinck AM, Martin JC et al (2012) Gut microbiota-derived propionate reduces cancer cell proliferation in the liver. Br J Cancer 107(8):1337–1344

Blumberg R, Powrie F (2012) Microbiota, disease, and back to health: a metastable journey. Sci Transl Med 4(137):137rv7

Bouskra D, Brézillon C, Bérard M, Werts C, Varona R, Boneca IG, Eberl G (2008) Lymphoid tissue genesis induced by commensals through NOD1 regulates intestinal homeostasis. Nature 456(7221):507–510

Brandl K, Plitas G, Mihu CN, Ubeda C, Jia T, Fleisher M et al (2008) Vancomycin-resistant enterococci exploit antibiotic-induced innate immune deficits. Nature 455(7214):804–807

Britton RA, Young VB (2014) Role of the intestinal microbiota in resistance to colonization by Clostridium difficile. Gastroenterology 146(6):1547–1553

Cani PD, Amar J, Iglesias MA, Poggi M, Knauf C, Bastelica D et al (2007) Metabolic endotoxemia initiates obesity and insulin resistance. Diabetes 56(7):1761–1772

Cerutti A, Rescigno M (2008) The biology of intestinal immunoglobulin a responses. Immunity 28(6):740–750

Chakraborti CK (2015) New-found link between microbiota and obesity. World J Gastrointest Pathophysiol 6(4):110

Charbonneau MR, Blanton LV, DiGiulio DB, Relman DA, Lebrilla CB, Mills DA, Gordon JI (2016) A microbial perspective of human developmental biology. Nature 535(7610):48–55

Clarke TB, Davis KM, Lysenko ES, Zhou AY, Yu Y, Weiser JN (2010) Recognition of peptidoglycan from the microbiota by Nod1 enhances systemic innate immunity. Nat Med 16(2):228–231

De La Cochetière MF, Durand T, Lepage P, Bourreille A, Galmiche JP, Dore J (2005) Resilience of the dominant human fecal microbiota upon short-course antibiotic challenge. J Clin Microbiol 43(11):5588–5592

Dethlefsen L, Relman DA (2011) Incomplete recovery and individualized responses of the human distal gut microbiota to repeated antibiotic perturbation. Proc Natl Acad Sci 108(Supplement 1):4554–4561

Dethlefsen L, Huse S, Sogin ML, Relman DA (2008) The pervasive effects of an antibiotic on the human gut microbiota, as revealed by deep 16S rRNA sequencing. PLoS Biol 6(11):e280

Dimmitt RA, Staley EM, Chuang G, Tanner SC, Soltau TD, Lorenz RG (2010) The role of postnatal acquisition of the intestinal microbiome in the early development of immune function. J Pediatr Gastroenterol Nutr 51(3):262

Dufour V, Millon L, Faucher JF, Bard E, Robinet E, Piarroux R et al (2005) Effects of a short-course of amoxicillin/clavulanic acid on systemic and mucosal immunity in healthy adult humans. Int Immunopharmacol 5(5):917–928

Eckburg PB, Bik EM, Bernstein CN, Purdom E, Dethlefsen L, Sargent M et al (2005) Diversity of the human intestinal microbial flora. Science 308(5728):1635–1638

Ferrer M, Méndez-García C, Rojo D, Barbas C, Moya A (2017) Antibiotic use and microbiome function. Biochem Pharmacol 134:114–126

Fouhy F, Ogilvie LA, Jones BV, Ross RP, Ryan AC, Dempsey EM et al (2014a) Identification of aminoglycoside and β-lactam resistance genes from within an infant gut functional metagenomic library. PLoS One 9(9):e108016

Fouhy F, Ross RP, Fitzgerald GF, Stanton C, Cotter PD (2014b) A degenerate PCR-based strategy as a means of identifying homologues of aminoglycoside and β-lactam resistance genes in the gut microbiota. BMC Microbiol 14(1):25

Francino MP (2016) Antibiotics and the human gut microbiome: dysbioses and accumulation of resistances. Front Microbiol 6:1543
Franzosa EA, Hsu T, Sirota-Madi A, Shafquat A, Abu-Ali G, Morgan XC, Huttenhower C (2015) Sequencing and beyond: integrating molecular 'omics' for microbial community profiling. Nat Rev Microbiol 13(6):360–372
Fung TC, Olson CA, Hsiao EY (2017) Interactions between the microbiota, immune and nervous systems in health and disease. Nat Neurosci 20(2):145
Garner CD, Antonopoulos DA, Wagner B, Duhamel GE, Keresztes I, Ross DA et al (2009) Perturbation of the small intestine microbial ecology by streptomycin alters pathology in a salmonella enterica serovar typhimurium murine model of infection. Infect Immun 77(7):2691–2702
Gopalakrishnan V, Helmink BA, Spencer CN, Reuben A, Wargo JA (2018) The influence of the gut microbiome on cancer, immunity, and cancer immunotherapy. Cancer Cell 33(4):570–580
Greer R, Dong X, Morgun A, Shulzhenko N (2016) Investigating a holobiont: microbiota perturbations and transkingdom networks. Gut Microbes 7(2):126–135
Hernández E, Bargiela R, Diez MS, Friedrichs A, Pérez-Cobas AE, Gosalbes MJ et al (2013) Functional consequences of microbial shifts in the human gastrointestinal tract linked to antibiotic treatment and obesity. Gut Microbes 4(4):306–315
Hill DA, Hoffmann C, Abt MC, Du Y, Kobuley D, Kirn TJ et al (2010) Metagenomic analyses reveal antibiotic-induced temporal and spatial changes in intestinal microbiota with associated alterations in immune cell homeostasis. Mucosal Immunol 3(2):148–158
Hill DA, Siracusa MC, Abt MC, Kim BS, Kobuley D, Kubo M et al (2012) Commensal bacteria–derived signals regulate basophil hematopoiesis and allergic inflammation. Nat Med 18(4):538–546
Hooper LV, Midtvedt T, Gordon JI (2002) How host-microbial interactions shape the nutrient environment of the mammalian intestine. Annu Rev Nutr 22(1):283–307
Human Microbiome Project Consortium (2012) Structure, function and diversity of the healthy human microbiome. Nature 486:207–214
Ivanov II, Atarashi K, Manel N, Brodie EL, Shima T, Karaoz U et al (2009) Th17 cells secrete interleukin-17 (IL-17), IL-17F, and IL-22 and have significant roles in protecting the host from bacterial and fungal infections, particularly at mucosal surfaces. Induction of intestinal Th17 cells by segmented filamentous bacteria. Cell 139:485–498
Jernberg C, Lofmark S, Edlund C, Jansson JK (2007) Long-term ecological impacts of antibiotic administration on the human intestinal microbiota. ISME J 1:56–66. https://doi.org/10.1038/ismej.2007
Karlsson FH, Tremaroli V, Nookaew I, Bergström G, Behre CJ, Fagerberg B et al (2013) Gut metagenome in European women with normal, impaired and diabetic glucose control. Nature 498(7452):99–103
Kline JN (2007) Eat dirt: CpG DNA and immunomodulation of asthma. Proc Am Thorac Soc 4(3):283–288
Knoop KA, McDonald KG, Kulkarni DH, Newberry RD (2016) Antibiotics promote inflammation through the translocation of native commensal colonic bacteria. Gut 65(7):1100–1109
Macfarlane GT, Macfarlane S (2011) Fermentation in the human large intestine: its physiologic consequences and the potential contribution of prebiotics. J Clin Gastroenterol 45:S120–S127
Maeda Y, Takeda K (2019) Host–microbiota interactions in rheumatoid arthritis. Exp Mol Med 51(12):1–6
Main BS, Minter MR (2017) Microbial immuno-communication in neurodegenerative diseases. Front Neurosci 11:151
Manuzak JA, Zevin AS, Cheu R, Richardson B, Modesitt J, Hensley-McBain T et al (2020) Antibiotic-induced microbiome perturbations are associated with significant alterations to colonic mucosal immunity in rhesus macaques. Mucosal Immunol 13(3):471–480
Maurice CF, Haiser HJ, Turnbaugh PJ (2013) Xenobiotics shape the physiology and gene expression of the active human gut microbiome. Cell 152(1–2):39–50

Mekonnen SA, Merenstein D, Fraser CM, Marco ML (2020) Molecular mechanisms of probiotic prevention of antibiotic-associated diarrhea. Curr Opin Biotechnol 61:226–234

Meropol SB, Edwards A (2015) Development of the infant intestinal microbiome: a bird's eye view of a complex process. Birth Defects Res C Embryo Today 105(4):228–239

Mikkelsen KH, Frost M, Bahl MI, Licht TR, Jensen US, Rosenberg J et al (2015) Effect of antibiotics on gut microbiota, gut hormones and glucose metabolism. PLoS One 10(11):e0142352

Murri M, Leiva I, Gomez-Zumaquero JM, Tinahones FJ, Cardona F, Soriguer F, Queipo-Ortuño MI (2013) Gut microbiota in children with type 1 diabetes differs from that in healthy children: a case-control study. BMC Med 11(1):46

Nobel YR, Cox LM, Kirigin FF, Bokulich NA, Yamanishi S, Teitler I et al (2015) Metabolic and metagenomic outcomes from early-life pulsed antibiotic treatment. Nat Commun 6(1):1–15

Palmer C, Bik EM, DiGiulio DB, Relman DA, Brown PO (2007) Development of the human infant intestinal microbiota. PLoS Biol 5(7):e177

Pérez-Cobas AE, Gosalbes MJ, Friedrichs A, Knecht H, Artacho A, Eismann K et al (2013) Gut microbiota disturbance during antibiotic therapy: a multi-omic approach. Gut 62(11):1591–1601

Qin J, Li Y, Cai Z, Li S, Zhu J, Zhang F et al (2012) A metagenome-wide association study of gut microbiota in type 2 diabetes. Nature 490(7418):55–60

Rautava S, Ruuskanen O, Ouwehand A, Salminen S, Isolauri E (2004) The hygiene hypothesis of atopic disease—an extended version. J Pediatr Gastroenterol Nutr 38(4):378–388

Rodrigues RR, Greer RL, Dong X, DSouza KN, Gurung M, Wu JY et al (2017) Antibiotic-induced alterations in gut microbiota are associated with changes in glucose metabolism in healthy mice. Front Microbiol 8:2306

Samuel BS, Shaito A, Motoike T, Rey FE, Backhed F, Manchester JK et al (2008) Effects of the gut microbiota on host adiposity are modulated by the short-chain fatty-acid binding G protein-coupled receptor, Gpr41. Proc Natl Acad Sci 105(43):16767–16772

Sanz Y, Olivares M, Moya-Pérez Á, Agostoni C (2015) Understanding the role of gut microbiome in metabolic disease risk. Pediatr Res 77(1–2):236–244

Schumann A, Nutten S, Donnicola D, Comelli EM, Mansourian R, Cherbut C et al (2005) Neonatal antibiotic treatment alters gastrointestinal tract developmental gene expression and intestinal barrier transcriptome. Physiol Genomics 23(2):235–245

Strzępa A, Majewska-Szczepanik M, Lobo FM, Wen L, Szczepanik M (2017) Broad spectrum antibiotic enrofloxacin modulates contact sensitivity through gut microbiota in a murine model. J Allergy Clin Immunol 140(1):121–133

Sudo N, Yu XN, Aiba Y, Oyama N, Sonoda J, Koga Y, Kubo C (2002) An oral introduction of intestinal bacteria prevents the development of a long-term Th2-skewed immunological memory induced by neonatal antibiotic treatment in mice. Clin Exp Allergy 32(7):1112–1116

Sun L, Zhang X, Zhang Y, Zheng K, Xiang Q, Chen N et al (2019) Antibiotic-induced disruption of gut microbiota alters local metabolomes and immune responses. Front Cell Infect Microbiol 9:99

Tan J, McKenzie C, Potamitis M, Thorburn AN, Mackay CR, Macia L (2014) The role of short-chain fatty acids in health and disease. Adv Immunol 121:91–119

Tanaka S, Kobayashi T, Songjinda P, Tateyama A, Tsubouchi M, Kiyohara C et al (2009) Influence of antibiotic exposure in the early postnatal period on the development of intestinal microbiota. FEMS Immunol Med Microbiol 56(1):80–87

Thomas C, Gioiello A, Noriega L, Strehle A, Oury J, Rizzo G et al (2009) TGR5-mediated bile acid sensing controls glucose homeostasis. Cell Metab 10(3):167–177

Tolhurst G, Heffron H, Lam YS, Parker HE, Habib AM, Diakogiannaki E et al (2012) Short-chain fatty acids stimulate glucagon-like peptide-1 secretion via the G-protein–coupled receptor FFAR2. Diabetes 61(2):364–371

Tremaroli V, Bäckhed F (2012) Functional interactions between the gut microbiota and host metabolism. Nature 489(7415):242–249

Ubeda C, Taur Y, Jenq RR, Equinda MJ, Son T, Samstein M et al (2010) Vancomycin-resistant enterococcus domination of intestinal microbiota is enabled by antibiotic treatment in mice and precedes bloodstream invasion in humans. J Clin Invest 120(12):4332–4341

Ubeda C, Lipuma L, Gobourne A, Viale A, Leiner I, Equinda M et al (2012) Familial transmission rather than defective innate immunity shapes the distinct intestinal microbiota of TLR-deficient mice. J Exp Med 209(8):1445–1456

Utzschneider KM, Kratz M, Damman CJ, Hullarg M (2016) Mechanisms linking the gut microbiome and glucose metabolism. J Clin Endocrinol Metab 101(4):1445–1454

Valitutti F, Cucchiara S, Fasano A (2019) Celiac disease and the microbiome. Nutrients 11(10):2403

Vijay-Kumar M, Aitken JD, Carvalho FA, Cullender TC, Mwangi S, Srinivasan S et al (2010) Metabolic syndrome and altered gut microbiota in mice lacking toll-like receptor 5. Science 328(5975):228–231

Virella G (ed) (1997) Microbiology and infectious diseases. Williams & Wilkins

Vollaard EJ, Clasener HA (1994) Colonization resistance. Antimicrob Agents Chemother 38(3):409

Vrieze A, Out C, Fuentes S, Jonker L, Reuling I, Kootte RS et al (2014) Impact of oral vancomycin on gut microbiota, bile acid metabolism, and insulin sensitivity. J Hepatol 60(4):824–831

Wang B, Yao M, Lv L, Ling Z, Li L (2017) The human microbiota in health and disease. Engineering 3(1):71–82

Wang S, Qu Y, Chang L, Pu Y, Zhang K, Hashimoto K (2020) Antibiotic-induced microbiome depletion is associated with resilience in mice after chronic social defeat stress. J Affect Disord 260:448–457

Wen L, Ley RE, Volchkov PY, Stranges PB, Avanesyan L, Stonebraker AC et al (2008) Innate immunity and intestinal microbiota in the development of type 1 diabetes. Nature 455(7216):1109–1113

Wlodarska M, Willing B, Keeney KM, Menendez A, Bergstrom KS, Gill N et al (2011) Antibiotic treatment alters the colonic mucus layer and predisposes the host to exacerbated Citrobacter rodentium-induced colitis. Infect Immun 79(4):1536–1545

Xu L, Surathu A, Raplee I, Chockalingam A, Stewart S, Walker L et al (2020) The effect of antibiotics on the gut microbiome: a metagenomics analysis of microbial shift and gut antibiotic resistance in antibiotic treated mice. BMC Genomics 21:1–18

Zhang M, Sun K, Wu Y, Yang Y, Tso P, Wu Z (2017) Interactions between intestinal microbiota and host immune response in inflammatory bowel disease. Front Immunol 8:942

Zhang J, Bi JJ, Guo GJ, Yang L, Zhu B, Zhan GF et al (2019) Abnormal composition of gut microbiota contributes to delirium-like behaviors after abdominal surgery in mice. CNS Neurosci Ther 25(6):685–696

Zheng D, Liwinski T, Elinav E (2020) Interaction between microbiota and immunity in health and disease. Cell Res 30:1–15

Nutritional Modulation of Gut Microbiota Alleviates Metabolic and Neurological Disorders

6

M. Rajeswari, Sudhakar Pola, and D. S. L. Sravani

Abstract

The food we consume is a reservoir of nutrients and bioactive compounds. Adherence to a prudent diet confers nutrition and health. In the light of existing research, it is evident that there is an association between nutrients, dietary patterns, and disease manifestation mediated through gut microbiome and microbial metabolites. The gut microbiome is an adverse consortium and is envisaged as more potent than the human genome owing to its complexity and the myriad of physiological functions it executes on the host along multiple axes, including the liver, brain, and heart. The gut microbiome's variation is linked to infection vulnerability and long-term sequelae of metabolic, immune-mediated, and neurological disorders. Excavation of the complexity of interactions of diet, nutrients, and microbiota with particular reference to the microbiota-accessible carbohydrates, plant proteins, glutamine, short-chain fatty acids, vitamin D, zinc, calcium, and other nutrients under the umbrella of increasing scientific evidence is the need of the hour. Thus, managing the microbiome through judicious dietary approaches appears to be promising in reconfiguring microbial aberrations. Diet engineering prowess based on individual genes, omic profiles, and nutribiotic algorithms paves the way for precision nutrition tailored to preclude chronic diseases.

M. Rajeswari
Department of Food, Nutrition and Dietetics, AUCST, Andhra University, Visakhapatnam, India

S. Pola (✉)
Department of Biotechnology, AUCST, Andhra University, Visakhapatnam, India

D. S. L. Sravani
Division of Nutritional Sciences, Cornell University, Ithaca, NY, USA

P. Veera Bramhachari (ed.), *Human Microbiome in Health, Disease, and Therapy*, https://doi.org/10.1007/978-981-99-5114-7_6

Keywords

Nutrients · Microbiome · Diet · Short-chain fatty acids · Diet engineering · Nutribiotics

6.1 Introduction

The discovery of the gut microbiome, not far short of two decades ago, ushered in a new microbiome era with promising results in the host health arena. A plethora of research studies in the past decade revealed profounding facts on the trillions of primordial microorganisms collectively termed the microbiota residing within the human gut (Cho and Blaser 2012). Proper maintenance of these microbes in early infancy aids in adult life. Therefore, it can be achieved through a healthy diet, exercise, and a clean environment, as it plays an essential role in shaping gut microbes (Pola and Padi 2021). Numerous factors may affect the gut microbiota, of which dietary factors are crucial (Zmora et al. 2019). Thus food-microbiome-host synergy is vital in comprehending the complexity of microbiome research (Ezra-Nevo et al. 2020).

Diet is a commixture of nutrients and a repository of health-promoting phytochemicals, which, when consumed in prudent form, builds the nutritional status coupled with coveted health benefits. A collegial relationship appears to exist between nutrients and the gut microbiome (Frame et al. 2020). Today's world is in a nutrition transition phase, facing the dual burden of pretransition diseases like under nutrition and infections, post-transition overnutrition, and lifestyle-affiliated diseases such as metabolic syndrome and obesity (Raj 2020). Diet therapy and clinical nutritionology are in vogue as procuring the right nutrition is essential to avert these lifestyle diseases (Covarrubias et al. 2020).

A comprehensive exploration of the concepts related to diet/nutrient–microbiota crosstalk in the gut and the potential health benefits and detrimental consequences is thoroughly appreciated. This knowledge will be of tremendous use in devising gut microbiota-targeted nutritional strategies to improve human health. Developing next-generation nutribiotics catering to individual needs is not far from reach with the ongoing vigorous research in this area. The role of diet, nutrient intake, pre- and probiotics, and non-nutrient polyphenols in maintaining healthy gut microbiota is reviewed in the light of new evidence.

6.2 Carbohydrates

Carbohydrates are perhaps the most widely researched essential nutrients to impact the gut microbiome. Digestible carbohydrates mainly comprise starches and sugars. Examples of beverages with added sugars include sodas, sports drinks, energy drinks, sweetened water, fruit juices, coffee, and tea. While WHO recommendations suggest, an added sugar intake be limited to less than 10% of an individual's total calorie consumption, sugar consumption worldwide has been devastatingly increasing (WHO 2015).

A diet rich in simple carbohydrates and low in dietary fiber is believed to cause detrimental effects on the gut microbiome (Sonnenburg et al. 2016). A highly compounded interaction appears to exist between dietary sugar and gut microbiota. Excessive sugar consumption is strongly implicated in regulating gut colonization and microbial dysbiosis besides the formation of glycoconjugates and biofilms. Subsequent maladaptive processes on the host include increased expression of sugar transporters in the gut, metabolic deregulation, oxidative stress, peripheral insulin resistance, increased inflammation, and increased immune modulation paving the pathway for multiple undesirable health outcomes and poor oral health (Haque et al. 2020).

6.3 Impact of Sugars on Microbiota

6.3.1 Sugar Intake and Regulation of Colonization

The presence of sugars in the gut could alter the gut microbiota by affecting microbial physiology. For example, it was observed that increased glucose and fructose concentration in the gut suppresses polysaccharide utilization genes in *Bacteroides thetaiotaomicron* by catabolite repression leading to its reduced colonization ability in mice (Townsend et al. 2019). Additionally, glucose hampers the utilization of other carbon sources by inhibiting the synthesis of the signaling molecule cAMP (Kremling et al. 2015). Research on the effect of these transcriptional changes caused by specific types of sugars on the gut microbial composition is sparse.

6.3.2 Formation of Glycoconjugates

Dietary sugars are vital components of several glycoconjugates of microbes. The versatility and diversity of bacterial glycoconjugates result in a species-specific glycan fingerprint, providing a range of ligands to interact with the host environment (Tytgat and De Vos 2016). Some microbial exopolysaccharides can promote anti-inflammatory response (Mazmanian et al. 2008). On the other hand, lipopolysaccharides (LPS) trigger the genesis of low-grade inflammation, thereby, insulin resistance which is mediated by the gut microbial impact on the innate immune system through toll-like receptor 4 (TLR-4) and CD14 signaling (Cani et al. 2007).

6.3.3 Formation of Biofilms and Flagella

Sugar availability could also influence the genesis of flagellar structures and biofilms (Di Rienzi and Britton 2019). Biofilm microbial populations were shown to significantly impair the intestinal epithelial barrier, affect cellular proliferation by altering polyamine metabolism, enhance pro-inflammatory/pro-carcinogenic response, and exacerbate gut dysbiosis in experimental models (Li et al. 2017). Bacterial flagella usually stimulate the host immune system by activating TLR5 signaling. Still, its glycosylation due to excess consumption of sugars reduces TLR5 recognition in opportunistic pathogens such as *Burkholderia cenocepacia* and enhances *Clostridium difficile* adhesion to epithelial cells (Hanuszkiewicz et al. 2014; Valiente et al. 2016). The presence of various dietary sugars affects metabolism inside the gut bacterial cells, and the metabolites secreted into the gut are also altered.

6.3.4 High Sugar Consumption and Microbial Dysbiosis

High carbohydrate diets could promote *Ruminococcaceae* and *Lachnospiraceae* growth at the cost of *Enterobacteriaceae* and *Bifidobacteria*. Several animal studies have reported adverse health effects of high-dose fructose or glucose intake concerning gut microbiota (Takahashi et al. 2015). Analysis of the microbiomes of lactose-intolerant individuals or subjects with similar genetic conditions might illuminate dietary sugar and gut microbiome interactions (Goodrich et al. 2016; Blekhman et al. 2015).

6.3.5 Sugars, Gut Microbiome, and Host Health

Sugars act as essential substrates for the survival of small intestine microbes as they possess more carbohydrate utilization genes than microbes of large intestine (Zoetendal et al. 2012). Repeated sugar consumption increases sugar transporters and carbohydrate metabolizing enzymes such as hydrolases in the host gut. Metagenomic studies in mice strengthened these claims fed a high-sugar/high-fat diet (Turnbaugh et al. 2009; Carmody et al. 2015). In an elegant study, germ-free (GF) mice that consumed high concentration (8%) sucrose solution had heightened expression of type 1 taste receptor 3 (T1R3) and SGLT1 than control mice in the epithelial cells of the intestine. The potential role of gut microbiota modulations in the carbohydrate metabolism pathways, leading to altered food consumption, energy homeostasis, and weight gain, was explored by researchers in the light of these transcriptional changes in the small intestine (Swartz et al. 2012).

Moreover, evidence shows that a diet high in sugars increases the abundance of *Akkermansia* spp., intestinal permeability, metabolic endotoxemia, inflammation (Jena et al. 2014), and hepatic fat accumulation, ultimately leading to hepatic steatosis without affecting body weight. This is in congruence with the normal-weight obesity observed in Asians that may play a vital role in developing metabolic disorders (Do et al. 2018). Excessive intake of fructose also may predispose an individual to nonalcoholic fatty liver disease (NAFLD) by increasing uric acid generation via its metabolism by fructokinase. The pro-oxidative and pro-inflammatory uric acid

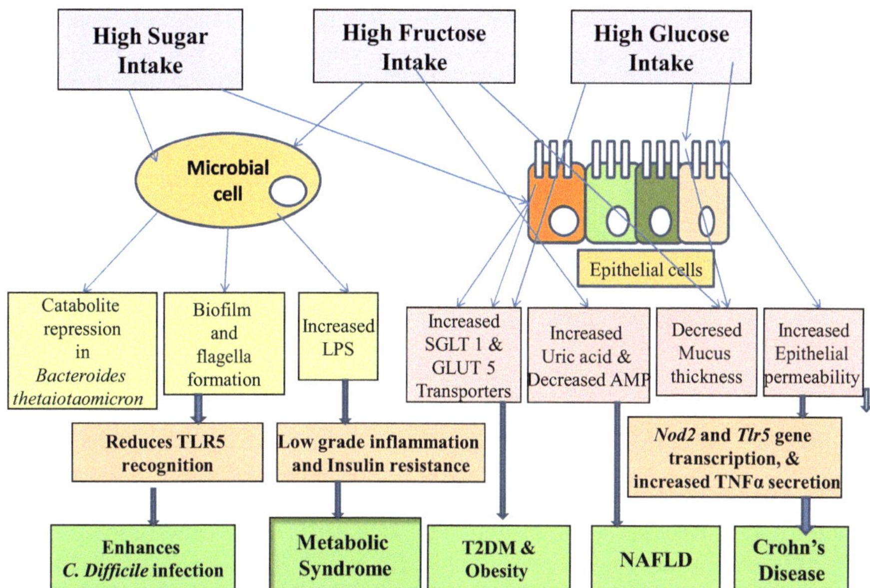

Fig. 6.1 Schematic representation of effect of sugar intake on gut microbiota and gut epithelium and consequent health outcomes

can increase gut permeability and endotoxemia, thus stimulating lipogenesis in the liver, progressing to NAFLD (Jensen et al. 2018).

Mechanistic studies established the role of high-sugar/high-fat (HS/HF) diet in the multifactorial etiology of Crohn's disease (CD) through microbial dysbiosis. Human carcinoembryonic antigen-related cell adhesion molecule 6 (CEACAMs) are abnormally expressed in individuals with CD. When transgenic mice expressing human CEACAMs were challenged with HS/HF diet, increased *E. coli* abundance was observed in gut microbiota with subsequent decrease in mucus layer thickness, intestinal permeability, and TNFα secretion, and induction of *Nod2* and *Tlr5* gene transcription. These modifications resulted in a higher ability of adherent-invasive *E. coli* (AIEC) bacteria to colonize the gut mucosal layer and to induce inflammation, potentiating CD pathogenesis and meriting the role of imprudent HS/HF diet (Martinez-Medina et al. 2014) (Fig. 6.1).

6.4 Proteins

Some dietary peptides and amino acids remain undigested in the host digestive system and are metabolized by the microbiota. Excessive protein intake and dispersal constraints of brush-border enzymes might be responsible (Smith and Macfarlane 1998). About 5–15 g of protein passes into the proximal colon daily and is digested by microbial populations (Conlon and Bird 2015). Remnants of protein digestion are acted up on by bacteria via bacterial proteinases and peptidases from species in particular *Clostridia*, *Propionibacterium*, *Prevotella*, *Bifidobacterium*, and *Bacteroides* (Maukonen and Saarela 2015; Jandhyala et al. 2015) subsequently used to form microbial cell components. Therefore, dietary protein, proteolytic fermentation, and metabolites converge immensely to affect host health (Smith and Macfarlane 1998).

6.4.1 Source of Protein and Gut Microbial Framework

Several investigations promulgated the influence of the dietary protein source on gut microbiota (Lazar et al. 2019). *Roseburia* and *Eubacterium rectale* educed in abundance on high-protein and low-carbohydrate diet with a subsequent decrease in the proportion of butyrate in feces (Russell et al. 2011). Specifically, animal protein consumption was associated with an increased abundance of bile-tolerant anaerobes, for example, *Alistipes*, *Bacteroides*, and *Bilophila*, which have been allied with increased risk of IBD (Cotillard et al. 2013). Additionally, higher intake of red and processed meat has been linked to increased levels of trimethylamine *N*-oxide (TMAO), choline, betaine, lecithin, and L-carnitine, all of which are associated with cardiovascular diseases and T2DM. It is also associated with more significant DNA damage in the colonic mucosa, especially when the diet is deprived of fermentable carbohydrates (Toden et al. 2007).

L-carnitine associated with red meat intake needs a special mention in the imbroglio of the diet, microbial metabolite, and host health interactions. L-carnitine is metabolized by the gut microbiota to trimethylamine (TMA) and is converted to TMAO in the liver by flavin monooxygenases. Available evidence in mice and humans revealed that the ability to transform l-carnitine to TMA or TMAO was associated with the abundance of bacterial genera such as *Prevotella*, *Deferribacteraceae*, *Anaeroplasmataceae*, and *Enterobacteriaceae* (Koeth et al. 2013). Processed meat consumption may increase colorectal cancer risk in humans via the production of heterocyclic amines (Butler et al. 2003), colonic cytotoxicity, hyperproliferation, and bloom of mucin-degrading bacteria (Ijssennagger et al. 2015). Conversely, specific gut microbial species, for instance, lactic-acid-producing bacteria, can protect from host DNA damage and neoplasia by binding the heterocyclic amines (Zsivkovits et al. 2003).

Evidence has shown that plant-derived proteins are linked with increased microbial diversity (Singh et al. 2017) and lower mortality than animal-derived proteins (Levine et al. 2014). *Bifidobacterium* and *Lactobacillus* abundance and intestinal SCFA levels have increased, augmenting mucosal barrier integrity (Kim et al. 2014) with pea protein extract supplementation (Świątecka et al. 2011). Similarly, mung bean protein was observed to alter the HFD-induced F/B ratio and to increase the abundance of *Ruminococcacea* in an HFD mice model (Nakatani et al. 2015). Evidence is also emerging that the quality and quantity of protein, cooking methods, and intake timing may affect the composition and diversity of gut microbiota (Yang et al. 2020a). For instance, fried meat was found to have *C. hidtolyticum perfringens*, a common food-borne pathogen compared to boiled meat in batch fermentations (Shen et al. 2010). Recent work highlighted that soy protein intake potentiated microbiota diversity and SCFA production. Observed effects were relatively more substantial when soy protein was consumed in the morning compared to the evening deliberating the effect of timing of exposure on microbiota (Tamura et al. 2020).

6.4.2 Amino Acids Influencing Gut Microbiota and Host Health

Functional amino acids such as tryptophan and glutamine might positively influence the gut microbiome-associated immune system—specifically, *Lactobacilli* spp. Utilize tryptophan as an energy source to generate aryl hydrocarbon receptor (AhR) ligands. This AhR agonist was found to limit central nervous system inflammation offering protection against multiple sclerosis (Rothhammer et al. 2018). On the contrary, increased indole levels in a rat model enhanced the probability of brain malfunctions (Jaglin et al. 2018). Indole is a metabolite of tryptophan from gut microbiota and a precursor of the AhR agonist indoxyl-3-sulfate (Schroeder et al. 2010).

Taurine being a necessary amino acid, besides nutritional roles, regulates neuroendocrine functions, has strong immunity-enhancing activities, and ameliorates intestinal inflammation. In immune-suppressed mice, taurine intervention reversed the reduction of *Lachnospiraceae* and *Ruminococcaceae* groups along with an increase in $CD3^+$ cells (T cells), $CD19^+$ cells (B cells), proving its immune potentiating competency (Fang et al. 2019). Taurine activates nucleotide-binding and oligomerization domain-like receptors (NLRs), especially NLRP6, an essential mediator of intestinal immunity and, therefore, can intimidate colitis severity (Levy et al. 2015). Taurine (165 mg/kg) in mice could regulate gut microecology by inhibiting the growth of pathogenic *Helicobacter*. This effect was shown to accelerate the production of SCFA and reduce LPS concentration (Yu et al. 2016). Similarly, 0.2% of taurine supplementation could restore the rice-field eel gut microbial dysbiosis induced by dietary oxidized fish oil (Peng et al. 2019).

6.4.3 Protein Metabolites in the Gut and Host Health

High-protein diets generally decreased the total SCFA production, in particular, of acetate and propionate, while branched-chain fatty acids (BCFA) such as isovalerate and isobutyrate were increased in concentration. The enhanced concentrations of BCFA were correlated with increased *Alistipes* and *Bacteroides* relative abundance. BCFAs are exclusively produced when gut bacteria ferment branched-chain amino acids. So, they are reliable biomarkers of protein fermentation. Interestingly, *Bacteroidetes* and *Firmicutes* species produce propionate and butyrate from peptide and amino acid fermentation (Shortt et al. 2018).

Spermidine/spermine, cadaverine, and putrescine are examples of polyamines exhibiting the governance of gene transcription and protein translation. Especially the spermidine-elicited hyphenation modification of eukaryotic translation initiation factor 5A (EIF5A) plays a crucial role. Besides, polyamines regulate metabolic functions and increase mitochondrial substrate oxidation. Enterocytes in the small intestine may utilize putrescine as an energy source. Polyamines protect the gut barrier function and induce gut maturation and longevity, thus supporting gut physiology. Abnormally increased concentrations of ammonia may lead to the development of malignant growths. Moreover, ammonia release incites inflammation, decreasing butyrate transporter expression and affecting butyrate uptake by colonocytes (Villodre et al. 2015). High ammonia concentration also decreases butyrate oxidation in colonocytes, affecting intestinal barrier function (Anand et al. 2016).

Intestinal uptake of p-cresol was associated with cardiovascular risk in patients with chronic kidney disease (Gryp et al. 2017). Similarly, phenylacetylglutamine is another microbial metabolite from phenylalanine associated with cardiovascular disease risk (Poesen et al. 2016). Notably, tyrosine metabolite, 4-ethyl phenyl sulfate (4-EPS), can induce autism spectrum disorder (ASD)-like behaviors. However, on a positive note, *Bacteroides fragilis* administration can reduce this neurotoxic metabolite, thus mitigating anxiety-like behavior (Hsiao et al. 2013).

6.5 Fats

About 5–10 gm of lipid enters the proximal colon daily, mostly of dietary origin (Conlon and Bird 2015). Several human and animal studies have suggested that a high-fat diet considerably decreases the abundance of *Lactobacillus* and *Bifidobacterium* spp. and increases *Clostridiales*, *Bacteroides*, and *Enterobacteriales* (Drasar et al. 2007). Concerning the type of fat, a high monounsaturated fat (MUFA) intake did not affect any bacterial genus abundance, albeit reduced total bacterial abundance and plasma total and LDL cholesterol (Fava et al. 2013). Moreover, n-3 PUFAs appear to alter gut microbial composition with a unique increase in *Akkermansia* spp. in favor of a lean phenotype (Bellenger et al. 2019).

6.5.1 High Intake of Fat, Microbial Dysbiosis, and Host Health

Araújo et al. (2017) reviewed the link between HFD and metabolic complications mediated by gut dysbiosis. Based on the randomized clinical trials, the pre-obesogenic mechanisms induced by HFD include intestinal dysbiosis, increased circulating levels of LPS in humans, decreased expression of antimicrobial peptides and gap junction proteins, and reduction in gut barrier integrity, ultimately leading to metabolic syndrome. The altered gut microbiota due to HFD is associated with reduced panteth antimicrobial peptides lysozyme, Reg IIIγ, elevated circulating inflammatory cytokines IFNγ, and TNF-α (Guo et al. 2017). These insults are caused by HFD within a short period impacting gut microbiota composition (Kim et al. 2019).

Saturated fat facilitates the conjugation of bile acids, expanding the abundance of *Bilophila wadsworthia*, which produces secondary bile acids, initiating barrier disruption and leading to colitis (Devkota et al. 2012). High-fat diet and gut dysbiosis might be a dual burden to T2DM who are already at risk for developing tuberculosis (Arias et al. 2019). Furthermore, high-fat-induced microbial dysbiosis is hypothesized in neuroinflammation and cognitive decline, while the rationale elicits further exploration (Deshpande et al. 2019).

6.5.2 Impact of the Source of Fat on Gut Microbiota

The impact of the source of fat on gut microbiota was deliberated with divergent revelations. In animals, administration of lard fat increased *Bacteroides* and *Bilophila*, while fish-oil increased *Bifidobacterium*, *Adlercreutzia*, *Lactobacillus*, *Streptococcus*, and *Akkermansia muciniphila*. Further, lard expanded the systemic TLR activation and reduced insulin sensitivity compared to fish oil (Caesar et al. 2015). A recent investigation noticed that taking virgin coconut oil (20%) for 16 weeks did not decrease hyperglycemia in individuals with diabetes (Djurasevic et al. 2018). Novel fat sources are explored to alleviate the ill effects of saturated fats. On a promising note, rats supplemented with fullerene C_{60} olive/coconut oil solution for 12 weeks demonstrated gut microbiota compositional alterations toward that which could potentially improve lipid homeostasis, causing a reduction in serum triglycerides concentration (Đurašević et al. 2020).

6.6 Vitamins

6.6.1 Water-Soluble Vitamins

Expanded scientific evidence indicated that the gut *Bifidobacterium* and *Bacteroidetes* can synthesize several B vitamins (Hill 1997). Specifically, the phyla *Fusobacteria*, *Bacteroidetes*, and *Proteobacteria* possess essential pathways for synthesizing riboflavin and biotin, while *Fusobacteria* is equipped with B12 production.

The gut microbiota utilizes dietary vitamins for their physiological functions and grants promising health benefits to the host. For example, *F. prausnitzii* utilizes riboflavin (vitamin B2) for extracellular electron transfer (Khan et al. 2012). Antioxidant vitamins are now being investigated as new intervention means for treating dysbiosis. In a mechanistic human study, 100 mg/day of supplemental riboflavin reduced fecal *E. coli* (Steinert et al. 2016). The bacterial synthesis of vitamins is affected by numerous factors. For example, during *S. typhimurium* infection, lectin RegIIIb, an antimicrobial peptide of host mucosa, kills *Bacteroidetes* leading to decreased levels of vitamin B6, affecting remission of the disease (Sperandio 2017).

6.6.2 Fat-Soluble Vitamins

Growing evidence indicates that vitamin A deficiency is linked to increased susceptibility to infection due to the disruption of the mucosal barrier. The rationale implied a reduction in the relative proportion of *Lactobacillus* spp. Increase in *E. coli*, a bloom of *Bacteroides vulgates*, coupled with diminished MUC2 expression in the gut, downregulation of defensin expression, and upregulation of toll-like receptors and expression (Amit-Romach et al. 2009). Furthermore, the inhibitory effects of retinol on the *Bacteroides vulgatus* are potentially mediated by a reduction in levels of bile acid, such as deoxycholic acid, that inhibit its growth (Hibberd et al. 2017).

Prospective studies have helped elucidate other pathways by which vitamin A offers resistance to infections in the host. Vitamin deficiency is deemed responsible for the heightened susceptibility to *Citrobacter* infection, which induces the depletion of vitamin A reserves in the liver (Spencer et al. 2014). Norovirus is a major cause of nonbacterial gastroenteritis in industrialized countries. On the other hand, excess of RA *Salmonella* increases in abundance owing to its resistance to antimicrobials, whereas protective commensal species such as *Clostridia* perish. Interestingly, commensals like *Clostridium* can suppress RA synthesis by intestinal epithelial cells (IECs). Thus, by reducing RA synthesis, commensal species limit *Salmonella*'s ability to alter the host immune response (Grizotte-Lake et al. 2018). Vitamin A and gut dysbiosis have recently been commonly associated with autism spectrum disorders (Liu et al. 2017).

The vitality of vitamin D for gut mucosal immune protection is elicited from a study in mice deficient in vitamin D, which exhibited diminished expression of Paneth cell defensins, mucin 2, and tight junction genes (Su et al. 2016). Furthermore, recommended vitamin D uptake in humans diminished the abundance of *Coprococcus* and *Bifidobacterium* and increased the *Prevotella* abundance along with a depletion in circulatory LPS levels (Luthold et al. 2017). A recent study showed a dose-dependent increase in *Akkermansia* and *Bacteroides* and reduced relative abundance of *Porphyromonas* after vitamin D_3 supplementation. These changes were associated with decreased inflammatory bowel disease incidence (Charoenngam et al. 2020). Zuo et al. (2019) showed that vitamin D intake was associated with a high prevalence of antihypertensive bacteria, including *Pseudoflavonifractor*, *Paenibacillus*, *Ruminiclostridium*, and *Marvinbryantia*. Recent studies have established the role of autoimmune diseases (Yamamoto and Jørgensen 2020) and radiation therapy in causing vitamin D deficiency by altering the gut microbiome and vitamin D receptor signaling pathways, indicating the need for supplementation (Huang et al. 2019).

Vitamin E could impact gut microbiota, but a considerable dearth of research in this area exists. In a recent study involving mice, alpha-tocopherol supplementation for 34 days revealed that a lower dose increased spleen and body weight. It reduced the ratio of *Firmicutes* to *Bacteroidetes* in the gut (Choi et al. 2020). Similarly, scarcity also exists in explaining vitamin K and gut microbiota interactions.

6.7 Minerals

Minerals are present abundantly in many food sources and play important roles in many biological processes of living organisms, including bacteria (Lordan et al. 2020). In food grains, these minerals are complexed with phytic acid and degraded by bacterial phytases in the gut (Sandberg and Andlid 2002).

High calcium (Ca) intake has a profound influence on epithelial cell proliferation and differentiation (Mariadason et al. 2001), inversely associated with colon cancer incidence (Keum et al. 2014) and obesity (Zhang et al. 2019). Diets rich in Ca favor the growth of *Lactobacilli* and maintain intestinal integrity (Gomes et al. 2015).

A recent mechanistic study established the role of adequate Ca intake in improving colon health by strengthening the gut barrier. When human coronoid cultures were maintained at1.5–3.0 mM of calcium, along with tight junction protein expression, a concomitant increase in the expression of desmosomal proteins cadherin-17 and desmoglein-2 and increased desmosome formation was noticed, revealing the role of calcium in barrier integrity (McClintock et al. 2020). Aquamin, calcium, magnesium, and the multi-mineral-rich natural product obtained from red marine algae has polyp prevention efficacy based on preclinical studies. In a recent elegant study, 30 healthy human participants (both males and females) were given an Aquamin (800 mg of calcium per day) for 90 days. Results revealed a change in microbial diversity, a reduction in total bacterial DNA load, bile acid levels, and an increase in short-chain fatty acids (SCFA), especially acetate. This study extols the safety and tolerance of Aquamin in healthy human participants as an effective dietary strategy for preventing colon polyp chemopreventive agents (Aslam et al. 2020).

Magnesium is involved in various physiological processes, for example, the relaxation of smooth muscle (Uberti et al. 2020). Magnesium deficiency is associated with decreased gut microbial diversity and increased incidence of chronic disease (Grober et al. 2015), including anxiety-like behavior in mice (Jørgensen et al. 2015). A magnesium-rich marine mineral blend was associated with increased microbial diversity in adult male rats (Crowley et al. 2018). Inulin was associated with a higher abundance of *Bifidobacterium* in rats, increasing calcium and magnesium absorption in the gut. Proton pump inhibitors (PPIs) cause decreased microbial diversity, defective mucus formation, and enhanced permeability in tight junctions and adversely affect ion reabsorption, especially magnesium ions. However, supplementation with prebiotic inulin fibers (20 g/d) can attenuate PPI-induced hypomagnesemia by stimulating gut microbes (Thursby and Juge 2017; Gommers et al. 2019). Phosphorus supplementation (1000 mg/day) in humans improved gut microbial diversity and fecal SCFA concentration (Trautvetter et al. 2018).

Over the past few decades, there has been immense interest in iron supplementation in malnourished children, focusing on its impact on anemia (Fischbach et al. 2006). Various mechanisms are hypothesized in explaining the symbiotic relationship between host and native gut microbes during times of iron deficiency. It is proposed that commensal bacteria can facilitate iron uptake in the host via the secretion of enterobactin (Ent), a siderophore (Qi and Han 2018). However, another established mechanism observes that Ent, with a high affinity for iron, scavenges iron from host mitochondria (Yilmaz and Li 2018), leading to the secretion of siderocalin in the host, which binds and sequesters Ent to fight back this "iron piracy" (Golonka et al. 2019). These mechanisms warrant further elucidation. Recent work provides evidence of the cross talk between metabolism and gut microbiota, resulting in systemic iron homeostasis (Das et al. 2020).

Zinc is vital for conserving epithelial integrity (Ohashi and Fukada,2019), and deficiencies of zinc transporters were reported to alter the gut microbial structure (Mao et al. 2019). Mice's neurobehavioral dysfunctions were brought on by oral exposure to zinc oxide nanoparticles (ZnONPs), which are often employed in the food industry and produce memory and spatial learning deficits and suppress locomotor activity.

The antimicrobial properties of copper (Cu) have been widely studied (O'Gorman and Humphreys 2012) and the use of Cu-coated surfaces to control nosocomial infections (Grass et al. 2011). A recent prospective observational study highlighted the importance of Cu in reducing the incidence of hand-transmitted healthcare-associated infections (Zerbib et al. 2020).

High salt intake is associated with high fecal salinity and linked to decreased diversity and depletion in anti-obese microbial members such as *Akkermansia* and *Bifidobacterium*, specifically *B. longum* and *B. adolescentis* (Seck et al. 2019). When rats received 20% fructose in drinking water, and 4% sodium chloride, gut dysbiosis and a decreased ratio of F/B contributed to high blood pressure. Furthermore, high fructose and salt intake (HFS) increased serum triacylglycerol, renin, and angiotensin II, activating the intrarenal renin-angiotensin system. Therefore, gut microbiota-targeted therapy could effectively improve HFS-induced hypertension (Chen et al. 2020b).

6.8 Dietary Fiber

Dietary fibers are classified as fermentable and non-fermentable according to their fermentability by microbes in the gut (Dhingra et al. 2012). Examples of poorly fermented fibers are cellulose, lignin, and psyllium, which are insoluble fibers. Pectins and β-glucan are highly fermentable, soluble, and viscous. The gut microbiota quickly ferments nonviscous soluble fibers such as resistant starches, inulin, maltodextrins, fructo, and galactooligosaccharides (Holscher 2017). A novel hierarchical classification of dietary fibers based on their selectivity to gut microorganisms was recently proposed by Cantu-Jungles. Fibers such as FOS and inulin which are highly accessible and can be utilized by many colonic microbes are named low-specificity fibers and are placed at the top of the hierarchy. On the other hand, insoluble glucans, which have structural properties that only a few bacteria can use, would serve as an example of high-specificity fibers positioned at the bottom of the hierarchy (Cantu-Jungles and Hamaker 2020).

Previous literature showed that several intestinal bacterial taxa, including *Firmicutes*, *Bacteroides*, and *Bifidobacterium*, utilize fructan (Chijiiwa et al. 2020). The growth of *Bacteroides* was found to be promoted by soluble dietary fiber from *Lentinula edodes* (LESDF) fraction 3 with branched-chain structure, while fraction two increased production of propionic and butyric acid, demonstrating the impact of fiber structure on microbial diversity and metabolite production (Xue et al. 2020).

Colonic microorganisms ferment fiber and prebiotics resistant to digestion in the small intestine to create short-chain fatty acids (SCFAs) butyrate, propionate, and acetate (Zmora et al. 2019). SCFAs produced by gut bacteria mediate most health benefits conferred on dietary fiber.

6.9 Biological Effects of SCFAs

6.9.1 SCFAs and Host Mucus Production

Dietary fibers and SCFAs stimulate mucus production, increasing mucosal thickness and reducing bacterial translocation and infection (Schroeder et al. 2018). Hypoxia-inducible factor (HIF) promotes mucus production, and SCFAs are essential for maintaining the stability of this transcription factor (Kelly et al. 2015). *Faecalibacterium prausnitzii* reduces the impact of acetate on mucus, thus preventing the overproduction of mucus and helping to maintain the integrity of the gut epithelium (Wrzosek et al. 2013). Furthermore, fiber deprivation in the diet facilitates some gut bacteria to use host mucin glycans (Sonnenburg et al. 2005). A changed gut flora brought on by a reduced mucus layer can increase vulnerability to infections and the onset of chronic diseases (Zou et al. 2018).

6.9.2 SCFAs and Histone Deacetylase

Propionate and butyrate manifest inhibition of histone deacetylases (HDAC), a molecule that could suppress pro-inflammatory effects. HDAC deregulation is linked to several diseases, including cancers, neurological disorders, and diabetes. Therefore, HDAC inhibition is one mechanism by which SCFAs impact gene expression regulation with important implications for health (Koh et al. 2016).

6.9.3 SCFAs in Glucose Homeostasis and Appetite Control

SCFAs, notably butyrate and propionate, tether to GPR-43 and GPR-41 on the enteroendocrine L cells, which can stimulate the secretion of peptide YY (PYY) and glucagon-like peptide-1 (GLP-1) resulting in reduced appetite and improved glucose homeostasis in the host. Amplified expression of GPR41 in white adipose tissue stimulates leptin production in adipocytes mediated through propionic acid. Leptin, an anorexigenic hormone, suppresses food (Allin et al. 2015).

SCFAs are absorbed via the portal vein and metabolized in the liver. Predominantly, acetate enters the peripheral circulation and mediates energy homeostasis mechanisms. In white adipose tissue, SCFAs decrease fat storage and reduce lipolysis, leading to reduced plasma-free fatty acids levels. In brown adipose tissue, SCFAs enhance thermogenicity. In skeletal muscle, SCFAs increase the capacity for lipid oxidation by enhancing the development of type 1 muscle fibers. These facts clarify the immense role of SCFAs in energy homeostasis (Sukkar et al. 2019).

6.9.4 SCFAs and Tight Junction Proteins

A loss of intestinal epithelial integrity called "leaky gut" may have severe consequences for health. Circulating SCFAs from gut microbiota enhance the production of gut epithelial proteins of tight junction occludin and claudin-5, thus strengthening the blood-brain barrier (BBB), thereby limiting the entry of undesirable metabolites into brain tissue and thus bestowing neuroimmune integrity (Sampson and Mazmanian 2015).

6.9.5 SCFAs and Protection from Susceptibility to Infections

Infections such as influenza appear to affect gut microbiota structure and amend the production of SCFAs. The altered microbiota could favor superinfection post-influenza. Administration of acetate reduces an individual's proneness to secondary bacterial infection by restoring the alveolar macrophage activity, and receptor FFAR2 protects against bacterial superinfection (Sencio et al. 2020).

6.10 Types of Fiber and Effects on Gut Microbial Diversity

Recent investigations on different sources of fiber proved their prebiotic effects. Fiber from sweet potato residue (SPDF), *Hibiscus sabdariffa* (*Hb*) calyces, Agave fructans (AF) and oligofructans (OF), and carrot dietary fiber (CDF) in in vitro and in vivo studies increased the abundance of *Bifidobacterium* and *Lactobacillus*, with an associated increase in the concentration of butyrate and propionate while decreasing *Enterobacillus*, *Clostridium perfringens* and *Bacteroides*. Specifically, SPDF supplementation resulted in a higher villus length to the fossa deepness ratio, suggesting enhanced utilization in the GI tract. Additionally, bound phenols present in *Hibiscus sabdariffa* (*Hb*) calyces and CDF were found to be instrumental in the fermentation and add to the antioxidant properties of these fibers (Liu et al. 2020a; Sáyago-Ayerdi et al. 2020).

Besides soluble fibers, some insoluble fibers also seem to exhibit prebiotic effects. Fiber from soya hulls dietary fiber (SHDF) markedly altered the structure of the fecal microbiota community, specifically beneficial microbes, which supports SHDF as a novel gut microbiota modulator for beneficial health effects (Yang et al.

2020b). The symbiotic effect was explored in a recent human intervention trial involving 24 healthy volunteers, which evaluated the impact of fermented salami (30 g) with a probiotic *Lactobacillus rhamnosus HN001* with added citrus fiber. After intervention for 4 weeks, higher abundances of butyrate-producing gut microbes were observed, along with a significant decrease in the inflammatory markers CRP and TNFα (Pérez-Burillo et al. 2020).

6.11 Fiber and Host Health

6.11.1 Type2 DM

Various fermentable fibers impact gut microbiota and glycemic control (Adeshirlarijaney and Gewirtz 2020). T2DM is generally associated with a reduced abundance of *A. muciniphilia* and *E. rectale*. Inulin fiber has the potential to increase levels of *A. muciniphilia* and restore the protection against inflammation and hyperglycemia. Similarly, resistant starch type 4 from maize might help correct the dysbiosis exhibited in T2DM (Deehan et al. 2020).

Several fibers from foods are extensively studied for their effect on T2DM mediated through gut microbes. Some recently investigated ones include medicinal mushroom *Phellinus linteus polysaccharide extract* (PLPE), barley soluble and insoluble fibers, and polysaccharide-rich extracts from *Apocynum venetum* leaves and rice bran dietary fiber (RBDF), all proved to reverse insulin resistance. The mediated mechanisms are increased SCFAs by enhancing the abundance of SCFA-producing bacteria, improved intestinal barrier function, and reduced systemic inflammation (Liu et al. 2020b; Li et al. 2020; Yuan et al. 2020).

6.11.2 Obesity and Metabolic Syndrome

Arabinoxylan oligosaccharides (AXOS) intake increases the proportion of butyrate producers in the gut microbiota, which could modulate parameters related to metabolic syndrome (Kjølbæk et al. 2020). Recently, processed fibers are gaining popularity owing to their potential to reduce obesity and related disease occurrence. The unique attribute of cellulose nanofiber (CN) to form highly viscous dispersions in water echoes its similarity with soluble dietary fibers (DFs) having glycemic control. It was observed that 0.2% CN intake reduced obesity induced by a high-fat diet (HFD) along with a shift of gut microbiota composition with pronounced *Lactobacillaceae* (Nagano and Yano 2020a). Additionally, CN intake and exercise suppressed the weight gain, increased fat mass, and improved blood glucose control. Exercise alone increased *Ruminococcaceae* prevalence, whereas a combination of exercise and CN intake increased *Eubacteriaceae*, key butyrate producers (Nagano and Yano 2020b). Gut microbiota could protect against triptolide (TP) through propionate production. Supplementing with propionate significantly reduced the expression of genes involved in the biosynthesis pathway of fatty acid

(*Srebp1c*, *Fasn*, and *Elovl6*), which results in decreased long-chain fatty acids in the liver. Therefore, propionate supplementation could be a plausible clinical strategy to minimize toxicity induced by drugs (Huang et al. 2020).

6.11.3 Musculoskeletal Health

Tendons, bones, cartilage, and joints may be impaired by early changes in muscle integrity (Collins et al. 2018), leading to musculoskeletal-related conditions. A high-fat or high-sucrose diet, a risk factor for obesity, could cause similar musculoskeletal damage (Collins et al. 2016). A combination of prebiotic oligofructose supplementation, aerobic exercise, and separate administration of these interventions prevented knee damage associated with obesity in rats (Rios et al. 2019).

6.11.4 Colonic Health and Colon Cancer

According to data from earlier studies, gut microbiota may play a causal role in developing and spreading colorectal cancer. Jujube polysaccharides (JP) and the gut microbiota composition were the subjects of a recent study. In animal models, the fiber in jujube fruits has been found to inhibit cancer development. By easing colitis, challenging with JP prevented colon cancer and significantly decreased the Firmicutes/Bacteroidetes ratio (Ji et al. 2020).

6.11.5 Secondary Responses to Fiber Consumption

Some studies have also shown the flip side of fiber and gut microbiota interactions. One report has indicated that fiber-derived butyrate promoted tumorigenesis by inducing stem cell generation (Belcheva et al. 2014). Similarly, refined inulin due to processing errors may increase the risk of hepatocellular carcinoma. However, these observations need validation in humans (Vijay-Kumar 2020). Eating more fiber for beneficial health effects has encouraged manufacturers to fortify processed food that is otherwise nutritionally deprived with refined dietary fibers (Singh and Vijay-Kumar 2020). Due to vigorous food processing methods, the so-called nutritious fiber is converted to refined fibers with a nullified impact on gut biotics (Payling et al. 2020).

6.12 Probiotics

Probiotics are beneficial microbes that benefit the host when consumed sufficiently. Fermented foods like cultured milk products and sauerkraut appear to influence the gut microbiota by producing cytokines such as IL-10 (Foligné et al. 2016) (Fig. 6.2).

6.12.1 Probiotics and Metabolic Disorders

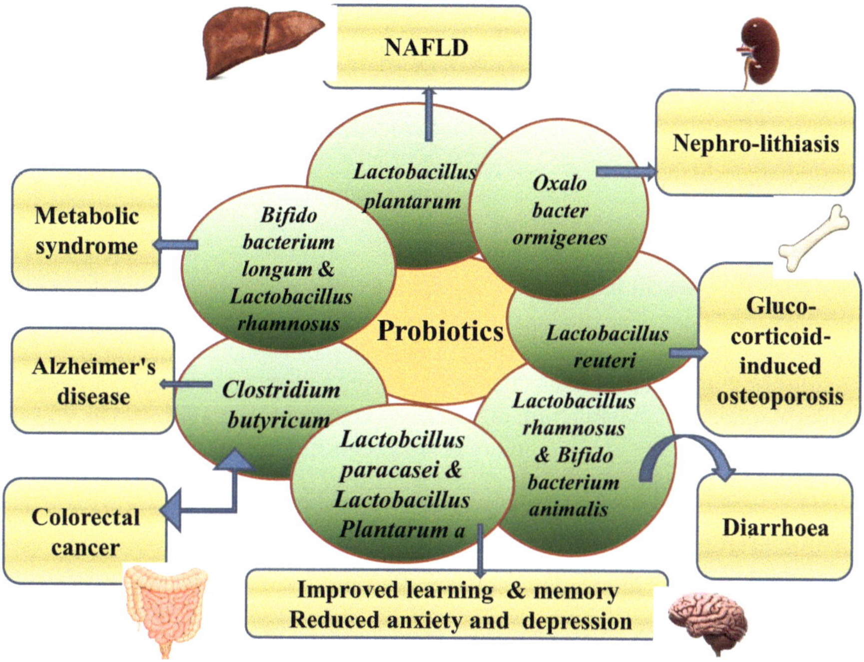

Fig. 6.2 Schematic representation of the impact of different probiotics and their combinations on health conditions

A well-appreciated study promulgated the impact of 3–12 weeks-long probiotic supplementation in obese subjects (Borgeraas et al. 2018). The mechanism of action for the weight reduction properties of probiotic supplementation was hypothesized as remodeling of energy metabolism, altered glucose and lipid metabolism genes, reduced endotoxin release, lowered inflammation, and change in the parasympathetic nerve activity. A recent elegant study revealed decreased *Bacteroidetes-to-Firmicutes* ratio and obesity parameters up on supplementation of Mediterranean diet (MD) along with *Bifidobacterium longum* and *Lactobacillus rhamnosus* as probiotics indicating the efficacy of probiotics (Pellegrini et al. 2020).

6.12.2 Probiotics and Gut-Brain Axis Imbalance and Cognitive Decline

Probiotic intervention counteracts Alzheimer's disease AD progression by affecting glucose homeostasis by causing restoration of the brain levels of the glucose transporters, reducing Tau phosphorylation by modulating protein kinase B (pAkt) and by decreasing advanced glycation end products (Bonfili et al. 2020).

A recent study in mice exemplified the benefits of ProBiotic-4 composed of *Lactobacillus casei*, *Bifidobacterium lactis*, *Lactobacillus acidophilus*, and *Bifidobacterium bifidum* administration in improving memory deficit and fecal microbiota composition. Additionally, a decrease in histone-2AX phosphorylation and abrogation of RIG-I oxidative DNA damage markers were observed, indicating the role of ProBiotic-4 in brain function (Yang et al. 2020c).

6.12.3 Probiotics and Colon Cancers

Probiotic Clostridium butyricum was shown to inhibit excessive fat-induced cancers in the intestine. The profound effects of *C. butyricum* on intestinal tumor cells include decreased proliferation and increased apoptosis, decreased pathogenic and biotransforming bacteria, and increased SCFA-producing bacteria, thus proving the potential efficacy of butyrate-producing bacteria against cancers (Chen et al. 2020a). In a unique study, the gut microbiota of mice undergoing chemotherapy exhibited bacterial dysbiosis leading to intestinal mucositis. When supplemented with mice, probiotic strains were found to increase *Enterobacteriales* and *Turicibacterales*, along with reduced bacterial translocation, thus proving efficacy in intestinal mucositis (Yeung et al. 2020).

6.12.4 Probiotics and Other Diseases

Mice were used to examine the effects of probiotics on the makeup of the gut microbiota and glucocorticoid-induced osteoporosis (GIO). *Lactobacillus reuteri* (LR), a probiotic, reduced trabecular bone loss in mice treated with prednisolone while reversing the inhibition of Wnt10b in bone. Finally, it was determined that GIO was prevented by bone-specific Wnt10b overexpression (Schepper et al. 2020). A recent systematic review revealed that there is still no evidence to recommend probiotics to treat constipation in children and adolescents despite probiotics' positive effects on specific characteristics of the intestinal habitat.

6.13 Polyphenols

The association between dietary polyphenols and human health has been receiving attention due to their illustrious antioxidant capacity, the protection offered to mucosa architecture, and the enrichment of the environment for commensal gut microbial blossom. Dietary polyphenols are conjugated to various lipids, organic acids, and sugars hydrolyzed by the colonic microbiota into absorbable metabolites with various health benefits (Marín et al. 2015). A supplemental dose of 396 mg/d of polyphenols was recommended to stimulate the growth of probiotics and decrease possible pathogens in the human gut without affecting other significant microbes (Ma and Chen 2020).

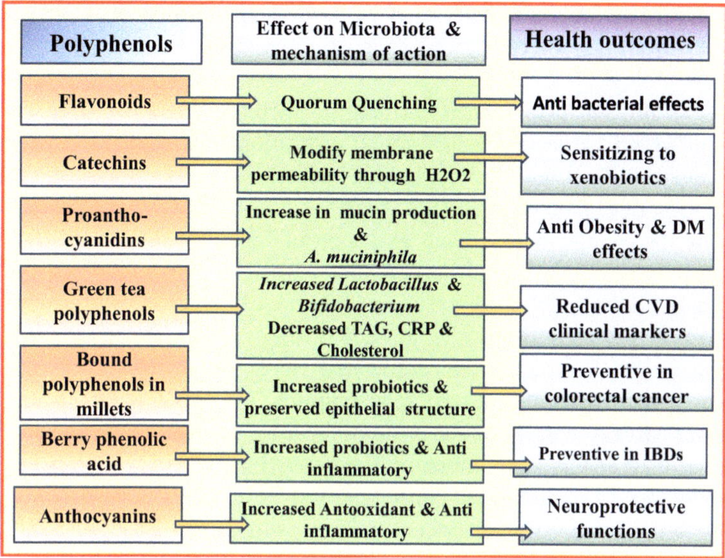

Fig. 6.3 Summary of health effects of different polyphenols mediated through gut microbiota

6.13.1 Dietary Polyphenols and Gut Microbiota

Dietary polyphenols were shown to hinder bacterial quorum sensing, improve membrane permeability, and increase the susceptibility of bacteria to xenobiotics (Di Meo et al. 2020) (Fig. 6.3).

6.13.2 Quorum Quenching (QQ) Mechanism

Flavonoid compounds can interfere with microbial cross-communication, causing an anti-biofilm effect. Quorum quenching (QQ) interferes with the microbial communication process of quorum sensing (QS), which involves the constant release of signaling molecules into the environment. Well applauded for its anti-inflammatory and anticancer properties, curcumin also exerts quorum quenching, portraying antibacterial effects (Kali et al. 2016; Packiavathy et al. 2014). Other dietary polyphenols that exhibited decreased QS were flavonoids from *Centella asiatica* (Vasavi et al. 2016), *Ananas comosus* extract, Musa paradisiacal water extract (Musthafa et al. 2010), grapefruit extract containing furocoumarins (Girennavar et al. 2008), and orange extract that is rich in flavones (Vikram et al. 2010). Polyphenols bind bacterial cell membranes and thereby modify membrane permeability affecting their growth.

6.13.3 Dietary Polyphenols and Gut Microbiota-Associated Health Benefits

Polyphenols prevent using macronutrients to increase energy uptake, and many have been found to decrease nutritional absorption in the gastrointestinal tract. Other important functions of polyphenols concerning energy homeostasis include the regulation of glucose homeostasis, repression of lipid synthesis, increased thermogenesis, fat oxidation, and fecal excretion of lipids (Van Hul and Cani 2019). Resveratrol is now considered as potential AMPK activator. Several preclinical investigations have established that *A. muciniphila* affluence is inversely associated with lifestyle diseases (Pierre et al. 2013). Berry phenolic acid consumption increased probiotic microbiota associated with anti-inflammatory functions (Lavefve et al. 2020). Long-term consumption of anthocyanins from *Lycium ruthenicum* Murray (ACN) could promote healthy microflora in the gut coupled with antioxidant and anti-inflammatory functions (Peng et al. 2020). The glycosidic form in plants is transformed into resveratrol in the gut by the action of *Bifidobacterium infantis* and *Lactobacillus acidophilus* (Basholli-Salihu et al. 2016). Akin to this, *E. coli* converts curcumin to tetrahydrocurcumin, an active metabolite with greater antioxidant activity than curcumin and potent anti-inflammatory and neuroprotective functions (Di Meo et al. 2019) (Fig. 6.4).

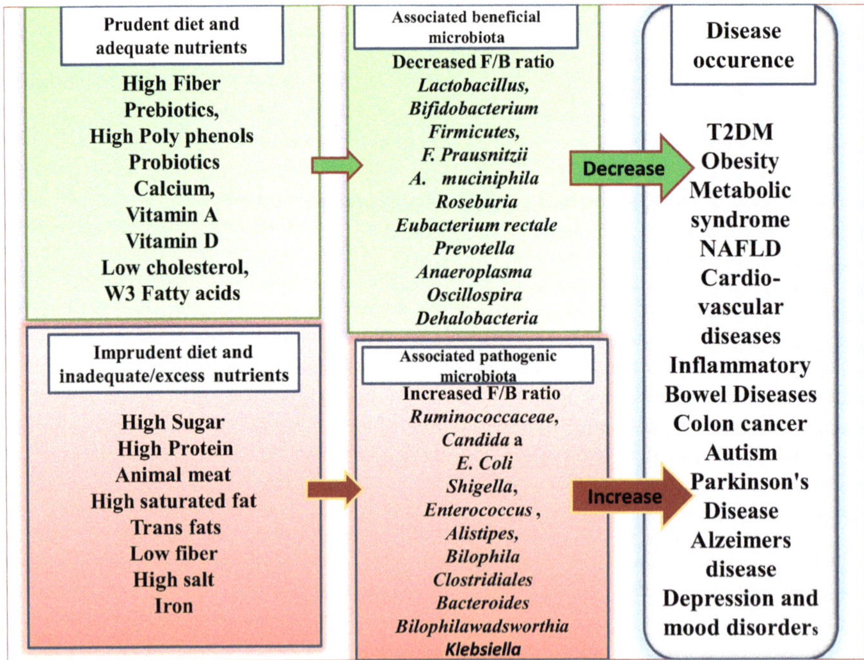

Fig. 6.4 Interaction of nutrients, other dietary factors with microbiota, and impact on host health

6.14 Conclusions and Future Perspectives

Nutrient-host-microbe interactions are complex, and this novel area has emerged as one of the notable areas of microbiomics yielding enticing contributions to illness and health. Now, escalating evidence advocates that diet impacts the abundance of gut microbes and their synthesis of metabolic by-products. Research in this field has expanded beyond individual nutrients to whole foods, dietary patterns, diet quality, and food processing methods. Recently, dietary patterns, including Mediterranean, vegan, Nordic, and low FODMAP, have been widely studied, but divergent outcomes anticipate further exploration. With the advent of food technology, along with the food, we also partake in various nonnutritive sweeteners, colors, preservatives, pesticides, and nano-sized foods. The influence of these factors on gut microbes needs further exploration. The analysis of the metabolic impact of dietary patterns presents unique challenges, as it encompasses several food items, diet-derived compounds, and numerous individual-based endogenous metabolic processes. The influential roles of the human microbiota should be probed from nadir to zenith to facilitate microbiome-based dietary treatment strategies scalable for personalized preventive medical nutrition therapy.

References

Adeshirlarijaney A, Gewirtz AT (2020) Considering gut microbiota in treatment of type 2 diabetes mellitus. Gut Microbes 11(3):1–12

Allin KH, Nielsen T, Pedersen O (2015) Mechanisms in endocrinology: gut microbiota in patients with type 2 diabetes mellitus. Eur J Endocrinol 172(4):R167–R177

Amit-Romach E, Uni Z, Cheled S, Berkovich Z, Reifen R (2009) Bacterial population and innate immunity-related genes in rat gastrointestinal tract are altered by vitamin A-deficient diet. J Nutr Biochem 20(1):70–77

Anand S, Kaur H, Mande SS (2016) Comparative in silico analysis of butyrate production pathways in gut commensals and pathogens. Front Microbiol 7:1945

Araújo JR, Tomas J, Brenner C, Sansonetti PJ (2017) Impact of high-fat diet on the intestinal microbiota and small intestinal physiology before and after the onset of obesity. Biochimie 141:97–106

Arias L, Goig GA, Cardona P, Torres-Puente M, Díaz J, Rosales Y, Garcia E, Tapia G, Comas I, Vilaplana C, Cardona P-J (2019) Influence of gut microbiota on progression to tuberculosis generated by high fat diet-induced obesity in C3HeB/FeJ mice. Front Immunol 10:2464

Aslam MN, Bassis CM, Bergin IL, Knuver K, Zick SM, Sen A et al (2020) A calcium-rich multimineral intervention to modulate colonic microbial communities and metabolomic profiles in humans: results from a 90-day trial. Cancer Prev Res 13(1):101–116

Basholli-Salihu M, Schuster R, Mulla D, Praznik W, Viernstein H, Mueller M (2016) Bioconversion of piceid to resveratrol by selected probiotic cell extracts. Bioprocess Biosyst Eng 39:1879–1885

Belcheva A et al (2014) Gut microbial metabolism drives transformation of MSH2-deficient colon epithelial cells. Cell 158:288–299

Bellenger J, Bellenger S, Escoula Q, Bidu C, Narce M (2019) N-3 polyunsaturated fatty acids: an innovative strategy against obesity and related metabolic disorders, intestinal alteration and gut microbiota dysbiosis. Biochimie 159:66–71

Blekhman R, Goodrich JK, Huang K, Sun Q, Bukowski R, Bell JT et al (2015) Host genetic variation impacts microbiome composition across human body sites. Genome Biol 16(1):191

Bonfili L, Cecarini V, Gogoi O, Berardi S, Scarpona S, Angeletti M et al (2020) Gut microbiota manipulation through probiotics oral administration restores glucose homeostasis in a mouse model of Alzheimer's disease. Neurobiol Aging 87:35–43

Borgeraas H, Johnson LK, Skattebu J, Hertel JK, Hjelmesæth J (2018) Effects of probiotics on body weight, body mass index, fat mass and fat percentage in subjects with overweight or obesity: a systematic review and meta-analysis of randomized controlled trials. Obes Rev 19(2):219–232

Butler LM et al (2003) Heterocyclic amines, meat intake, and association with colon cancer in a population-based study. Am J Epidemiol 157:434–445

Caesar R, Tremaroli V, Kovatcheva-Datchary P, Cani PD, Bäckhed F (2015) Crosstalk between gut microbiota and dietary lipids aggravates WAT inflammation through TLR signaling. Cell Metab 22:658–668

Cani PD, Amar J, Iglesias MA et al (2007) Metabolic endotoxemia initiates obesity and insulin resistance. Diabetes 56:1761–1772

Cantu-Jungles TM, Hamaker BR (2020) New view on dietary fiber selection for predictable shifts in gut microbiota. MBio 11(1):e02179–e02119

Carmody RN, Gerber GK, Luevano JM, Gatti DM, Somes L, Svenson KL, Turnbaugh PJ (2015) Diet dominates host genotype in shaping the murine gut microbiota. Cell Host Microbe 17:72–84

Charoenngam N, Shirvani A, Kalajian TA, Song A, Holick MF (2020) The effect of various doses of oral vitamin D3 supplementation on gut microbiota in healthy adults: a randomized, double-blinded, Dose-response Study. Anticancer Res 40(1):551–556

Chen D, Jin D, Huang S, Wu J, Xu M, Liu T et al (2020a) Clostridium butyricum, a butyrate-producing probiotic, inhibits intestinal tumor development through modulating Wnt signaling and gut microbiota. Cancer Lett 469:456–467

Chen Y, Zhu Y, Wu C, Lu A, Deng M, Yu H et al (2020b) Gut dysbiosis contribute to high fructose induced salt sensitive hypertension in Sprague-Dawley rats. Nutrition 75–76:110766

Chijiiwa R, Hosokawa M, Kogawa M et al (2020) Single-cell genomics of uncultured bacteria reveals dietary fiber responders in the mouse gut microbiota. Microbiome 8:5

Cho I, Blaser MJ (2012) The human microbiome: at the interface of health and disease. Nat Rev Genet 13:260–270

Choi Y, Lee S, Kim S, Lee J, Ha J, Oh H et al (2020) Vitamin E (α-tocopherol) consumption influences gut microbiota composition. Int J Food Sci Nutr 71(2):221–225

Collins KH, Paul HA, Hart DA, Reimer RA, Smith IC, Rios JL et al (2016) A high-fat high-sucrose diet rapidly alters muscle integrity, inflammation and gut microbiota in male rats. Sci Rep 6:37278

Collins KH, Herzog W, MacDonald GZ, Reimer RA, Rios JL, Smith IC et al (2018) Obesity, metabolic syndrome, and musculoskeletal disease: common inflammatory pathways suggest a central role for loss of muscle integrity. Front Physiol 9:112

Conlon MA, Bird AR (2015) The impact of diet and lifestyle on gut microbiota and human health. Nutrients 7(1):17–44

Cotillard A, Kennedy SP, Kong LC, Prifti E, Pons N, Le Chatelier E et al (2013) Dietary intervention impact on gut microbial gene richness. Nature 500:585–588

Covarrubias AO, Escalera AM, Romero A, Sánchez C, Salinas H, Solís A (2020) Bioethics and the relationship with human nutrition. Mex Bioeth Rev ICSA 1(2):16–19

Crowley E, Long-Smith C, Murphy A, Patterson E, Murphy K, O'Gorman D, Stanton C, Nolan Y (2018) Dietary supplementation with a magnesium-rich marine mineral blend enhances the diversity of gastrointestinal microbiota. Mar Drugs 16:216

Das NK, Schwartz AJ, Barthel G, Inohara N, Liu Q, Sankar A et al (2020) Microbial metabolite signaling is required for systemic iron homeostasis. Cell Metab 31(1):115–130

Deehan EC, Yang C, Perez-Muñoz ME, Nguyen NK, Cheng CC, Triador L et al (2020) Precision microbiome modulation with discrete dietary Fiber structures directs short-chain fatty acid production. Cell Host Microbe 27(3):389–404.e6

Deshpande NG, Saxena J, Pesaresi TG, Carrell CD, Ashby GB, Liao MK, Freeman LR (2019) High fat diet alters gut microbiota but not spatial working memory in early middle-aged Sprague Dawley rats. PLoS One 14(5):e0217553

Devkota S et al (2012) Dietary- fat-induced taurocholic acid promotes pathobiont expansion and colitis in Il10−/−mice. Nature 487:104–108

Dhingra D, Michael M, Rajput H, Patil RT (2012) Dietary fibre in foods: a review. J Food Sci Technol 49(3):255–266

Di Meo F, Margarucci S, Galderisi U, Crispi S, Peluso G (2019) Curcumin, gut microbiota, and neuroprotection. Nutrients 11:2426

Di Meo F, Valentino A, Petillo O, Peluso G, Filosa S, Crispi S (2020) Bioactive polyphenols and neuromodulation: molecular mechanisms in neurodegeneration. Int J Mol Sci 21(7):2564

Di Rienzi SC, Britton RA (2019) Adaptation of the gut microbiota to modern dietary sugars and sweeteners. Adv Nutr 11(3):616–629

Djurasevic S, Bojic S, Nikolic B, Dimkic I, Todorovic Z, Djordjevic J, Mitic-Culafic D (2018) Beneficial effect of virgin coconut oil on alloxan-induced diabetes and microbiota composition in rats. Plant Foods Hum Nutr 73(4):295–301

Do MH, Lee E, Oh M-J, Kim Y, Park H-Y (2018) High-glucose or -fructose diet cause changes of the gut microbiota and metabolic disorders in mice without body weight change. Nutrients 10:761. 11–14

Drasar BS, Crowther JS, Goddard P, Hawksworth G, Hill MJ, Peach S et al (2007) The relation between diet and the gut microflora in man. Proc Nutr Soc 32:49–52

Đurašević S, Nikolić G, Todorović A, Drakulić D, Pejić S, Martinović V, Mitić-Ćulafić D, Milić D, Kop TJ, Jasnić N, Đorđević J, Todorović Z (2020) Effects of fullerene C_{60} supplementation on gut microbiota and glucose and lipid homeostasis in rats. Food Chem Toxicol 140:111302

Ezra-Nevo G, Henriques SF, Ribeiro C (2020) The diet-microbiome tango: how nutrients lead the gut brain axis. Curr Opin Neurobiol 62:122–132

Fang H, Meng F, Piao F, Jin B, Li M, Li W (2019) Effect of taurine on intestinal microbiota and immune cells in Peyer's patches of immunosuppressive mice. In: Taurine 11. Springer, Singapore, pp 13–24

Fava F, Gitau R, Griffin BA, Gibson GR, Tuohy KM, Lovegrove JA (2013) The type and quantity of dietary fat and carbohydrate alter faecal microbiome and short-chain fatty acid excretion in a metabolic syndrome "at-risk" population. Int J Obes 37:216–223

Fischbach MA, Lin H, Liu DR, Walsh CT (2006) How pathogenic bacteria evade mammalian sabotage in the battle for iron. Nat Chem Biol 2:132–138

Foligné B, Parayre S, Cheddani R, Famelart MH, Madec MN, Plé C et al (2016) Immunomodulation properties of multi-species fermented milks. Food Microbiol 53:60–69

Frame LA, Costa E, Jackson SA (2020) Current explorations of nutrition and the gut microbiome: a comprehensive evaluation of the review literature. Nutr Rev 78(10):798–812

Girennavar B, Cepeda ML, Soni KA, Vikram A, Jesudhasan P, Jayaprakasha GK, Pillai SD, Patil BS (2008) Grapefruit juice and its furocoumarins inhibits autoinducer signaling and biofilm formation in bacteria. Int J Food Microbiol 125(2):204–208

Golonka R, Yeoh BS, Vijay-Kumar M (2019) The iron tug-of-war between bacterial siderophores and innate immunity. J Innate Immun 11(3):249–262

Gomes JMG, Costa JA, Alfenas RC (2015) Could the beneficial effects of dietary calcium on obesity and diabetes control be mediated by changes in intestinal microbiota and integrity? Br J Nutr 114:1756–1765

Gommers LMM, Ederveen THA, van der Wijst J, Overmars-Bos C, Kortman GAM, Boekhorst J, Bindels RJM, de Baaij JHF, Hoenderop JGJ (2019) Low gut microbiota diversity and dietary magnesium intake are associated with the development of PPI-induced hypomagnesemia. FASEB J 33:11235–11246

Goodrich JK, Davenport ER, Beaumont M, Jackson MA, Knight R, Ober C et al (2016) Genetic determinants of the gut microbiome in UK twins. Cell Host Microbe 19(5):731–743

Grass G, Rensing C, Solioz M (2011) Metallic copper as an antimicrobial surface. Appl Environ Microbiol 77:1541–1547

Grizotte-Lake M, Zhong G, Duncan K, Kirkwood J, Iyer N, Smolenski I et al (2018) Commensals suppress intestinal epithelial cell retinoic acid synthesis to regulate interleukin-22 activity and prevent microbial dysbiosis. Immunity 49(6):1103–1115

Gröber U, Schmidt J, Kisters K (2015) Magnesium in prevention and therapy. Nutrition 7:8199–8226

Gryp T, Vanholder R, Vaneechoutte M, Glorieux G (2017) P-Cresyl sulfate. Toxins 9(2):52

Guo X, Li J, Tang R, Zhang G, Zeng H, Wood RJ, Liu Z (2017) High fat diet alters gut microbiota and the expression of paneth cell-antimicrobial peptides preceding changes of circulating inflammatory cytokines. Mediat Inflamm 2017:9474896

Hanuszkiewicz A, Pittock P, Humphries F, Moll H, Rosales AR, Molinaro A et al (2014) Identification of the flagellin glycosylation system in Burkholderia cenocepacia and the contribution of glycosylated flagellin to evasion of human innate immune responses. J Biol Chem 289(27):19231–19244

Haque M, McKimm J, Sartelli M, Samad N, Haque SZ, Bakar MA (2020) A narrative review of the effects of sugar-sweetened beverages on human health: a key global health issue. J Popul Ther Clin Pharmacol 27(1):e76–e103

Hibberd MC, Wu M, Rodionov DA, Li X, Cheng J, Griffin NW et al (2017) The effects of micronutrient deficiencies on bacterial species from the human gut microbiota. Sci Transl Med 9(390):eaal4069

Hill MJ (1997) Intestinal flora and endogenous vitamin synthesis. Eur J Cancer Prev 6:S43–S45

Holscher HD (2017) Dietary fiber and prebiotics and the gastrointestinal microbiota. Gut Microbes 8(2):172–184

Hsiao EY, McBride SW, Hsien S, Sharon G, Hyde ER, McCue T, CodelliJA CJ, Reisman SE, Petrosino JF et al (2013) Microbiota modulate behavioralandphysiological abnormalities associated with neurodevelopmental disorders. Cell 155(7):1451–1463

Huang R, Xiang J, Zhou P (2019) Vitamin D, gut microbiota, and radiation-related resistance: a love-hate triangle. J Exp Clin Cancer Res 38(1):1–10

Huang JF, Zhao Q, Dai MY, Xiao XR, Zhang T, Zhu WF, Li F (2020) Gut microbiota protects from triptolide-induced hepatotoxicity: key role of propionate and its downstream signalling events. Pharmacol Res 155:104752

Ijssennagger N et al (2015) Gut microbiota facilitates dietary heme- induced epithelial hyperproliferation by opening the mucus barrier in colon. Proc Natl Acad Sci U S A 112:10038–10043

Jaglin M, Rhimi M, Philippe C, Pons N, Bruneau A, Goustard B, Dauge V, Maguin E, Naudon L, Rabot S (2018) Indole, a signaling molecule droduced by the gut microbiota, negatively impacts emotional behaviors in rats. Front Neurosci 12:216

Jandhyala SM, Talukdar R, Subramanyam C, Vuyyuru H, Sasikala M, Nageshwar Reddy D (2015) Role of the normal gut microbiota. World J Gastroenterol 21:8787–8803

Jena PK, Singh S, Prajapati B, Nareshkumar G, Mehta T, Seshadri S (2014) Impact of targeted specific antibiotic delivery for gut microbiota modulation on high-fructose-fed rats. Appl Biochem Biotechnol 172(8):3810–3826

Jensen T, Abdelmalek MF, Sullivan S, Nadeau KJ, Green M, Roncal C et al (2018) Fructose and sugar: a major mediator of non-alcoholic fatty liver disease. J Hepatol 68(5):1063–1075

Ji X, Hou C, Gao Y, Xue Y, Yan Y, Guo X (2020) Metagenomic analysis of gut microbiota modulatory effects of jujube (Ziziphus jujuba mill.) polysaccharides in a colorectal cancer mouse model. Food Funct 11:163–173

Jørgensen BP, Winther G, Kihl P, Nielsen DS, Wegener G, Hansen AK, Sørensen DB (2015) Dietary magnesium deficiency affects gut microbiota and anxiety-like behaviour in c57bl/6n mice. Acta Neuropsychiatr 27:307–311

Kali A, Bhuvaneshwar D, Charles PM et al (2016) Antibacterial synergy of curcumin with antibiotics against biofilm producing clinical bacterial isolates. J Basic Clin Pharm 7(3):93–96

Kelly CJ, Zheng L, Campbell EL, Saeedi B, Scholz CC, Bayless AJ, Wilson KE, Glover LE, Kominsky DJ, Magnuson A, Weir TL, Ehrentraut SF, Pickel C, Kuhn KA, Lanis JM, Nguyen V, Taylor CT, Colgan SP (2015) Crosstalk between microbiota-derived short-chain fatty acids and intestinal epithelial HIF augments tissue barrier function. Cell Host Microbe 17:662–671

Keum NN, Aune D, Greenwood DC, Ju W, Givannucci E (2014) Calcium intake and cancer risk: dose-response meta-analysis of prospective observational studies. Int J Cancer 135:1940–1948

Khan MT, Browne WR, van Dijl JM, Harmsen HJ (2012) How can Faecalibacteriumprausnitzii employ riboflavin for extracellular electron transfer? Antioxid Redox Signal 17(10):1433–1440

Kim CH, Park J, Kim M (2014) Gut microbiota-derived short-chain fatty acids, T cells, and inflammation. Immune Netw 14:277

Kim SJ, Kim SE, Kim AR, Kang S, Park MY, Sung MK (2019) Dietary fat intake and age modulate the composition of the gut microbiota and colonic inflammation in C57BL/6J mice. BMC Microbiol 19(1):193

Kjølbæk L, Benítez-Páez A, Del Pulgar EMG, Brahe LK, Liebisch G, Matysik S et al (2020) Arabinoxylan oligosaccharides and polyunsaturated fatty acid effects on gut microbiota and metabolic markers in overweight individuals with signs of metabolic syndrome: a randomized cross-over trial. Clin Nutr 39(1):67–79

Koeth RA et al (2013) Intestinal microbiota metabolism of l-carnitine, a nutrient in red meat, promotes atherosclerosis. Nat Med 19:576–585

Koh A, De Vadder F, Kovatcheva-Datchary P, Backhed F (2016) From dietary fiber to host physiology: short- chain fatty acids as key bacterial metabolites. Cell 165:1332–1345. 39

Kremling A, Geiselmann J, Ropers D, de Jong H (2015) Understanding carbon catabolite repression in *Escherichia coli* usingquantitative models. Trends Microbiol 23:99–109

Lavefve L, Howard LR, Carbonero F (2020) Berry polyphenols metabolism and impact on human gut microbiota and health. Food Funct 11:45–65

Lazar V, Ditu L-M, Pircalabioru GG, Picu A, Petcu L, Cucu N, Chifiriuc MC (2019) Gut microbiota, host organism, and diet trialogue in diabetes and obesity. Front Nutr 6:21

Levine ME, Suarez JA, Brandhorst S, Balasubramanian P, Cheng CW, Madia F et al (2014) Low protein intake is associated with a major reduction in IGF-1, cancer, and overall mortality in the 65 and younger but not older population. Cell Metab 19:407–417

Levy M, Thaiss CA, Zeevi D, Dohnalova L, Zilberman-Schapira G, Mahdi JA et al (2015) Microbiota-modulated metabolites shape the intestinal microenvironment by regulating NLRP6 inflammasome signaling. Cell 163(6):1428–1443

Li S, Konstantinov SR, Smits R, Peppelenbosch MP (2017) Bacterial biofilms in colorectal cancer initiation and progression. Trends Mol Med 23(1):18–30

Li Y, Rahman SU, Huang Y, Zhang Y, Ming P, Zhu L et al (2020) Green tea polyphenols decrease weight gain, ameliorate alteration of gut microbiota, and mitigate intestinal inflammation in canines with high-fat-diet-induced obesity. J Nutr Biochem 78:108324

Liu J, Liu X, Xiong XQ, Yang T, Cui T, Hou NL et al (2017) Effect of vitamin a supplementation on gut microbiota in children with autism spectrum disorders-a pilot study. BMC Microbiol 17(1):204

Liu, M., Li, X., Zhou, S., Wang, T. T., Zhou, S., Yang, K., et al. (2020a). Dietary fiber isolated from sweet potato residues promote healthy gut microbiome profile. Food Funct, 11(1): 689–699

Liu Y, Wang C, Li J, Li T, Zhang Y, Liang Y, Mei Y (2020b) Phellinus linteus polysaccharide extract improves insulin resistance by regulating gut microbiota composition. FASEB J 34(1):1065–1078

Lordan C, Thapa D, Ross RP, Cotter PD (2020) Potential for enriching next-generation health-promoting gut bacteria through prebiotics and other dietary components. Gut Microbes 11(1):1–20

Luthold RV, Fernandes GR, Franco- de-Moraes AC, Folchetti LG, Ferreira SR (2017) Gut microbiota interactions with the immunomodulatory role of vitamin D in normal individuals. Metabolism 69:76–86

Ma G, Chen Y (2020) Polyphenol supplementation benefits human health via gut microbiota: a systematic review via meta-analysis. J Funct Foods 66:103829

Mao Z, Lin H, Su W, Li J, Zhou M, Li Z et al (2019) Deficiency of ZnT8 promotes adiposity and metabolic dysfunction by increasing peripheral serotonin production. Diabetes 68:1197–1209
Mariadason JM, Bordonaro M, Aslam F, Shi L, Kuraguchi M, Velcich A et al (2001) Downregulation of beta-catenin TCF signaling is linked to colonic epithelial cell differentiation. Cancer Res 61:3465–3471
Marín L, Miguélez EM, Villar CJ, Lombó F (2015) Bioavailability of dietary polyphenols and gut microbiota metabolism: antimicrobial properties. Biomed Res Int 2015:905215
Martinez-Medina M, Denizot J, Dreux N et al (2014) Western diet induces dysbiosis with increased *E coli* in CEABAC10 mice, alters host barrier function favouring AIEC colonisation. Gut 63:116–124
Maukonen J, Saarela M (2015) Human gut microbiota: does diet matter? Proc Nutr Soc 74:23–36
Mazmanian SK, Round JL, Kasper DL (2008) A microbial symbiosis factor prevents intestinal inflammatory disease. Nature 453:620–625
McClintock SD, Attili D, Dame MK, Richter A, Silvestri SS, Berner MM et al (2020) Differentiation of human colon tissue in culture: effects of calcium on trans-epithelial electrical resistance and tissue cohesive properties. PLoS One 15(3):e0222058
Musthafa KS, Ravi AV, Annapoorani A, Packiavathy IS, Pandian SK (2010) Evaluation of anti-quorum-sensing activity of edible plants and fruits through inhibition of the N-acyl-homoserine lactone system in Chromobacterium violaceum and Pseudomonas aeruginosa. Chemotherapy 56:333–339
Nagano T, Yano H (2020a) Dietary cellulose nanofiber modulates obesity and gut microbiota in high-fat-fed mice. Bioact Carbohydr Diet Fibre 22:100214
Nagano T, Yano H (2020b) Effect of dietary cellulose nanofiber and exercise on obesity and gut microbiota in mice fed a high-fat-diet. Biosci Biotechnol Biochem 84(3):613–620
Nakatani A, Li X, Miyamoto J, Igarashi M, Watanabe H, Sutou A, Watanabe K, Motoyama T, Neis E, Dejong C, Rensen S (2015) The role of microbial amino acid metabolism in host metabolism. Nutrients 7:2930–2946
O'Gorman J, Humphreys H (2012) Application of copper to prevent and control infection. Where are we now? J Hosp Infect 81:217e223
Ohashi W, Fukada T (2019) Contribution of zinc and zinc transporters in the pathogenesis of inflammatory bowel diseases. J Immunol Res 2019:8396878
Packiavathy IA, Priya S, Pandian SK et al (2014) Inhibition of biofilm development of uropathogens by curcumin - an anti-quorum sensing agent from *Curcuma longa*. Food Chem 148:453–460
Payling L, Fraser K, Loveday SM, Sims I, Roy N, McNabb W (2020) The effects of carbohydrate structure on the composition and functionality of the human gut microbiota. Trends Food Sci Technol 97:233–248
Pellegrini M, Ippolito M, Monge T, Violi R, Cappello P, Ferrocino I et al (2020) Gut microbiota composition after diet and probiotics in overweight breast cancer survivors: a randomized open-label pilot intervention trial. Nutrition 74:110749
Peng M, Luo H, Kumar V, Kajbaf K, Hu Y, Yang G (2019) Dysbiosis of intestinal microbiota induced by dietary oxidized fish oil and recovery of diet-induced dysbiosis via taurine supplementation in rice field eel (Monopterus albus). Aquaculture 512:734288
Peng Y, Yan Y, Wan P, Dong W, Huang K, Ran L et al (2020) Effects of long-term intake of anthocyanins from Lycium ruthenicum Murray on the organism health and gut microbiota in vivo. Food Res Int 130:108952
Pérez-Burillo S, Pastoriza S, Gironés A, Avellaneda A, Francino MP, Rufián-Henares JA (2020) Potential probiotic salami with dietary fiber modulates metabolism and gut microbiota in a human intervention study. J Funct Foods 66:103790
Pierre JF, Heneghan AF, Feliciano RP, Shanmuganayagam D, Roenneburg DA, Krueger CG et al (2013) Cranberry proanthocyanidins improve the gut mucous layer morphology and function in mice receiving elemental enteral nutrition. JPEN J Parenter Enteral Nutr 37(3):401–409
Poesen R, Claes K, Evenepoel P, De Loor H, Augustijns P, Kuypers D, Meijers B (2016) Microbiota-derived phenylacetylglutamine associates with overall mortality and cardiovascular disease in patients with CKD. J Am Soc Nephrol 27:3479–3487

Pola S, Padi DL (2021) Overview on human gut microbiome and its role in immunomodulation. In: Bramhachari PV (ed) Microbiome in human health and disease. Springer, Singapore, pp 69–82
Qi B, Han M (2018) Microbial siderophore enterobactin promotes mitochondrial iron uptake and development of the host via interaction with ATP synthase. Cell 175(2):571–582. e11
Raj S (2020) Influences of the nutrition transition on chronic disease. In: Integrative and functional medical nutrition therapy. Humana, Cham, pp 17–29
Rios JL, Bomhof MR, Reimer RA, Hart DA, Collins KH, Herzog W (2019) Protective effect of prebiotic and exercise intervention on knee health in a rat model of diet-induced obesity. Sci Rep 9:3893
Rothhammer V, Borucki DM, Tjon EC, Takenaka MC, Chao CC, Ardura-Fabregat A, de Lima KA, Gutierrez-Vazquez C, Hewson P, Staszewski O et al (2018) Microglial control of astrocytes in response to microbial metabolites. Nature 557(7707):724–728
Russell WR, Gratz SW, Duncan SH, Holtrop G, Ince J, Scobbie L et al (2011) Highprotein, reduced-carbohydrate weight-loss diets promote metabolite profiles likely to be detrimental to colonic health. Am J Clin Nutr 93:1062–1072
Sampson TR, Mazmanian SK (2015) Control of brain development, function, and behavior by the microbiome. Cell Host Microbe 17(5):565–576
Sandberg AS, Andlid T (2002) Phytogenic and microbial phytases in human nutrition. Int J Food Sci Technol 37:823–833
Sáyago-Ayerdi SG, Zamora-Gasga VM, Venema K (2020) Changes in gut microbiota in predigested Hibiscus sabdariffa L calyces and agave (Agave tequilana weber) fructans assessed in a dynamic in vitro model (TIM-2) of the human colon. Food Res Int 132:109036
Schepper JD, Collins F, Rios-Arce ND, Kang HJ, Schaefer L, Gardinier JD et al (2020) Involvement of the gut microbiota and barrier function in glucocorticoid-induced osteoporosis. J Bone Miner Res 35(4):801–820
Schroeder JC, Dinatale BC, Murray IA, Flaveny CA, Liu Q, Laurenzana EM, Lin JM, Strom SC, Miecinski CJ, Amin S et al (2010) The uremic toxin 3-indoxyl sulfate is a potent endogenous agonist for the human aryl hydrocarbon receptor. Biochemistry 49(2):393–400
Schroeder BO, Birchenough G, Stahlman M et al (2018) Bifidobacteria or fiber protects against diet-induced microbiota-mediated colonic mucus deterioration. Cell Host Microbe 23:27–40.e7
Seck EH, Senghor B, Merhej V, Bachar D, Cadoret F, Robert C et al (2019) Salt in stools is associated with obesity, gut halophilic microbiota and Akkermansia muciniphila depletion in humans. Int J Obes 43(4):862–871
Sencio V, Barthelemy A, Tavares LP, Machado MG, Soulard D, Cuinat C et al (2020) Gut dysbiosis during influenza contributes to pulmonary pneumococcal superinfection through altered short-chain fatty acid production. Cell Rep 30(9):2934–2947
Shen Q, Chen YA, Tuohy KM (2010) A comparative in vitro investigation into the effects of cooked meats on the human faecal microbiota. Anaerobe 16:572–577
Shortt C, Hasselwander O, Meynier A, Nauta A, Fernández EN, Putz P et al (2018) Systematic review of the effects of the intestinal microbiota on selected nutrients and non-nutrients. Eur J Nutr 57(1):25–49
Singh V, Vijay-Kumar M (2020) Beneficial and detrimental effects of processed dietary fibers on intestinal and liver health: health benefits of refined dietary fibers need to be redefined! Gastroenterol Rep 8(2):85–89
Singh RK, Chang HW, Yan D, Lee KM, Ucmak D, Wong K et al (2017) Influence of diet on the gut microbiome and implications for human health. J Transl Med 15:73
Smith E, Macfarlane G (1998) Enumeration of amino acid fermenting bacteria in the human large intestine: effects of pH and starch on peptide metabolism and dissimilation of amino acids. FEMS Microbiol Ecol 25:355–368
Sonnenburg JL, Xu J, Leip DD, Chen CH, Westover BP, Weatherford J, Buhler JD, Gordon JI (2005) Glycan foraging in vivo by an intestine adapted bacterial symbiont. Science 307:1955–1959
Sonnenburg ED, Smits SA, Tikhonov M, Higginbottom SK, Wingreen NS, Sonnenburg JL (2016) Diet-induced extinctions in the gut microbiota compound over generations. Nature 529:212–215

Spencer SP, Wilhelm C, Yang Q, Hall JA, Bouladoux N, Boyd A et al (2014) Adaptation of innate lymphoid cells to a micronutrient deficiency promotes type 2 barrier immunity. Science 343(6169):432–437
Sperandio V (2017) Take your pick: vitamins and microbiota facilitate pathogen clearance. Cell Host Microbe 21(2):130–131
Steinert RE, SadaghianSadabad M, Harmsen HJ, Weber P (2016) The prebiotic concept and human health: a changing landscape with riboflavin as a novel prebiotic candidate? Eur J Clin Nutr 70(12):1461
Su D et al (2016) Vitamin D signaling through induction of paneth cell defensins maintains gut microbiota and improves metabolic disorders and hepatic steatosis in animal models. Front Physiol 7:498
Sukkar AH, Lett AM, Frost G, Chambers E (2019) Regulation of energy expenditure and substrate oxidation by short chain fatty acids. J Endocrinol 242(2):R1–R8
Swartz TD, Duca FA, De Wouters T, Sakar Y, Covasa M (2012) Up-regulation of intestinal type 1 taste receptor 3 and sodium glucose luminal transporter-1 expression and increased sucrose intake in mice lacking gut microbiota. Br J Nutr 107(5):621–630
Świątecka D, Dominika Ś, Narbad A, Arjan N, Ridgway KP, Karyn RP et al (2011) The study on the impact of glycated pea proteins on human intestinal bacteria. Int J Food Microbiol 145:267–272
Takahashi Y, Sugimoto K, Soejima Y, Kumagai A, Koeda T, Shojo A, Nakagawa K, Harada N, Yamaji R, Inui H (2015) Inhibitory effects of eucalyptus and banaba leaf extracts on non-alcoholicsteatohepatitisinduced by a high-fructose/high-glucose diet in rats. Biomed Res Int 2015:296207
Tamura K, Sasaki H, Shiga K, Miyakawa H, Shibata S (2020) The timing effects of soy protein intake on mice gut microbiota. Nutrients 12(1):87
Thursby E, Juge N (2017) Introduction to the human gut microbiota. Biochem J 474:1823–1836
Toden S, Bird AR, Topping DL, Conlon MA (2007) High red meat diets induce greater numbers of colonic DNA double-strand breaks than white meat in rats: attenuation by high-amylose maize starch. Carcinogenesis 28:2355–2362
Townsend GE, Han W, Schwalm ND, Raghavan V, Barry NA, Goodman AL, Groisman EA (2019) Dietary sugar silences a colonization factor in a mammalian gut symbiont. Proc Natl Acad Sci 116:233–238
Trautvetter U, Camarinha-Silva A, Jahreis G, Lorkowski S, Glei M (2018) High phosphorus intake and gut-related parameters–results of a randomized placebo-controlled human intervention study. Nutr J 17(1):23
Turnbaugh PJ, Ridaura VK, Faith JJ, Rey FE, Knight R, Gordon JI (2009) The effect of diet on the human gut microbiome: a metagenomic analysis in humanized gnotobiotic mice. Sci Transl Med 1:6ra14
Tytgat HL, De Vos WM (2016) Sugar coating the envelope: glycoconjugates for microbe–host crosstalk. Trends Microbiol 24(11):853–861
Uberti F, Morsanuto V, Ruga S, Galla R, Farghali M, Notte F et al (2020) Study of magnesium formulations on intestinal cells to influence myometrium cell relaxation. Nutrients 12(2):573
Valiente E, Bouché L, Hitchen P, Faulds-Pain A, Songane M, Dawson LF et al (2016) Role of glycosyltransferases modifying type B flagellin of emerging hypervirulent Clostridium difficile lineages and their impact on motility and biofilm formation. J Biol Chem 291(49):25450–25461
Van Hul M, Cani PD (2019) Targeting carbohydrates and polyphenols for a healthy microbiome and healthy weight. Curr Nutr Rep 8(4):307–316
Vasavi HS, Arun AB, Rekha PD (2016) Anti-quorum sensing activity of flavonoid-rich fraction from Centella asiatica L. against Pseudomonas aeruginosa PAO1. J Microbiol Immunol Infect 49(1):8–15
Vijay-Kumar M (2020) Abstract IA20: gut microbiota dysbiosis and dietary fermentable fibers in a pickle: a brew for liver cancer. Cancer Res 80(8_Supplement):IA20

Vikram A, Jayaprakasha GK, Jesudhasan PR, Pillai SD, Patil BS (2010) Suppresion of bacterial cell-cell signalling, biofilm formation and type III secretion system by citrus flavonoids. J Appl Microbiol 109:515–527

Villodre TC, Boudry C, Stumpff F, Aschenbach JR, Vahjen W, Zentek J, Pieper R (2015) Down-regulation of monocarboxylate transporter 1 (MCT1) gene expression in the colon of piglets is linked to bacterial protein fermentation and pro-inflammatory cytokine-mediated signalling. Br J Nutr 113:610–617

World Health Organization (2015) Information note about intake of sugars recommended in the WHO guideline for adults and children. WHO Press, World Health Organization, Geneva

Wrzosek L, Miquel S, Noordine ML, Bouet S, Joncquel Chevalier-Curt M, Robert V, Philippe C, Bridonneau C, Cherbuy C, Robbe-Masselot C et al (2013) *Bacteroides thetaiotaomicron* and *Faecalibacterium prausnitzii* influence the production of mucus glycans and the development ofgoblet cells in the colonic epithelium of a gnotobiotic model rodent. BMC Biol 11:61

Xue Z, Ma Q, Chen Y, Lu Y, Wang Y, Jia Y et al (2020) Structure characterization of soluble dietary fiber fractions from mushroom Lentinula edodes (Berk.) Pegler and the effects on fermentation and human gut microbiota in vitro. Food Res Int 129:108870

Yamamoto EA, Jørgensen TN (2020) Relationships between vitamin D, gut microbiome, and systemic autoimmunity. Front Immunol 10:3141

Yang R, Shan S, Zhang C, Shi J, Li H, Li Z (2020a) Inhibitory effects of bound polyphenol from foxtail millet bran on colitis-associated carcinogenesis by the restoration of gut microbiota in a mice model. J Agric Food Chem 68(11):3506–3517

Yang C, Xu Z, Deng Q, Huang Q, Wang X, Huang F (2020b) Beneficial effects of flaxseed polysaccharides on metabolic syndrome via gut microbiota in high-fat diet fed mice. Food Res Int 131:108994

Yang L, Zhao Y, Huang J, Zhang H, Lin Q, Han L et al (2020c) Insoluble dietary fiber from soy hulls regulates the gut microbiota in vitro and increases the abundance of bifidobacteriales and lactobacillales. J Food Sci Technol 57(1):152–162

Yeung CY, Chiang Chiau JS, Cheng ML, Chan WT, Chang SW, Chang YH et al (2020) Modulations of probiotics on gut microbiota in a 5-fluorouracil-induced mouse model of mucositis. J Gastroenterol Hepatol 35(5):806–814

Yilmaz B, Li H (2018) Gut microbiota and iron: the crucial actors in health and disease. Pharmaceuticals (Basel) 11(4):98

Yu H, Guo Z, Shen S, Shan W (2016) Effects of taurine on gut microbiota and metabolism in mice. Amino Acids 48(7):1601–1617

Yuan Y, Zhou J, Zheng Y, Xu Z, Li Y, Zhou S, Zhang C (2020) Beneficial effects of polysaccharide-rich extracts from Apocynum venetum leaves on hypoglycemic and gut microbiota in type 2 diabetic mice. Biomed Pharmacother 127:110182

Zerbib S, Vallet L, Muggeo A, de Champs C, Lefebvre A, Jolly D, Kanagaratnam L (2020) Copper for the prevention of outbreaks of health care–associated infections in a long-term care facility for older adults. J Am Med Dir Assoc 21(1):68–71

Zhang F, Ye J, Zhu X, Wang L, Gao P, Shu G, Jiang Q, Wang S (2019) Anti-obesity effects of dietary calcium: the evidence and possible mechanisms. Int J Mol Sci 20:3072

Zmora N, Suez J, Elinav E (2019) You are what you eat: diet, health and the gut microbiota. Nat Rev Gastroenterol Hepatol 16(1):35–56

Zoetendal EG, Raes J, Van den Bogert B, Arumugam M, Booijink CCGM, Troost FJ, Bork P, Wels M, de Vos WM, Kleerebezem M (2012) The human small intestinal microbiota is driven by rapid uptake and conversion of simple carbohydrates. ISME J 6:1415–1426

Zou J et al (2018) Fiber- mediated nourishment of gut microbiota protects against diet- induced obesity by restoring IL-22-mediated colonic health. Cell Host Microbe 23:41–53

Zsivkovits M et al (2003) Prevention of heterocyclic amine induced DNA damage in colon and liver of rats by different lactobacillus strains. Carcinogenesis 24:1913–1918

Zuo K, Li J, Xu Q, Hu C, Gao Y, Chen M et al (2019) Dysbiotic gut microbes may contribute to hypertension by limiting vitamin D production. Clin Cardiol 42(8):710–719

7 Hepatocellular Carcinoma and Human Gut Microbiome: Association with Disease and Scope for Therapeutic Intervention

Ishfaq Hassan Mir, Saqib Hassan, Joseph Selvin, and Chinnasamy Thirunavukkarasu

Abstract

Hepatocellular carcinoma (HCC) is the world's third most prevalent cause of cancer-related mortality. HCC frequently occurs in patients with chronic liver diseases, and it is triggered by a vicious cycle of liver damage, inflammation, and regeneration. Current research showcases that the bacterial microbiome has an indispensable part in fostering the development of HCC and associated liver disorders. This chapter will explore the mechanisms by which the gut microbiota triggers the progression of hepatocarcinogenesis and associated liver disorders, with a particular emphasis on obesity, alcoholic liver disease, metabolic-associated alcoholic fatty liver disease, cirrhosis, and HCC. The pertinent mechanisms, encompassing bile acids, Toll-like receptors, mycotoxicosis, and immune checkpoint inhibitors, facilitating the progression of such maladies are covered as well. Furthermore, several prospective highlights for the diagnosis and treatment interventions are presented, which may be used in future clinical settings for combating HCC. Based on preclinical accomplishments, we highlight the gut-microbiota-liver axis as an intriguing target for the concurrent prevention of chronic liver disease progression and HCC induction.

I. H. Mir · C. Thirunavukkarasu (✉)
Department of Biochemistry and Molecular Biology, School of Life Sciences, Pondicherry University, Puducherry, India

S. Hassan
Future Leaders Mentoring Fellow, American Society for Microbiology, Washington, DC, USA

Department of Biotechnology, School of Bio and Chemical Engineering, Sathyabama Institute of Science and Technology, Chennai, Tamil Nadu, India

J. Selvin
Department of Microbiology, School of Life Sciences, Pondicherry University, Puducherry, India

P. Veera Bramhachari (ed.), *Human Microbiome in Health, Disease, and Therapy*, https://doi.org/10.1007/978-981-99-5114-7_7

Keywords

Hepatocellular · Carcinoma · Liver · Bile · Microbiome

7.1 Introduction

Hepatocellular carcinoma (HCC) is one of the most lethal malignancies in developing countries. On an annual basis, these primary liver tumors inflict roughly 600,000 deaths worldwide (Mir et al. 2021). It is a highly destructive malignancy with a dismal prognosis and survivability. Among all cancer types, HCC is the sixth most prevalent type of cancer and the third largest cause of cancer-related mortality across the globe. HCC incidences are triggered by long-term liver complications such as cirrhosis, endemic HBV/HCV infections, metabolic-associated fatty liver disease(MAFLD), aflatoxin exposure, and alcohol-related liver diseases (Chakraborty and Sarkar 2022). HCC is more frequently brought on by HCV infection in the United States, but HBV-associated liver cancer is more commonly seen in Asia and developing nations (Mir et al. 2021, 2022). From the molecular perspective, HCC is characterized by improperly coordinated signal transduction mechanisms that encourage tumor growth, progression, and metastasis upon interacting with the tumor microenvironment. Compared to women, men experience it more frequently. According to research, gender disparities in the development and progression of HCC are influenced by the reinforcing effect of androgens and the protective effect of estrogen (Li et al. 2019). Despite recent advancements in its treatment, the prognosis of HCC patients remains uncertain. Early diagnosis, prognosis, and therapy boost the likelihood of survival, whereas later phases only have access to palliative care. The life span of HCC patients relies on the stage of the tumor at diagnosis. A 5-year survival rate is attainable with early diagnosis and effective therapy, even though some months in the advanced stage are anticipated.

If HCC is diagnosed early, it can be managed more efficiently (Mir et al. 2022). There are several treatment options available, including liver transplantation, ablation, resection, chemotherapy, and radiation therapy. However, at the advanced stage, sorafenib is implemented as a prototype therapy that affects the receptors of vascular endothelial growth factor (VEGF) and platelet-derived growth factor (PDGF), thereby inhibiting angiogenesis and metastasis (Mir et al. 2021). The FDA approved cabozantinib in the year 2019 for use as an alternate therapy for HCC. Cabozantinib is a multi-tyrosine kinase inhibitor that targets the tyrosine-protein kinase Met (c-MET), the vascular endothelial growth factor receptor 2 (VEGFR2), and the tyrosine kinase receptors AXL and RET (Zhang et al. 2020). Several therapeutic agents that block the kinase activity and immunoglobulins targeted at various sites of interest are now being studied for HCC treatment. However, considerable research is needed to comprehend the molecular mechanisms underlying therapeutic benefits and escape or resistance mechanisms in HCC (Mir et al. 2021). Given the complex pathophysiology of HCC, existing therapies continue to fall short of patient expectations.

The microbiome impacts key biochemical, inflammatory, and immunological processes, and it is vital in many gastrointestinal and liver pathologies. Recent experimental studies have demonstrated that the microbiome serves a significant part in the progression of hepatocarcinogenesis. Dysfunctions of the gut bacterial flora have a profound impact on liver damage. There is mounting evidence that dysbiosis has a role in the emergence of obesity, metabolic illness, chronic liver disease (CLD), and hepatocarcinogenesis (Schwabe and Greten 2020). The microbiome does not actively communicate with the liver. However, the liver and the gut are anatomically interconnected (Anstee et al. 2019). It has been demonstrated that the gut microbiota and the metabolites of the gut microbiota contribute significantly to the development of hepatocarcinogenesis and its intervention. For example, deoxycholic acid (DCA), a secondary bile acid produced by bacteria, and lipoteichoic acid (LTA), a component of Gram-negative bacteria's cell walls, co-activated formulation of prostaglandin-endoperoxide synthase 2 or cyclooxygenase-2 (COX-2) in senescent hepatic stellate cells (HSCs) to enhance PGE2-mediated blockade of antitumor immunity, inducing hepatocarcinogenesis (Loo et al. 2017). It has been established that products originating from the gut microbiota can alter hepatic immunity and inflammation to influence the progression of nonalcoholic steatohepatitis (NASH) and virus-induced HCC (Schwabe and Greten 2020). To prevent the passage of luminal contents, such as intestinal microorganisms, inside the body, a solitary sheet of epithelial cells in the intestines acts as a physical barrier. It is noteworthy that individuals with CLDs comprising alcoholic hepatitis, cirrhosis, and hepatocellular carcinoma have greater serum lipopolysaccharide (LPS) concentration than normal individuals, indicating enhanced gut epithelial barrier permeability. Additionally, a study on animals revealed that chemical alteration of this barrier stimulates the development of hepatic tumors (Komiyama et al. 2021). As a result, increased intestinal permeability has been linked to tumor development in people with CLDs. In another study, the taxonomic richness in fecal samples was higher in liver cancer individuals that responded to anti-programmed cell death protein 1 (PD-1) treatment than in nonresponders (Zheng et al. 2019). Apart from that, patients who responded to treatment had higher concentrations of the *Ruminococcaceae* spp. and *Akkermansia muciniphila*, whereas nonresponders had higher levels of *Proteobacteria*.

A recurring trend in numerous animal models examining the impact of dysbiosis on hepatocellular carcinoma is that treatment of a combined approach of a wide range of antibiotics resulted in gut sterilization and a decline in tumor volume, as well as the prevention of HCC progression (Rattan et al. 2020). In order to combat dysbiosis and demonstrate a reduction in HCC growth, probiotics have also been employed in murine HCC models. To improve the patient's prognosis, it is crucial to diagnose HCC as soon as possible. Because of the noninvasiveness, high efficacy, and accuracy of gut microbiota, it is advantageous in the diagnosis of disease. In addition, more and more research points to the potential use of gut microbiota as a biomarker for a broad range of diseases, including liver cirrhosis (Kang et al. 2021). The gut microbiome and the HCC tumor microenvironment are two complicated systems that may be explored with greater precision because of advances in human microbiome research.

The current chapter first outlines existing findings that explored the role of intestinal microbiota in the incidence and progression of HCC. Then, we focused on gut microbiota-mediated HCC treatment and early detection. Some potential highlights for diagnosis and treatment are presented, which might be employed in future clinical applications.

7.2 The Gut Microbiota

The intestine is a vital organ of the body that facilitates digestion and absorption. It is one of the primary immune organs involved in the management of normal bodily functions. The accomplishment of intestinal operation is greatly aided by the occurrence of diverse microbes in it. The "gut microbiota" is a collection of microorganisms that populate the intestine, comprising bacteria, archaea, eukarya, viruses, and parasites. According to studies, the GI tract is home to more than 10^{14} microorganisms, having approximately 100 times as much genomic material as the human genome and about ten times more bacterial cells than human cells (Thursby and Juge 2017). Because of the occurrence of a large number of bacterial cells living synergistically within the human body, humans are also referred to as superorganisms (Gill et al. 2006). The host receives several perks from the microbiota through a variety of physiological mechanisms, such as improved gut integrity, safeguarding against infections, and modulation of host immunity. However, these mechanisms may be impaired as a result of a change in microbial balance, a condition referred to as dysbiosis. As more advanced methods for evaluating and analyzing complex ecosystems emerge, a role for the microbiome in a wide range of intestinal and extraintestinal ailments has become abundantly evident (Thursby and Juge 2017).

Firmicutes and Bacteroidetes are the most prevalent phyla of gut bacteria, accounting for 90% of the gut microbiota, followed by Actinobacteria, Proteobacteria, Fusobacteria, and Verrucomicrobia. The Firmicutes phylum has about 200 distinct genera, including *Lactobacillus*, *Bacillus*, *Clostridium*, *Enterococcus*, and *Ruminococcus*. *Clostridium* genera account for 95% of the Firmicutes phyla. Bacteroidetes include prominent genera such as *Bacteroides* and *Prevotella* (Rinninella et al. 2019). A steady increase in microbiota concentration can be observed throughout the gastrointestinal tract, with low concentrations in the stomach and exceptionally high concentrations in the colon. Only 10^1 bacteria per gram are present in the stomach, but higher densities and more diverse bacterial populations can be found in the duodenum (10^3/g), jejunum (10^4/g), ileum (10^7/g), and colon (10^{12} bacteria/gram) (Dieterich et al. 2018). The diversity of bacteria in the small intestine is less extensive than in the colon. The small intestine's bacterial density is constrained by the O_2 gradient, antibacterial proteins, bile acids, and hydrogen ion concentration. As a result, rapidly developing facultative anaerobes, primarily *Lactobacillaceae* and *Enterobacteriaceae*, predominate in this scenario (Luo et al. 2022). Given the presence of bile and digestive enzymes along with the swift passage of food, the duodenum is an adverse environment for microbial survival, and only a minimal number of distinct bacteria may be located therein. The

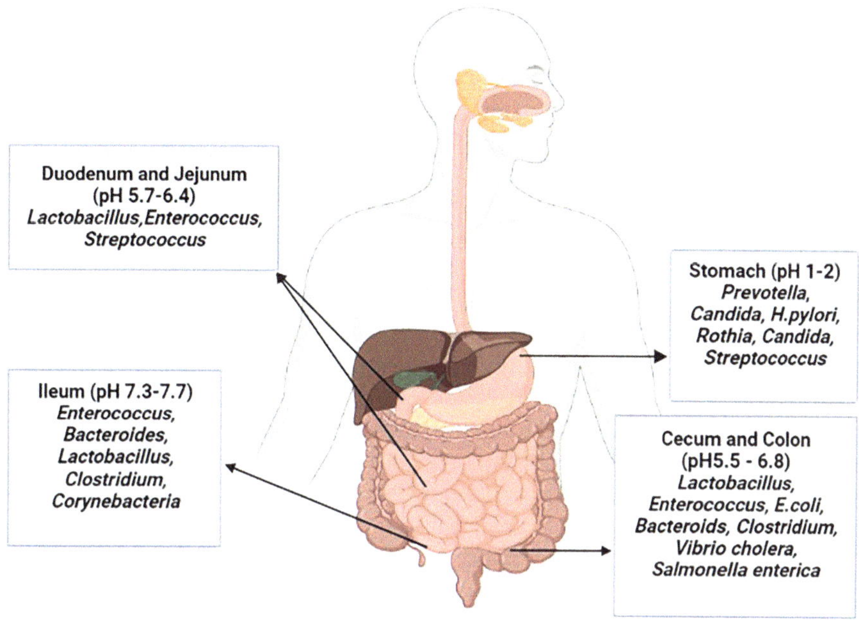

Fig. 7.1 The localization and richness of bacteria in the human gastrointestinal tract

duodenum predominately comprises microorganisms from three phyla: Firmicutes, Proteobacteria, and Actinobacteria. The jejunum is inhabited by a diverse and dense colony of Gram-positive aerobic and facultative anaerobic bacteria. The ileum, on the other hand, has an ileal bacterial density of up to 109 CFU/mL and a predominance of anaerobic bacteria. These bacteria include *Enterococcus*, *Bacteroides*, *Lactobacillus*, *Clostridium*, and *Corynebacteria* (Luo et al. 2022). The most affluent microbial community is by far found in the large intestine (Fig. 7.1), which has sluggish flow rates and a pH range of neutral to mildly acidic. It is dominated by obligate anaerobic microorganisms (Rinninella et al. 2019).

7.2.1 The Intestinal Epithelial Barrier

A symbiotic interaction between the host and microbiota is established on the strict segregation of bacterial entities from the host chamber. A properly maintained, multilayer barrier facilitates this partitioning in the intestines (Pradere et al. 2010). This barrier depends on a healthy epithelial lining consisting of a mucus layer, Paneth and goblet cells, lymphatic tissue, and several released substances like antibodies and defensins. Because of the frequent variations in the luminal constituents of the intestines and the fast epithelial cell turnover, the gut barrier is a quick-reacting, highly changing structure. Paneth cells secrete antimicrobial peptides that control the microbiota through the frequent sampling of gut microbes by specialistic epithelial cells known as M cells; conversely, the microbiome regulates the intestinal barrier and epithelial cell proliferation (Peterson and Artis 2014).

In this intricate system, bile acids play a crucial role in modulating the function of the epithelial barrier and the proliferation of intestinal epithelial cells through the signal transduction pathways controlled by the farnesoid X-activated receptor (FXR) and the epidermal growth factor receptor (EGFR) (Dossa et al. 2016). A vital interaction between the liver, bacteria, and the gut is maintained via bile acids. Bile acids are generated in the hepatic system and then processed by bacteria. IECs, which are intestinal epithelial cells, express the FXR receptor, which senses the bile acids and communicates the information to the liver through the FGF19 receptor (Modica et al. 2012). The diversity of the intestinal microbiome and the integrity of the intestinal barrier are significantly impacted by acute and CLDs, respectively, contributing to dysbiosis and a leaky gut. Numerous studies evaluating intestinal permeability in patients with chronic liver abnormalities that could potentially lead to cirrhosis have been published (Pijls et al. 2013). Several mechanisms that lead to the breakdown of the intestinal barrier and the emergence of a leaky gut are most likely multifaceted. Some of the mechanisms that promote this aberration include bile acid secretion, dysbiosis mediated by bacteria, and a surge in the secretion of inflammatory cytokines.

Recent research has revealed significant changes in the intestinal microbiota of cirrhotic patients, including an increase in *Enterobacteriaceae* as well as strains commonly found in the oral microbiota, such as *Veillonellaceae* and *Streptococcaceae*. Simultaneously, the prevalence of useful bacteria in the gut, such as *Lachnospiraceae*, is diminishing. The cirrhosis stage is favorably correlated with *Enterobacteriaceae* and negatively correlated with *Lachnospiraceae* (Yu and Schwabe 2017). The present knowledge of the mechanisms regarding the changes in the gut microbiota in patients with liver disorders is still lacking and hampered by a number of factors. To understand how dysbiotic microbes contribute to liver disease, well-designed functional investigations are required. Not only must dysbiosis be validated as a factor of liver disease development and progression, but it must also be determined whether dysbiosis contributes to gut leakiness in CLD patients.

7.2.2 Effect of Intestinal Microbiome on HCC Progression

The advancement of HCC is aided by aberrations such as cirrhosis, nonalcoholic fatty liver disease, adiposity, and nonalcoholic fatty liver disease (Fig. 7.2). The following section will go through what is currently known about gut flora in these circumstances.

7.2.3 Obesity

Intestinal microbiomes are essential for controlling how much energy is extracted from the diet, which in turn impacts obesity. Studies utilizing GF mice, which are

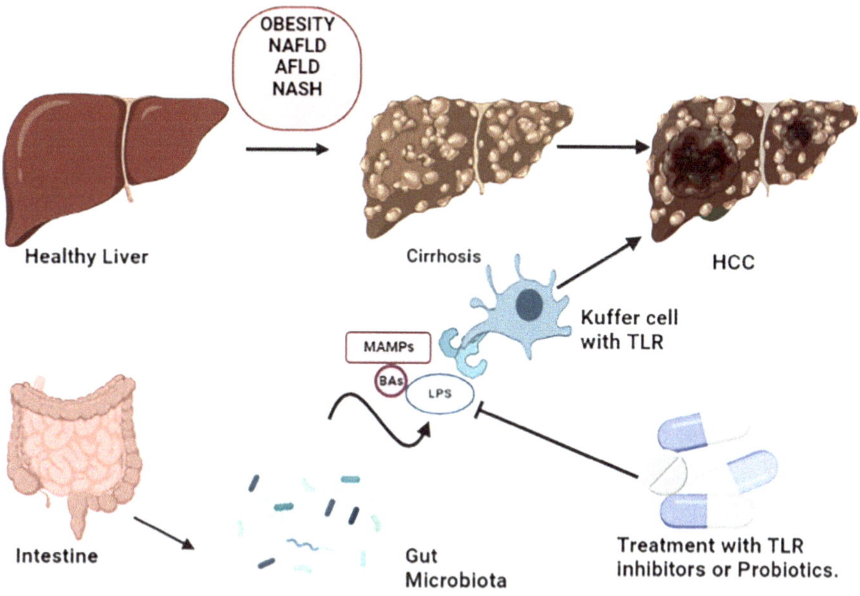

Fig. 7.2 The progression of HCC and its management using the gut microbiome. Hepatocarcinogenesis is triggered by long-term liver comorbidities such as ALD, MAFLD, NASH, and cirrhosis. This is facilitated by gut dysbiosis, which yields microbial byproducts, notably bile acids and lipopolysaccharides. The gut barrier can be strengthened and HCC progression impeded by altering the gut microbiota with probiotics, prebiotics, antibiotics, or bacterial antagonists

maintained in sterile facilities, suggested the association between gut flora and obesity in the beginning. It has been observed that germ-free mice have lower levels of body fat despite consuming more calories than typical specific pathogen-free (SPF) mice (Bäckhed et al. 2004). Through a variety of methods, calorie intake may rise due to intestinal microbes, including the metabolism of complex sugars and polysaccharides that the host ordinarily cannot digest. Furthermore, germ-free circumstances that favor catabolism alter several metabolic processes, such as fatty acid oxidation in the liver and glycogen breakdown (Bäckhed et al. 2007).

It has been proposed that the nature of the intestinal flora serves a significant role in the occurrence of obesity. When compared to their lean counterparts, naturally obese animal models exhibit a more considerable proportion of intestinal *Firmicutes* and a relative enrichment of microbial traits for polysaccharide breakdown (Davis 2016). According to some fecal transplantation investigations, germ-free animal models that acquire microbiomes from obese individuals accumulate more body fat than controls. Rodent models have helped us comprehend how well the gut microbiota contributes to obesity, but they are constrained by human physiological and metabolic characteristics. A potent study tool is the gnotobiotic pig model, which has been created using pigs inoculated with human microbiota (Wang and Donovan 2015)

7.2.4 Alcoholic Liver Disease (ALD)

ALD triggers about 50% of all cases of cirrhosis, and it is a cofactor in liver abnormalities driven by viral infection and nonalcoholic steatohepatitis. Multiple mechanisms that contribute to alcohol-induced HCC have been identified, albeit they are not adequately addressed. Alcohol is primarily transformed into acetaldehyde in the cytosol of hepatic cells, which is then metabolized to acetate in the mitochondrion. Acetaldehyde plays a crucial role in tumorigenesis by interacting with DNA and proteins to generate adducts. Acetaldehyde, a poisonous byproduct of ethanol, accumulates in hepatocytes and induces inflammation and liver fibrogenesis, which is the principal cause of ALD (Takase et al. 2021).

In the past 20 years, it has been evident how crucial the gut microbiota is in the early phases of ALD, and there is mounting evidence that LPS and ALD are closely related. An elevation in LPS in the portal circulation from undetectable quantities following ethanol administration in rats demonstrated that even a solitary alcohol binge is enough to promote microbial translocation. Consequently, individuals experiencing prolonged alcoholism had higher serum LPS levels (Yu and Schwabe 2017). The potential of alcohol as well as its derivative acetaldehyde to impair tight junctions adds to the increased rates of bacterial translocation during ALD. Microbial proliferation and significant alterations in the gut microbial population are triggered by prolonged ethanol usage. After alcohol consumption, a tiny portion of it is absorbed in the oral cavity. Around 20% of it is progressively absorbed in the stomach, and a significant quantity is absorbed in the intestinal tract (Levitt et al. 1997). Several studies have demonstrated that drinking alcohol could significantly affect the diversity of intestinal microbiota. When contrasted with the nonexposed placebo group, rodents that consumed alcohol for 13 weeks displayed decreased and increased frequency in *Lactobacilli* and *Bacteroidetes*, respectively (Kosnicki et al. 2019). Alcohol intake causes gastrointestinal hyperpermeability, which facilitates the ability of microbial species and their byproducts to penetrate the portal and systemic circulation. As a consequence, alterations in the gut microbiota triggered by alcohol might have an effect on the different tissues and organs of the body (Stärkel et al. 2018).

7.2.5 Metabolic-Associated Fatty Liver Disease (MAFLD)

Despite being identified as a disorder only about 20 years ago, MAFLD is currently the most pervasive hepatic disorder worldwide and is expected to emerge as the primary source of CLDs, including hepatocellular carcinoma. When compared to other chronic hepatic disorders, MAFLD has a minimal proportion of risk in HCC development, but it contributes significantly to HCC development at the population level because of its high incidence (Michelotti et al. 2013). Research has demonstrated that the microbiome of obese people is far more effective at extracting energy and, as a result, induces obesity. Antibiotic use, therefore, cuts down a high-fat diet (HFD)-influenced MAFLD in animal models (Jiang et al. 2015).

The gut microbiota changes related to MAFLD vary according to the clinical phase of the disorder. Disease progression is accompanied by reduced microbiota diversity, a rise in Gram-negative bacteria, primarily *Proteobacteria*, and a decline in Gram-positive bacteria, mainly *Firmicutes*. Physiologically, a switch from valuable to pathogenic organisms that induce the formation of an inflammatory and biochemically toxic gut environment, which in turn causes gut barrier impairment, exposes the liver to nutritional and microbiota-derived elements and accelerates the advancement of MAFLD (Hrncir et al. 2021). In some studies, it was revealed that compared to healthy controls, MAFLD patients have greater levels of *Prevotella* and *Porphyromonas* species. However, a reduced proportion of *Bacteroidetes* is also observed (Albhaisi and Bajaj 2021). Despite an extensive amount of preclinical evidence exploring and pointing to a connection between dysbiosis and MAFLD, the role of the gut microbiota in MAFLD has only been evaluated in a modest number of cross-sectional human studies. According to one investigation, the magnitude of MAFLD is linked to gut microbiota dysbiosis and changes in its metabolic activity (Schwenger et al. 2019). Pediatric research revealed that patients with NASH had higher concentrations of *E. coli* than control participants, which was correlated with increased blood alcohol content (Zhu et al. 2013).

7.2.6 Gut Microbiota, PCOS, and MAFLD

Increased prevalence of MAFLD is reported in patients with polycystic ovary syndrome (PCOS) (Vassilatou 2014). Hepatic steatosis (HS) has been reported in PCOS patients previously (Gambarin-Gelwan et al. 2007). Alcohol-producing bacteria such as *Bifidobacterium* may contribute to the pathogenesis of MAFLD in PCOS (Zhu et al. 2013). Jobira et al. (2021) found higher % RA of *Bifidobacterium* in adolescents with HS, which suggests that bacterial taxa involved in ethanol production may contribute to endogenous ethanol production in NALFD in PCOS. They conclude that there is a link between the gut microbiome and metabolic disease in adolescents with HS and PCOS. Hassan et al. (2022) observed an enrichment of *Bifidobacterium* in the gut microbiome of women with PCOS. The relationship among bile acids, gut microbiota, and metabolic diseases has been explored in earlier studies (Jia et al. 2019; Wahlström et al. 2016). Primary bile acids are used as substrates by gut microbial enzymes, which produce secondary bile acids, which are then circulated between the gut and liver via enterohepatic circulation (Wahlström et al. 2016). Bile acid metabolism begins in the gastrointestinal tract involving microbiota possessing bile salt hydrolase activity (Ridlon et al. 2006). Bile salt hydrolase activity is common in *Bifidobacterium* and *Lactobacillus* (Tanaka et al. 1999).

7.2.7 Cirrhosis

Cirrhosis, which is characterized by extensive fibrosis and the depletion of hepatocytes, is regarded as an end-stage liver disorder. A cirrhotic liver can be the consequence of any of the liver conditions explained above (Bhat et al. 2016). A large percentage of liver malignancies occur in patients with cirrhosis of the liver. Prominent features of liver cirrhosis include dysbiosis and leaky gut. In the case of patients suffering from liver cirrhosis, dysbiosis and leaky gut are considered to have a role in the progression of hepatocarcinogenesis (Akkız 2021). In the initial phases of CLD, there is high microbial translocation and dysbiosis. Such mechanisms lead to fibrosis and cirrhosis advancement. As a result, dysbiosis and leaky gut are prominent attributes of all phases of CLD, promoting the gradual induction of HCC (Akkız 2021). At the moment, liver cirrhosis has no better treatment available. The sole option is to control its symptoms while also reducing its progression. One and only intervention, a liver transplant, may be performed if the liver is badly damaged. Depending on the etiology of the disease, the financial burden of managing cirrhosis varies by around 2 billion dollars (Lee and Suk 2020).

Peptides essential for matrix disintegration include matrix metalloproteinases (MMPs) and tissue antagonists of metalloproteinases, which are MMP inhibitors. The discovery that MMPs are produced in liver damage suggests that a possible cause of liver fibrosis is the destruction of the regular extracellular matrix. According to a prior study, fibrosis in pulmonary disorders is correlated to MMPs and the microbiota (Taylor et al. 2015). MMPs are linked to several hepatic injury phases, including cirrhosis and hepatocarcinogenesis (Lee and Suk 2020).

Cirrhosis has been linked to altered gut microbiota function and organization. In terms of design, the fecal microbiota in cirrhotic individuals exhibits a decline in diversification and an elevation in microorganisms that stimulate the immune system. For example, a decrease of potentially advantageous *Firmicutes* (e.g., *Lachnospiraceae* and *Ruminococcaceae*) has been observed in cirrhotic individuals. Comparable alterations can indeed be observed inside the oral cavity, blood serum, and other tissues of such patients (Wang et al. 2021). In contrast to the structural breakdown of the gastrointestinal barrier, cirrhosis is characterized by gut penetration with lymphocytes, as demonstrated by the growth of TNF-alpha and IFN-gamma-expressing lymphocytes and the reduction of Th17 cells. As the disease progresses, cirrhosis is also associated with decreased bile flow and compromised FXR signal transduction mechanisms (Wang et al. 2021).

As of now, certain scientific evidence from human research as well as animal studies suggest that the prevalence of HCC is associated with the gut microbiome (Table 7.1). More research is required to determine how initial phases of liver disorders exhibit distinct intestinal microbiota profiles from different etiologies and whether these differences may fade in the final phases.

Table 7.1 Alterations in gut microbiome attributed to HCC progression in human and animal research

Models	Disorder	Gut microbiome diversity	References
Rodent	HCC	Altered	Yu et al. (2010)
Rodent	HCC	Altered	Dapito et al. (2012)
Rodent	NASH/ HCC	*Atopobium* spp.↑, *Bacteroides* spp.↑, *Bacteroides vulgatus* ↑, *Bacteroides acidifaciens*↑, *Bacteroides uniformis*↑, *Clostridium cocleatum*↑, *Desulfovibrio* spp.↑	Xie et al. (2016)
Rodent	HCC	*Lactobacillus* spp.↓, *Bifidobacterium* spp.↓ *Enterococcus* spp.↓	Zhang et al. (2012)
Rodent	MAFLD/ HCC	*Mucispirillum*↑, *Desulfovibrio*↑, *Anaerotruncus*↑, *Desulfovibrionaceae*↑, *Bifidobacterium*↓, *Bacteroides*↓	Zhang et al. (2020)
Rodent	HCC	Altered	Yoshimoto et al. (2013)
Human	HCC	*E. coli* ↑	Grąt et al. (2016)
Human	HCC	*Proteobacteria* ↑, *Desulfococcus* ↑, *Enterobacter* ↑, *Prevotella* ↑ *Veillonella* ↑, *Cetobacterium* ↓	Ni et al. (2019)
Human	HCC	*Bacteroides* ↑, *Akkermansia* ↓, *Bifidobacterium* ↓	Ponziani et al. (2019)
Human	HCC	*Klebsiella* ↑, *Haemophilus* ↑, *Alistipes* ↓, *Phascolarctobacterium* ↓, *Ruminococcus* ↓	Ren et al. (2019)
Human	HCC	*Neisseria* ↑, *Enterobacteriaceae* ↑, *Veillonella* ↑, *Limnobacter* ↑, *Enterococcus* ↓, *Phyllobacterium* ↓, *Clostridium* ↓, *Ruminococcus* ↓, *Coprococcus*↓	Zheng et al. (2020)

7.3 Strategies by Which the Gut Microbiome Mediates the Progression of Hepatocarcinogenesis

7.3.1 Bile Acids

Bile acids (BAs) comprise a class of H_2O-soluble steroids produced by the liver cells as a result of cholesterol breakdown. The gut microbiota dehydroxylates the two primary BAs and converts them into secondary BAs. Deoxycholic acid and lithocholic acid are produced through the conversion of cholic acid and chenodeoxycholic acid, respectively (Wang et al. 2013). It is imperative to strictly regulate BA concentrations because the pathological effects of deregulated BAs encompass cholestasis and malignancy. Research findings have demonstrated that BAs may have cancer-causing propensity (Liao et al. 2011).

The liver is the body's primary metabolic and detoxifying organ. As a result, most toxicants and endobiotics, such as BAs, have the ability to inflict liver damage. BAs are beneficial for boosting biliary outflow in the liver and getting rid of xenobiotic compounds, steroid hormone metabolites, triglycerides, and bilirubin. Yet, extremely high BAs concentrations induce DNA alteration in the hepatocytes, which may substantially accelerate the rate at which tumor suppressor genes and oncogenes mutate. Apart from that, they can also aggravate cellular damage and inflammation to stimulate hepatocarcinogenesis (Wang et al. 2013). According to a study, switching from primary to secondary BAs influenced the natural killer cell intrusion in the liver and mediation of liver cancer in mice (Ma et al. 2018). According to in vitro research, BAs might cause immediate hepatic cell destruction by ROS-mediated programmed cell death (Yerushalmi et al. 2001).

Gut microbes, including the species *Clostridium*, influenced the bioconversion of BAs. It has been observed that *Clostridium* clusters are implicated in the production of deoxycholic acid by the induction of 7α-dehydroxylation of primary bile acids, which has a role in HCC progression. Additionally, it has been demonstrated that *C. scindens* significantly decreases the amount of hepatic natural killer T lymphocytes (Luo et al. 2022). Receptors activated by BAs, such as GPCRs and FXRs, are the potential regulators of BA homeostasis and have a role in hepatocarcinogenesis (Wang et al. 2013).

7.3.2 Toll-Like Receptors (TLRs)

TLRs are pattern-recognition receptors (PRRs) that mainly serve as microbial detectors and, thus, are vital for the commencement of immunologic and inflammatory processes. TLR4 is thought to be the more prominent TLR to mediate HCC among the other TLRs, and it can be expressed by various cells in the hepatic tissues, such as Kupffer cells, lymphocytes, and natural killer cells. Kupffer cells are essential phagocytic cells that are liver based. Hepatic stellate cell (HSC) stimulation and fibrogenesis are reported to be facilitated by Kupffer cells. Cirrhosis, or the impairment of hepatic functioning, is a long-term consequence of fibrosis. TLR is principally responsible for recognizing lipopolysaccharides (Le Noci et al. 2021).

LPS is a constituent of the bacterial cell wall (Gram-negative). The high-fat diet was observed to elevate the concentration of LPS in the mice serum along with the decline in *Lactobacillus* and *Bifidobacterium* populations (Luo et al. 2022). In one study, it was revealed that the microbiome might trigger HCC by stimulating TLR4 via diethylnitrosamine and CCL4 toxins (Dapito et al. 2012). Apart from TLR4, TLR9 has also been implicated in colon cancer and hepatocarcinogenesis (Gao et al. 2018). TLR stimulation is known to play a key role in the inflammation-fibrosis-HCC circuit. TLR4 enhances the TGF-beta signal transduction pathway and liver fibrogenesis, both of which are implicated in the HCC progression. TLR9 signal transduction was observed to induce steatohepatitis and fibrosis in mice via the activation of IL1. Furthermore, it is critical for cell cycle progression in HCC. TLR4 and TLR9, on the other hand, have also been shown to possess antitumorigenic characteristics (Song et al. 2018). TLR analogs trigger the host immune response and generate an augmented lymphocyte response, which has led to greater therapeutic success when administered as adjuvants in conjunction with radiotherapy and immunotherapy. Thus, altering TLR activity by regulating the microbiome could be a therapeutic option for HCC. Most importantly, the pathways of TLR4 activation and their interaction with other signal transduction pathways in the HCC microenvironment will undoubtedly offer a viable novel approach for treating hepatocarcinogenesis.

7.3.3 Mycotoxicosis

Prolonged mycotoxicosis, including hepatocarcinogenesis, is triggered by the contamination of a multitude of mycotoxins. Such underlying mechanism typically entails changes to gene regulation, epigenetic modification management, and DNA adduct generation. Apparently, one of the variables driving mycotoxin-stimulated liver cancer is gut microbiome disruption (Liew and Mohd-Redzwan 2018). *Aspergillus flavus* as well as *Aspergillus parasiticus* yield the mycotoxin referred to as aflatoxin (AF). The most typical mycotoxin discovered in both human and animal food is AFB1. AFB1 is possibly the most powerful liver cancer carcinogen known to exist in animals and is categorized as a Group I carcinogenic agent, and it predominantly affects the liver (Muhammad et al. 2017). Considering AFB1's impact on the intestinal microbiome, there is scant evidence available. The possibility that AFB1 might modulate the microbiome in a dose-dependent fashion has been investigated earlier. It has been estimated that AFB1 reduced the phylogenetic diversity and enhanced even distribution within the bacterial communities of male F344 rats (Wang et al. 2016).

Through the disruption of microflora, toxic trace elements could also have an impact on facilitating HCC. According to research, the consumption of arsenic enhanced gut permeability and the population of Gram-negative microorganisms, leading to LPS translocation in the liver and inducing hepatocarcinogenesis (Choiniere and Wang 2016). Extensive research is required to ascertain the relationship between intestinal flora and mycotoxins as well as the implications of this association for the prevention and treatment of hepatocellular carcinoma.

7.3.4 Immune Checkpoint Inhibitors

Immune checkpoint blockade has become a promising strategy in the treatment of HCC. Tumor growth is related to immune suppression because malignant cells stimulate various immunological regulatory mechanisms to block therapeutic interventions (Darvin et al. 2018). Presently, inhibition of cytotoxic T-lymphocyte-associated protein 4 (CTLA-4), programmed cell death protein 1 (PD-1), and its ligand (PD-L1) are widely recognized and have been authorized for treating a variety of cancers. T lymphocytes exhibit CTLA-4 and PD-1 as glycoproteins on their membrane. After being triggered by their ligands, T-cell activity is suppressed, which may potentially result in T-cell programmed cell death. As a result, they are unable to produce the antitumor effects needed to eradicate tumor cells. This makes T cells a reasonable target in cancer immunotherapy (Ribas and Wolchok 2018).

Individuals suffering from advanced HCC and infected with HCV underwent initial testing for the monoclonal antibody tremelimumab, which targets CTLA-4. According to the observations, tremelimumab improved anti-HCV immunity in addition to its anticancer activity (Sangro et al. 2013). Apart from CTLA4, nivolumab has been used for blocking PD-1 in hepatocellular carcinoma (van Doorn et al. 2020).

Checkpoint inhibitors offer opportunities in the treatment of HCC. Another strategy is being carefully investigated right now. Clinical trials are being conducted to examine combinations of immune checkpoint inhibitors and tyrosine kinase inhibitors. Recent studies in melanoma patients demonstrated improved outcomes when checkpoint inhibitors were combined (Larkin et al. 2015). Furthermore, mounting research suggests that the gut microbiome impacts immune checkpoint potency. For instance, cyclophosphamide improves the passage of the upper digestive tract, allowing the buildup of *Barnesiella intestinihominis* in the colon as well as the remobilization of *Enterococcus hirae*, which normally resides in the intestinal tract to the spleen. These two events work in concert to stimulate antitumor effects (Daillère et al. 2016). *Bifidobacterium* species, which stimulate antigen-presenting cells, were found to be correlated with the treatment response of PD-1/PD-L1 inhibition. In order to enhance the anticancer potential of immune checkpoint inhibitors, altering gut microbial composition is thus a prospective therapy option for combating hepatocarcinogenesis.

7.3.5 HCC and Bacterial Metabolites

There is evidence that bacterial metabolites have a role in the impact of dysbiosis on the emergence of HCC and liver disease, presumably in a disease-specific manner (Caussy et al. 2019). Most tumors can synthesize, elongate, and desaturate fatty acids to enhance proliferation because they have an aberrantly active lipid metabolism (Röhrig and Schulze 2016). However, strategies that target fatty acid metabolism and, in particular, fatty acid desaturation, are only effective against specific subsets of cancer cells. This implies that many cancer cells have a flexibility in their fatty acid metabolism that has not been fully explored. Recent research has demonstrated that some cancer cells can utilize a different fatty acid desaturation process. It was discovered that palmitate desaturases to the uncommon fatty acid sapienate in primary human liver carcinomas and mouse hepatocellular carcinomas, supporting membrane production during proliferation. As a result, stearoyl-CoA desaturase 88-dependent known fatty acid desaturation route is circumvented by sapienate production in cancer cells. Treatment with the mixture of 1,3-dimethylbutylamin (DMBA) and ferredoxin (FD) led to the development of HCC in a mouse model for MAFLD. Additionally, a significant rise in Gram-positive bacterial species, particularly distinct *Clostridium* clusters, was reported (Yoshimoto et al. 2013).

7.4 Early Diagnosis and Possible Future Therapeutic Interventions for HCC

Although there is not a clear link between human research and the processes discovered in animal studies, they do seem to demonstrate dysbiosis patterns linked to increased inflammation, alteration of the intestinal barrier, and potential immune system impacts. Future therapies could therefore be focused on preserving a balanced microbiome, avoiding dysbiosis, and influencing downstream effector processes linked to the emergence of HCC. The use of a combination of broad-spectrum antibiotics resulted in sterilization of the gut, a decrease in tumor burden, and the prevention of HCC growth, which is a recurrent pattern seen in various animal HCC models used to explore the impact of dysbiosis on hepatocarcinogenesis (Yoshimoto et al. 2013; Xie et al. 2017). While it is not realistic to treat patients with broad-spectrum antibiotics continuously in clinical studies, there has been research into other antibiotics that can alter the gut flora to have a more favorable profile and have fewer adverse effects (Ponziani et al. 2017). It has been demonstrated that the non-absorbable antibiotic rifaximin causes the overgrowth of "beneficial" microorganisms like *Bifidobacterium*, *Faecalibacterium*, and *Lactobacillus* without significantly changing the microbiome composition. Rifaximin has been shown in at least one study to have a modest inhibitory effect on the development of HCC in DEN-exposed mice treated with rifaximin, despite not being extensively utilized in HCC animal model studies for gut decontamination (Dapito et al. 2012).

7.4.1 Probiotics and HCC

Probiotics are live microorganisms, which benefit the host if they are given in sufficient doses (Culligan et al. 2009). There are a very few studies on the use of probiotic supplements as a dietary strategy to lower the risk of HCC caused by aflatoxins. For instance, dietary treatment with probiotics like live *Lactobacillus rhamnosus* LC705 and *Propionibacterium freudenreichii subsp. Shermani*) could successfully reduce the excretion of aflatoxin-DNA adduct (AFB1-N7-guanine) in urine in a clinical investigation. Probiotic treatment decreased tumor incidence and levels of c-myc, bcl-2, cyclin D1, and rasp-21 in a rat study looking at the chemo-preventive effect of probiotic-fermented milk and chlorophyllin on AFB1-induced HCC (Kumar et al. 2011). This suggests that probiotics have the ability to protect against AFB1-induced hepatocarcinogenesis. In a different rat study, administration of VSL#3, which contains four *Lactobacilli*, three *Bifidobacteria*, and one *Streptococcus thermophilus subsp. Salivarius*, prevented the progression of cirrhosis to HCC by restoring gut homeostasis and reducing intestinal and hepatic inflammation. Probiotics slow down the development of HCC in mice (Li et al. (2016)). Giving mice with liver tumor injections, the probiotic cocktail Prohep, which contains *Lactobacillus rhamnosus* GG, *Escherichia coli Nissle* 1917, and heat-inactivated VSL#3, may change the composition of the gut microbiota and shrink liver tumors.

7.4.2 Prebiotics and HCC

Prebiotics are oligosaccharides that are not absorbed, such as lactulose. They work to promote the growth of helpful bacteria and inhibit the growth of harmful bacteria, so changing the balance of the gut microbiota, in addition, they can trigger the creation of SCFAs and control the immune response (Fotiadis et al. 2008). Prebiotics can therefore either treat or prevent HCC. Treatment with inulin-type fructans, according to Bindels et al. (2012), reduced the infiltration of hepatic BaF3 cells and reduced inflammation while raising the amount of portal propionate in mice given transplanting BaF3 cells that have been Bcr-Abl-transfected. In addition, propionate can prevent BaF3 cells from proliferating by the cAMP-dependent pathway in vitro. In addition to slowing down the growth of BaF3 cells, propionate also inhibits the proliferation of other human cancer cells via activating the Gi/Gq-protein-coupled receptor 2, also known as GPR43, which binds to propionate. Overall, these findings are favorable to prebiotics as a novel anticancer therapy approach.by the cAMP-dependent pathway in vitro.

7.4.3 HCC and Fecal Microbiota Transplantation (FMT)

FMT has recently been demonstrated to play a part in various cancer forms, including non-small cell lung cancer, colorectal cancer, and melanoma (Kang et al. 2017; Kang and Cai 2021). FMT in particular can improve the effectiveness of immunotherapy with checkpoint inhibitors against various kinds of cancer. FMT may also be used to treat or prevent HCC. In addition, FMT may be used to boost the anticancer effects of immune checkpoint inhibitors in the treatment of HCC.

7.4.4 Antibiotics and HCC

By preventing bacterial DNA transcription, protein synthesis, and other biological processes, antibiotics can stop the growth and translocation of intestinal flora and decrease the liver's uptake of pro-inflammatory signals from the leaky gut. Rifaximin is a semisynthetic, water-insoluble, nonsystemic antibiotic with minimal gastrointestinal absorption. Rifaximin is currently exclusively approved for the treatment of recurrent hepatic encephalopathy in people with liver cirrhosis (Bajaj et al. 2021; Caraceni et al. 2021). However, many hepatologists feel that rifaximin is a reasonable alternative to quinolones or other systemic antibiotics to prevent spontaneous bacterial peritonitis because of the broad-spectrum antibacterial effect (good effect on Gram-positive and Gram-negative aerobic and anaerobic bacteria). Future research will concentrate on using rifaximin to treat CLDs like liver cirrhosis, even if it is unclear how this may affect the development of HCC. Other antibiotics like norfloxacin and isoproterenol have also been shown to have effects. Isoproterenol can decrease the expression of carcinogenic gene products by decreasing STAT3 activation, but norfloxacin can prevent spontaneous bacterial peritonitis and lower mortality in patients with cirrhosis (Dai et al. 2015). Studies have also looked into using bacteriophages to control the microbiome for therapeutic objectives. Specific bacterial groups can be targeted by bacteriophages, which cause less disruption to symbiotic and untargeted bacteria (Budynek et al. 2010).

7.4.5 Gut Microbiome and Early Diagnosis of HCC

The noninvasiveness, high efficacy, and accuracy of the gut microbiota make it helpful for disease diagnosis. Given that they are linked to the emergence of liver diseases such cirrhosis/fibrosis and cancer, gut microbial changes may act as indicators of HCC (Meng et al. 2018). According to recent studies on the relationship between gut microbiota and HCC, it is critical to find microbiome biomarkers based on gut microbial changes in CLD to detect HCC at an early stage, according to current studies on the relationship between gut microbiota and HCC. Recently, in order to characterize the gut microbiome among HCC cases and evaluate the potential to use it as a noninvasive biomarker to diagnose HCC, the results in normal subjects were validated by characterizing the gut microbiota, identifying the biomarkers, and building the HCC classifiers among early HCC patients, cirrhosis patients, and normal subjects. As reported by Ren et al. (2019), early HCC with cirrhosis had a higher fecal microbial diversity than cirrhosis alone. Additionally, early HCC had a higher abundance of the phylum Actinobacteria than cirrhosis. As a result, early HCC had higher abundances of 13 taxa than cirrhosis, including *Parabacteroides* and *Gemmiger*. In early HCC patients compared to normal people, the abundances of butyrate-producing genera decreased, while those of LPS-producing genera rose. Additionally, the authors determined the ideal 30 microbiological markers between early HCC cases and non-HCC cases. Notably, it was confirmed that gut microbial indicators had a powerful potential for identifying early or even advanced HCC.

7.5 Conclusion and Future Directions

Over the past few decades, significant advancements have been achieved in our understanding of the gut microbiota. The gut microbiota has a significant role in regulating host physiological processes, including bile acid metabolism and immunological responses, and it can influence the onset of liver disorders via the gut-liver axis. Currently, it is believed that a long-term impact of CLDs led to the development of HCC. Growing evidence connects CLDs to particular dietary practices, changes in the gut microbiota's structure and function, and reductions in beneficial species and increases in dangerous species. Microbes will soon be used for both diagnosis and treatment. In the future, altering the gut microbiome will probably play a significant role in the detection and management of HCC. We need to fully grasp the structural and functional alterations in the gut microbiota in liver illnesses in order to fully appreciate their role to the development of HCC on the basis of the obvious related pathophysiology. However, it is important to note that the majority of our current knowledge of the gut microbiota is based on research using animal models and patient-provided fecal microbial samples. In conclusion, clinicians can create a more effective treatment plan for each patient based on gut microbial testing that also satisfies the criteria for "personalized medicine." In clinical settings, the microbiota is altered to control gut-liver signals that encourage the development of HCC, thereby improving patient survival and curative outcomes.

Acknowledgments The first author Ishfaq Hassan Mir acknowledges the Indian Council of Medical Research (ICMR), New Delhi, India, for financial assistance in the form of Senior Research Fellowship [ICMR-SRF; S.No. 45/17/2022-/BIO/BMS].

References

Akkız H (2021) The gut microbiome and hepatocellular carcinoma. J Gastrointest Cancer 52(4):1314–1319. https://doi.org/10.1007/s12029-021-00748-1

Albhaisi SAM, Bajaj JS (2021) The influence of the microbiome on NAFLD and NASH. Clin Liver Dis 17(1):15–18. https://doi.org/10.1002/cld.1010

Anstee QM, Reeves HL, Kotsiliti E, Govaere O, Heikenwalder M (2019) From NASH to HCC: current concepts and future challenges. Nat Rev Gastroenterol Hepatol 16(7):411–428. https://doi.org/10.1038/s41575-019-0145-7

Bäckhed F, Ding H, Wang T, Hooper LV, Koh GY, Nagy A, Semenkovich CF, Gordon JI (2004) The gut microbiota as an environmental factor that regulates fat storage. Proc Natl Acad Sci U S A 101(44):15718–15723. https://doi.org/10.1073/pnas.0407076101

Bäckhed F, Manchester JK, Semenkovich CF, Gordon JI (2007) Mechanisms underlying the resistance to diet-induced obesity in germ-free mice. Proc Natl Acad Sci U S A 104(3):979–984. https://doi.org/10.1073/pnas.0605374104

Bajaj JS, Sikaroodi M, Shamsaddini A, Henseler Z, Santiago-Rodriguez T, Acharya C, Fagan A, Hylemon PB, Fuchs M, Gavis E, Ward T, Knights D, Gillevet PM (2021) Interaction of bacterial metagenome and virome in patients with cirrhosis and hepatic encephalopathy. Gut 70(6):1162–1173. https://doi.org/10.1136/gutjnl-2020-322470

Bhat M, Arendt BM, Bhat V, Renner EL, Humar A, Allard JP (2016) Implication of the intestinal microbiome in complications of cirrhosis. World J Hepatol 8(27):1128–1136. https://doi.org/10.4254/wjh.v8.i27.1128

Bindels LB, Porporato P, Dewulf EM, Verrax J, Neyrinck AM, Martin JC, Scott KP, Buc Calderon P, Feron O, Muccioli GG, Sonveaux P, Cani PD, Delzenne NM (2012) Gut microbiota-derived propionate reduces cancer cell proliferation in the liver. Br J Cancer 107(8):1337–1344. https://doi.org/10.1038/bjc.2012.409

Budynek P, Dabrowska K, Skaradziński G, Górski A (2010) Bacteriophages and cancer. Arch Microbiol 192(5):315–320. https://doi.org/10.1007/s00203-010-0559-7

Caraceni P, Vargas V, Solà E, Alessandria C, de Wit K, Trebicka J, Angeli P, Mookerjee RP, Durand F, Pose E, Krag A, Bajaj JS, Beuers U, Ginès P, Liverhope Consortium (2021) The use of Rifaximin in patients with cirrhosis. Hepatology (Baltimore, Md.) 74(3):1660–1673. https://doi.org/10.1002/hep.31708

Caussy C, Ajmera VH, Puri P, Hsu CL, Bassirian S, Mgdsyan M, Singh S, Faulkner C, Valasek MA, Rizo E, Richards L, Brenner DA, Sirlin CB, Sanyal AJ, Loomba R (2019) Serum metabolites detect the presence of advanced fibrosis in derivation and validation cohorts of patients with non-alcoholic fatty liver disease. Gut 68(10):1884–1892. https://doi.org/10.1136/gutjnl-2018-317584

Chakraborty E, Sarkar D (2022) Emerging therapies for hepatocellular carcinoma (HCC). Cancers 14(11):2798. https://doi.org/10.3390/cancers14112798

Choiniere J, Wang L (2016) Exposure to inorganic arsenic can lead to gut microbe perturbations and hepatocellular carcinoma. Acta pharm Sin B 6(5):426–429. https://doi.org/10.1016/j.apsb.2016.07.011

Culligan EP, Hill C, Sleator RD (2009) Probiotics and gastrointestinal disease: successes, problems and future prospects. Gut Pathog 1(1):19. https://doi.org/10.1186/1757-4749-1-19

Dai X, Ahn KS, Kim C, Siveen KS, Ong TH, Shanmugam MK, Li F, Shi J, Kumar AP, Wang LZ, Goh BC, Magae J, Hui KM, Sethi G (2015) Ascochlorin, an isoprenoid antibiotic inhibits growth and invasion of hepatocellular carcinoma by targeting STAT3 signaling cascade through the induction of PIAS3. Mol Oncol 9(4):818–833. https://doi.org/10.1016/j.molonc.2014.12.008

Daillère R, Vétizou M, Waldschmitt N, Yamazaki T, Isnard C, Poirier-Colame V, Duong CPM, Flament C, Lepage P, Roberti MP, Routy B, Jacquelot N, Apetoh L, Becharef S, Rusakiewicz S, Langella P, Sokol H, Kroemer G, Enot D, Roux A et al (2016) Enterococcus hirae and Barnesiella intestinihominis facilitate cyclophosphamide-induced therapeutic immunomodulatory effects. Immunity 45(4):931–943. https://doi.org/10.1016/j.immuni.2016.09.009

Dapito DH, Mencin A, Gwak GY, Pradere JP, Jang MK, Mederacke I, Caviglia JM, Khiabanian H, Adeyemi A, Bataller R, Lefkowitch JH, Bower M, Friedman R, Sartor RB, Rabadan R, Schwabe RF (2012) Promotion of hepatocellular carcinoma by the intestinal microbiota and TLR4. Cancer Cell 21(4):504–516. https://doi.org/10.1016/j.ccr.2012.02.007

Darvin P, Toor SM, Sasidharan Nair V, Elkord E (2018) Immune checkpoint inhibitors: recent progress and potential biomarkers. Exp Mol Med 50(12):1–11. https://doi.org/10.1038/s12276-018-0191-1

Davis CD (2016) The gut microbiome and its role in obesity. Nutr Today 51(4):167–174. https://doi.org/10.1097/NT.0000000000000167

Dieterich W, Schink M, Zopf Y (2018) Microbiota in the gastrointestinal tract. Med Sci (Basel, Switzerland) 6(4):116. https://doi.org/10.3390/medsci6040116

Dossa AY, Escobar O, Golden J, Frey MR, Ford HR, Gayer CP (2016) Bile acids regulate intestinal cell proliferation by modulating EGFR and FXR signaling. Am J Physiol Gastrointest Liver Physiol 310(2):G81–G92. https://doi.org/10.1152/ajpgi.00065.2015

Fotiadis CI, Stoidis CN, Spyropoulos BG, Zografos ED (2008) Role of probiotics, prebiotics and synbiotics in chemoprevention for colorectal cancer. World J Gastroenterol 14(42):6453–6457. https://doi.org/10.3748/wjg.14.6453

Gambarin-Gelwan M, Kinkhabwala SV, Schiano TD, Bodian C, Yeh HC, Futterweit W (2007) Prevalence of nonalcoholic fatty liver disease in women with polycystic ovary syndrome. Clin Gastroenterol Hepatol 5(4):496–501. https://doi.org/10.1016/j.cgh.2006.10.010

Gao C, Qiao T, Zhang B, Yuan S, Zhuang X, Luo Y (2018) TLR9 signaling activation at different stages in colorectal cancer and NF-kappaB expression. Onco Targets Ther 11:5963–5971. https://doi.org/10.2147/OTT.S174274

Gill SR, Pop M, Deboy RT, Eckburg PB, Turnbaugh PJ, Samuel BS, Gordon JI, Relman DA, Fraser-Liggett CM, Nelson KE (2006) Metagenomic analysis of the human distal gut microbiome. Science (New York, N.Y.) 312(5778):1355–1359. https://doi.org/10.1126/science.1124234

Grąt M, Wronka KM, Krasnodębski M, Masior Ł, Lewandowski Z, Kosińska I, Grąt K, Stypułkowski J, Rejowski S, Wasilewicz M, Gałęcka M, Szachta P, Krawczyk M (2016) Profile of gut microbiota associated with the presence of hepatocellular cancer in patients with liver cirrhosis. Transplant Proc 48(5):1687–1691. https://doi.org/10.1016/j.transproceed.2016.01.077

Hassan S, Kaakinen MA, Draisma H, Zudina L, Ganie MA, Rashid A, Balkhiyarova Z, Kiran GS, Vogazianos P, Shammas C, Selvin J, Antoniades A, Demirkan A, Prokopenko I (2022) Bifidobacterium is enriched in gut microbiome of Kashmiri women with polycystic ovary syndrome. Genes 13(2):379. https://doi.org/10.3390/genes13020379

Hrncir T, Hrncirova L, Kverka M, Hromadka R, Machova V, Trckova E, Kostovcikova K, Kralickova P, Krejsek J, Tlaskalova-Hogenova H (2021) Gut microbiota and NAFLD: pathogenetic mechanisms, microbiota signatures, and therapeutic interventions. Microorganisms 9(5):957. https://doi.org/10.3390/microorganisms9050957

Jia ET, Liu ZY, Pan M, Lu JF, Ge QY (2019) Regulation of bile acid metabolism-related signaling pathways by gut microbiota in diseases. J Zhejiang Univ Sci B 20(10):781–792. https://doi.org/10.1631/jzus.B1900073

Jiang C, Xie C, Li F, Zhang L, Nichols RG, Krausz KW, Cai J, Qi Y, Fang ZZ, Takahashi S, Tanaka N, Desai D, Amin SG, Albert I, Patterson AD, Gonzalez FJ (2015) Intestinal farnesoid X receptor signaling promotes non-alcoholic fatty liver disease. J Clin Invest 125(1):386–402. https://doi.org/10.1172/JCI76738

Jobira B, Frank DN, Silveira LJ, Pyle L, Kelsey MM, Garcia-Reyes Y, Robertson CE, Ir D, Nadeau KJ, Cree-Green M (2021) Hepatic steatosis relates to gastrointestinal microbiota changes in obese girls with polycystic ovary syndrome. PLoS One 16(1):e0245219. https://doi.org/10.1371/journal.pone.0245219

Kang YB, Cai Y (2021) Faecal microbiota transplantation enhances efficacy of immune checkpoint inhibitors therapy against cancer. World J Gastroenterol 27(32):5362–5375. https://doi.org/10.3748/wjg.v27.i32.5362

Kang Y, Pan W, Cai Y (2017) Gut microbiota and colorectal cancer: insights into pathogenesis for novel therapeutic strategies. Darmflora und Darmkrebs: Einblicke in die Pathogenese für neue-Therapiestrategien. Z Gastroenterol 55(9):872–880. https://doi.org/10.1055/s-0043-116387

Kang Y, Cai Y, Yang Y (2021) The gut microbiome and hepatocellular carcinoma: implications for early diagnostic biomarkers and novel therapies. Liver Cancer 11(2):113–125. https://doi.org/10.1159/000521358

Komiyama S, Yamada T, Takemura N, Kokudo N, Hase K, Kawamura YI (2021) Profiling of tumour-associated microbiota in human hepatocellular carcinoma. Sci Rep 11(1):10589. https://doi.org/10.1038/s41598-021-89963-1

Kosnicki KL, Penprase JC, Cintora P, Torres PJ, Harris GL, Brasser SM, Kelley ST (2019) Effects of moderate, voluntary ethanol consumption on the rat and human gut microbiome. Addict Biol 24(4):617–630. https://doi.org/10.1111/adb.12626

Kumar M, Verma V, Nagpal R, Kumar A, Gautam SK, Behare PV, Grover CR, Aggarwal PK (2011) Effect of probiotic fermented milk and chlorophyllin on gene expressions and genotoxicity during AFB_1-induced hepatocellular carcinoma. Gene 490(1-2):54–59. https://doi.org/10.1016/j.gene.2011.09.003

Larkin J, Chiarion-Sileni V, Gonzalez R, Grob JJ, Cowey CL, Lao CD, Schadendorf D, Dummer R, Smylie M, Rutkowski P, Ferrucci PF, Hill A, Wagstaff J, Carlino MS, Haanen JB, Maio M, Marquez-Rodas I, McArthur GA, Ascierto PA, Long GV et al (2015) Combined Nivolumab and Ipilimumab or monotherapy in untreated melanoma. N Engl J Med 373(1):23–34. https://doi.org/10.1056/NEJMoa1504030

Le Noci V, Bernardo G, Bianchi F, Tagliabue E, Sommariva M, Sfondrini L (2021) Toll like receptors as sensors of the tumor microbial dysbiosis: implications in cancer progression. Front Cell Dev Biol 9:732192. https://doi.org/10.3389/fcell.2021.732192

Lee NY, Suk KT (2020) The role of the gut microbiome in liver cirrhosis treatment. Int J Mol Sci 22(1):199. https://doi.org/10.3390/ijms22010199

Levitt MD, Li R, DeMaster EG, Elson M, Furne J, Levitt DG (1997) Use of measurements of ethanol absorption from stomach and intestine to assess human ethanol metabolism. Am J Phys 273(4):G951–G957. https://doi.org/10.1152/ajpgi.1997.273.4.G951

Li J, Sung CY, Lee N, Ni Y, Pihlajamäki J, Panagiotou G, El-Nezami H (2016) Probiotics modulated gut microbiota suppresses hepatocellular carcinoma growth in mice. Proc Natl Acad Sci U S A 113(9):E1306–E1315. https://doi.org/10.1073/pnas.1518189113

Li Y, Xu A, Jia S, Huang J (2019) Recent advances in the molecular mechanism of sex disparity in hepatocellular carcinoma. Oncol Lett 17(5):4222–4228. https://doi.org/10.3892/ol.2019.10127

Liao M, Zhao J, Wang T, Duan J, Zhang Y, Deng X (2011) Role of bile salt in regulating Mcl-1 phosphorylation and chemoresistance in hepatocellular carcinoma cells. Mol Cancer 10:44. https://doi.org/10.1186/1476-4598-10-44

Liew WP, Mohd-Redzwan S (2018) Mycotoxin: its impact on gut health and microbiota. Front Cell Infect Microbiol 8:60. https://doi.org/10.3389/fcimb.2018.00060

Loo TM, Kamachi F, Watanabe Y, Yoshimoto S, Kanda H, Arai Y, Nakajima-Takagi Y, Iwama A, Koga T, Sugimoto Y, Ozawa T, Nakamura M, Kumagai M, Watashi K, Taketo MM, Aoki T, Narumiya S, Oshima M, Arita M, Hara E et al (2017) Gut microbiota promotes obesity-associated liver cancer through PGE_2-mediated suppression of antitumor immunity. Cancer Discov 7(5):522–538. https://doi.org/10.1158/2159-8290.CD-16-0932

Luo W, Guo S, Zhou Y, Zhao J, Wang M, Sang L, Chang B, Wang B (2022) Hepatocellular carcinoma: how the gut microbiota contributes to pathogenesis, diagnosis, and therapy. Front Microbiol 13:873160. https://doi.org/10.3389/fmicb.2022.873160

Ma C, Han M, Heinrich B, Fu Q, Zhang Q, Sandhu M, Agdashian D, Terabe M, Berzofsky JA, Fako V, Ritz T, Longerich T, Theriot CM, McCulloch JA, Roy S, Yuan W, Thovarai V, Sen SK, Ruchirawat M, Korangy F et al (2018) Gut microbiome-mediated bile acid metabolism regulates liver cancer via NKT cells. Science (New York, N.Y.) 360(6391):eaan5931. https://doi.org/10.1126/science.aan5931

Meng X, Li S, Li Y, Gan RY, Li HB (2018) Gut Microbiota's relationship with liver disease and role in hepatoprotection by dietary natural products and probiotics. Nutrients 10(10):1457. https://doi.org/10.3390/nu10101457

Michelotti GA, Machado MV, Diehl AM (2013) NAFLD, NASH and liver cancer. Nat Rev Gastroenterol Hepatol 10(11):656–665. https://doi.org/10.1038/nrgastro.2013.183

Mir IH, Guha S, Behera J, Thirunavukkarasu C (2021) Targeting molecular signal transduction pathways in hepatocellular carcinoma and its implications for cancer therapy. Cell Biol Int 45(11):2161–2177. https://doi.org/10.1002/cbin.11670

Mir IH, Jyothi KC, Thirunavukkarasu C (2022) The prominence of potential biomarkers in the diagnosis and management of hepatocellular carcinoma: current scenario and future anticipation. J Cell Biochem 123(10):1607–1623. https://doi.org/10.1002/jcb.30190

Modica S, Petruzzelli M, Bellafante E, Murzilli S, Salvatore L, Celli N, Di Tullio G, Palasciano G, Moustafa T, Halilbasic E, Trauner M, Moschetta A (2012) Selective activation of nuclear bile acid receptor FXR in the intestine protects mice against cholestasis. Gastroenterology 142(2):355–65.e654. https://doi.org/10.1053/j.gastro.2011.10.028

Muhammad I, Sun X, Wang H, Li W, Wang X, Cheng P, Li S, Zhang X, Hamid S (2017) Curcumin successfully inhibited the computationally identified CYP2A6 enzyme-mediated bioactivation of aflatoxin B1 in arbor acres broiler. Front Pharmacol 8:143. https://doi.org/10.3389/fphar.2017.00143

Ni J, Huang R, Zhou H, Xu X, Li Y, Cao P, Zhong K, Ge M, Chen X, Hou B, Yu M, Peng B, Li Q, Zhang P, Gao Y (2019) Analysis of the relationship between the degree of dysbiosis in gut microbiota and prognosis at different stages of primary hepatocellular carcinoma. Front Microbiol 10:1458. https://doi.org/10.3389/fmicb.2019.01458

Peterson LW, Artis D (2014) Intestinal epithelial cells: regulators of barrier function and immune homeostasis. Nat Rev Immunol 14(3):141–153. https://doi.org/10.1038/nri3608

Pijls KE, Jonkers DM, Elamin EE, Masclee AA, Koek GH (2013) Intestinal epithelial barrier function in liver cirrhosis: an extensive review of the literature. Liver Int 33(10):1457–1469. https://doi.org/10.1111/liv.12271

Ponziani FR, Zocco MA, D'Aversa F, Pompili M, Gasbarrini A (2017) Eubiotic properties of rifaximin: disruption of the traditional concepts in gut microbiota modulation. World J Gastroenterol 23(25):4491

Ponziani FR, Bhoori S, Castelli C, Putignani L, Rivoltini L, Del Chierico F, Sanguinetti M, Morelli D, ParoniSterbini F, Petito V, Reddel S, Calvani R, Camisaschi C, Picca A, Tuccitto A, Gasbarrini A, Pompili M, Mazzaferro V (2019) Hepatocellular carcinoma is associated with gut microbiota profile and inflammation in nonalcoholic fatty liver disease. Hepatology (Baltimore, Md.) 69(1):107–120. https://doi.org/10.1002/hep.30036

Pradere JP, Troeger JS, Dapito DH, Mencin AA, Schwabe RF (2010) Toll-like receptor 4 and hepatic fibrogenesis. Semin Liver Dis 30(3):232–244. https://doi.org/10.1055/s-0030-1255353

Rattan P, Minacapelli CD, Rustgi V (2020) The microbiome and hepatocellular carcinoma. Liver Transpl 26(10):1316–1327. https://doi.org/10.1002/lt.25828

Ren Z, Li A, Jiang J, Zhou L, Yu Z, Lu H, Xie H, Chen X, Shao L, Zhang R, Xu S, Zhang H, Cui G, Chen X, Sun R, Wen H, Lerut JP, Kan Q, Li L, Zheng S (2019) Gut microbiome analysis as a tool towards targeted non-invasive biomarkers for early hepatocellular carcinoma. Gut 68(6):1014–1023. https://doi.org/10.1136/gutjnl-2017-315084

Ribas A, Wolchok JD (2018) Cancer immunotherapy using checkpoint blockade. Science (New York, N.Y.) 359(6382):1350–1355. https://doi.org/10.1126/science.aar4060

Ridlon JM, Kang DJ, Hylemon PB (2006) Bile salt biotransformations by human intestinal bacteria. J Lipid Res 47(2):241–259. https://doi.org/10.1194/jlr.R500013-JLR200

Rinninella E, Raoul P, Cintoni M, Franceschi F, Miggiano GAD, Gasbarrini A, Mele MC (2019) What is the healthy gut microbiota composition? A changing ecosystem across age, environment, diet, and diseases. Microorganisms 7(1):14. https://doi.org/10.3390/microorganisms7010014

Röhrig F, Schulze A (2016) The multifaceted roles of fatty acid synthesis in cancer. Nat Rev Cancer 16(11):732–749. https://doi.org/10.1038/nrc.2016.89

Sangro B, Gomez-Martin C, de la Mata M, Iñarrairaegui M, Garralda E, Barrera P, Riezu-Boj JI, Larrea E, Alfaro C, Sarobe P, Lasarte JJ, Pérez-Gracia JL, Melero I, Prieto J (2013) A clinical trial of CTLA-4 blockade with tremelimumab in patients with hepatocellular carcinoma and chronic hepatitis C. J Hepatol 59(1):81–88. https://doi.org/10.1016/j.jhep.2013.02.022

Schwabe RF, Greten TF (2020) Gut microbiome in HCC - mechanisms, diagnosis and therapy. J Hepatol 72(2):230–238. https://doi.org/10.1016/j.jhep.2019.08.016

Schwenger KJ, Clermont-Dejean N, Allard JP (2019) The role of the gut microbiome in chronic liver disease: the clinical evidence revised. JHEP Rep 1(3):214–226. https://doi.org/10.1016/j.jhepr.2019.04.004

Song IJ, Yang YM, Inokuchi-Shimizu S, Roh YS, Yang L, Seki E (2018) The contribution of toll-like receptor signaling to the development of liver fibrosis and cancer in hepatocyte-specific TAK1-deleted mice. Int J Cancer 142(1):81–91. https://doi.org/10.1002/ijc.31029

Stärkel P, Leclercq S, de Timary P, Schnabl B (2018) Intestinal dysbiosis and permeability: the yin and yang in alcohol dependence and alcoholic liver disease. Clin Sci (London, England: 1979) 132(2):199–212. https://doi.org/10.1042/CS20171055

Takase T, Toyoda T, Kobayashi N, Inoue T, Ishijima T, Abe K, Kinoshita H, Tsuchiya Y, Okada S (2021) Dietary iso-α-acids prevent acetaldehyde-induced liver injury through Nrf2-mediated gene expression. PLoS One 16(2):e0246327. https://doi.org/10.1371/journal.pone.0246327

Tanaka H, Doesburg K, Iwasaki T, Mierau I (1999) Screening of lactic acid bacteria for bile salt hydrolase activity. J Dairy Sci 82(12):2530–2535. https://doi.org/10.3168/jds.S0022-0302(99)75506-2

Taylor SL, Rogers GB, Chen AC, Burr LD, McGuckin MA, Serisier DJ (2015) Matrix metalloproteinases vary with airway microbiota composition and lung function in non-cystic fibrosis bronchiectasis. Ann Am Thorac Soc 12(5):701–707. https://doi.org/10.1513/AnnalsATS.201411-513OC

Thursby E, Juge N (2017) Introduction to the human gut microbiota. Biochem J 474(11):1823–1836. https://doi.org/10.1042/BCJ20160510

van Doorn DJ, Takkenberg RB, Klümpen HJ (2020) Immune checkpoint inhibitors in hepatocellular carcinoma: an overview. Pharmaceuticals (Basel, Switzerland) 14(1):3. https://doi.org/10.3390/ph14010003

Vassilatou E (2014) Nonalcoholic fatty liver disease and polycystic ovary syndrome. World J Gastroenterol 20(26):8351–8363. https://doi.org/10.3748/wjg.v20.i26.8351

Wahlström A, Sayin SI, Marschall HU, Bäckhed F (2016) Intestinal crosstalk between bile acids and microbiota and its impact on host metabolism. Cell Metab 24(1):41–50. https://doi.org/10.1016/j.cmet.2016.05.005

Wang M, Donovan SM (2015) Human microbiota-associated swine: current progress and future opportunities. ILAR J 56(1):63–73. https://doi.org/10.1093/ilar/ilv006

Wang X, Fu X, Van Ness C, Meng Z, Ma X, Huang W (2013) Bile acid receptors and liver cancer. Curr Pathobiol Rep 1(1):29–35. https://doi.org/10.1007/s40139-012-0003-6

Wang J, Tang L, Glenn TC, Wang JS (2016) Aflatoxin B1 induced compositional changes in gut microbial communities of male F344 rats. Toxicol Sci 150(1):54–63. https://doi.org/10.1093/toxsci/kfv259

Wang R, Tang R, Li B, Ma X, Schnabl B, Tilg H (2021) Gut microbiome, liver immunology, and liver diseases. Cell Mol Immunol 18(1):4–17. https://doi.org/10.1038/s41423-020-00592-6

Xie G, Wang X, Liu P, Wei R, Chen W, Rajani C, Hernandez BY, Alegado R, Dong B, Li D, Jia W (2016) Distinctly altered gut microbiota in the progression of liver disease. Oncotarget 7(15):19355–19366. https://doi.org/10.18632/oncotarget.8466

Xie G, Wang X, Zhao A, Yan J, Chen W, Jiang R, Ji J, Huang F, Zhang Y, Lei S, Ge K, Zheng X, Rajani C, Alegado RA, Liu J, Liu P, Nicholson J, Jia W (2017) Sex-dependent effects on gut microbiota regulate hepatic carcinogenic outcomes. Sci Rep 7:45232. https://doi.org/10.1038/srep45232

Yerushalmi B, Dahl R, Devereaux MW, Gumpricht E, Sokol RJ (2001) Bile acid-induced rat hepatocyte apoptosis is inhibited by antioxidants and blockers of the mitochondrial permeability transition. Hepatology (Baltimore, Md.) 33(3):616–626. https://doi.org/10.1053/jhep.2001.22702

Yoshimoto S, Loo TM, Atarashi K, Kanda H, Sato S, Oyadomari S, Iwakura Y, Oshima K, Morita H, Hattori M, Honda K, Ishikawa Y, Hara E, Ohtani N (2013) Obesity-induced gut microbial metabolite promotes liver cancer through senescence secretome. Nature 499(7456):97–101. https://doi.org/10.1038/nature12347

Yu LX, Schwabe RF (2017) The gut microbiome and liver cancer: mechanisms and clinical translation. Nat Rev Gastroenterol Hepatol 14(9):527–539. https://doi.org/10.1038/nrgastro.2017.72

Yu LX, Yan HX, Liu Q, Yang W, Wu HP, Dong W, Tang L, Lin Y, He YQ, Zou SS, Wang C, Zhang HL, Cao GW, Wu MC, Wang HY (2010) Endotoxin accumulation prevents carcinogen-induced apoptosis and promotes liver tumorigenesis in rodents. Hepatology (Baltimore, Md.) 52(4):1322–1333. https://doi.org/10.1002/hep.23845

Zhang HL, Yu LX, Yang W, Tang L, Lin Y, Wu H, Zhai B, Tan YX, Shan L, Liu Q, Chen HY, Dai RY, Qiu BJ, He YQ, Wang C, Zheng LY, Li YQ, Wu FQ, Li Z, Yan HX et al (2012) Profound impact of gut homeostasis on chemically-induced pro-tumorigenic inflammation and hepatocarcinogenesis in rats. J Hepatol 57(4):803–812. https://doi.org/10.1016/j.jhep.2012.06.011

Zhang C, Yang M, Ericsson AC (2020) The potential gut microbiota-mediated treatment options for liver cancer. Front Oncol 10:524205. https://doi.org/10.3389/fonc.2020.524205

Zheng Y, Wang T, Tu X, Huang Y, Zhang H, Tan D, Jiang W, Cai S, Zhao P, Song R, Li P, Qin N, Fang W (2019) Gut microbiome affects the response to anti-PD-1 immunotherapy in patients with hepatocellular carcinoma. J Immunother Cancer 7(1):193. https://doi.org/10.1186/s40425-019-0650-9

Zheng R, Wang G, Pang Z, Ran N, Gu Y, Guan X, Yuan Y, Zuo X, Pan H, Zheng J, Wang F (2020) Liver cirrhosis contributes to the disorder of gut microbiota in patients with hepatocellular carcinoma. Cancer Med 9(12):4232–4250. https://doi.org/10.1002/cam4.3045

Zhu L, Baker SS, Gill C, Liu W, Alkhouri R, Baker RD, Gill SR (2013) Characterization of gut microbiomes in non-alcoholic steatohepatitis (NASH) patients: a connection between endogenous alcohol and NASH. Hepatology (Baltimore, Md) 57(2):601–609. https://doi.org/10.1002/hep.26093

Influence of Intestinal Microbiomes on COVID Progression and Its Effects by Immunotherapeutic Modulation

8

Niharikha Mukala, Sudhakar Pola, and Anusha Konatala

Abstract

The microbiota plays a crucial role in regulating various physiological functions and pathological conditions within the human body. An important aspect of COVID-19 pathogenesis is comprehending how various infections in the body, including COVID-19, affect and influence the microbiome. We may develop better diagnostics and strategies against COVID-19 infection by examining the association between the intestinal and respiratory microbiota. To take a broader scientific approach, we must answer several key questions, such as how microbiome diversity and composition vary from person to person, how accumulated microbiota can benefit individuals over time, and what factors contribute to microbiota development. Analyzing the signaling molecules that mediate biological mechanisms for immune responses between the host and microbiota and among microbiota may provide valuable insight. Several potential therapies to improve the microbiome or target specific microbiota include phage therapy, FMT, prebiotics, probiotics, and synbiotics. Modulating the neonatal microbiome has been a challenging goal to increase efficacy recently. However, much more research is required to engineer microbiome therapeutics. This chapter provides an overview of existing challenges and strategies to make the necessary modifications to restore the naive gut.

Keywords

COVID-19 · Microbiome · Viral infection · Dysbiosis · Microbiota

N. Mukala · S. Pola (✉)
Department of Biotechnology, AUCST, Andhra University, Visakhapatnam, India
e-mail: sudhakar@andhrauniversity.edu.in

A. Konatala
Department of Medical Biotechnology, University of Windsor, Windsor, ON, Canada

P. Veera Bramhachari (ed.), *Human Microbiome in Health, Disease, and Therapy*, https://doi.org/10.1007/978-981-99-5114-7_8

8.1 Microbiome

The word "microbiome" is the collection of genomes from all the microorganisms in an environment. The variation in the microbiota depends on the location of specific microbiota in the body. For instance, the gut microbiota differs from the skin microbiota (Sanapala and Pola 2021). The microbiome colonizes the human body in different roles and varies between individuals. The microbiome plays a major role in various physicochemical activities like forming tight junctions between the cells to maintain the integrity of the tissues, enhancing the production of T cells to boost the immune system, and controlling the body's metabolism (Pola and Padi 2021). The human microbiome is currently associated with several disorders, including inflammatory bowel disease, type 2 diabetes, Parkinson's disease, hypothyroidism, colorectal cancer, COPD, and rheumatoid arthritis. Other respiratory diseases are directly or indirectly associated with specific microorganism patterns (Yamamoto et al. 2021).

Microbiome formation is influenced by genetics (Mohan and Sudhakar 2022), delivery mode, infant feeding, nutrition, antibiotic administration, age, medication, vaccination (health complications), and geographical and seasonal differences. Diseased conditions in the human body impact microbiome composition. The areas where the microbiota plays a key role in humans are nutrition, immunity, behavior, and disease (Harper et al. 2021). The beneficial microbiota in the gut can help digest food that humans can't break down. The harmful microbiota can damage the immune system and make the body prone to various conditions like gastrointestinal diseases. As an example, the presence of the microbiome genus Pseudomonadales and Streptococcus is linked to human upper respiratory tract infections in patients.

Consuming certain live bacterial strains confers health benefits on the human body. These bacteria are known as probiotics, and the administration of probiotics helps reduce many bacterial infections. Some probiotics have also been engineered to kill pathogenic bacteria, known as "smart microbes" (Ronda et al. 2019). For instance, *Lactococcus lactis* was modified to create molecules that would attack *Enterococcus faecium*, a bacterium linked to the onset of meningitis in infants. Researchers have laid the groundwork for a future technique to alter microbes in an individual's gut.

The production of smart microbes as probiotics can be engineered using MAGIC (metagenomic alteration of gut microbiome by in situ conjugation). This method includes oral ingestion of bacteria that can transfer DNA with specific traits to the bacterial microbiome already present in the body (Baghbani et al. 2020). This method needs a lot of improvements to make use of it to treat infectious diseases in humans, increase the persistence of DNA in the gut, and ensure that the DNA can only be transferred to targeted, nonpathogenic strains of the microbiome.

8.2 Microbiomes of Infectious Disease

Microbiome research focuses on the microbial communities' behavior, interactions, and functions within a specified environment. The microbiota has enzymes that cannot be coded by the human genome but are necessary to fulfill some physiologic tasks, like the digestive enzymes that break down substances like polysaccharides and polyphenols and hydrolytic enzymes to regulate and balance cellular metabolism. The microbiota's function within a healthy host involves influencing diverse pathogens through colonization resistance, allowing the host's immune system to participate in immune cell differentiation, promoting the proliferation of granulocyte/monocyte progenitors, activating innate lymphoid cells and myeloid cells, triggering proinflammatory T- and B-cell responses, and initiating pre-inflammatory T- and B-cell secretion of SIGA (secretory IgA). Both secretory IgA and gut inflammation modify the microbiota composition, resulting in shifts in microbial proportions and increased pathogen growth.

Firstly, the microbiota induces alpha-defensin, beta-defensin, C-type lectins, secretory IgA, and other AMPs (antimicrobial peptides) that affect the immune system (innate and adaptive) through intestinal epithelia and paneth cell receptors; macrophage cell receptors such as TLRs or NLRs; and CCR6. TLRs are important for developing the mucosal and intestinal immune systems, decreasing inflammatory responses, and promoting immunological tolerance to the necessary microbial components. NLRs regulate IL-18 levels, immune responses (Konatala et al. 2021), dysbiosis, and intestinal hyperplasia. When antigens of the microbiota bind to these receptors, they start a chain reaction of signaling pathways that release antimicrobial compounds like defensins and stimulate T cells like T-helper 1 and 17 to make IL-1, IL-15, IL-17, IL-22, etc. They also stimulate B cells to produce antibodies (Harper et al. 2021) (Fig. 8.1).

8.3 Effects of the Microbiome in the COVID-19 Infection

COVID-19 is a respiratory disease ranging from mild (cough and/or fever) to severe pneumonia (ARDS and multiple organ failure). Angiotensin-converting enzyme 2 (ACE2) is the receptor to which the viral spike binds. This receptor is expressed on the respiratory and gastrointestinal epithelium, leading to changes in its microbiome composition during the infection (Liu et al. 2022).

SARS-CoV-2 is an enveloped, single-stranded RNA virus consisting of structural proteins like the nucleocapsid, membrane, envelope, and spike proteins. The viral particles mediated by the S glycoprotein attach and fuse to the host cell membrane and get inserted into the virion membrane in multiple copies with a crown-like appearance. The S protein of coronaviruses is cleaved into S1 and S2 subunits by proprotein convertases during their biosynthesis or after reaching their target site.

ACE2 is an 805-amino acid carboxypeptidase, and the downregulation of ACE2 leads to a severe disturbance in the renin-angiotensin-aldosterone system. As shown in Fig. 8.2, the virus bound to the ACE2 receptor induces conformational changes in

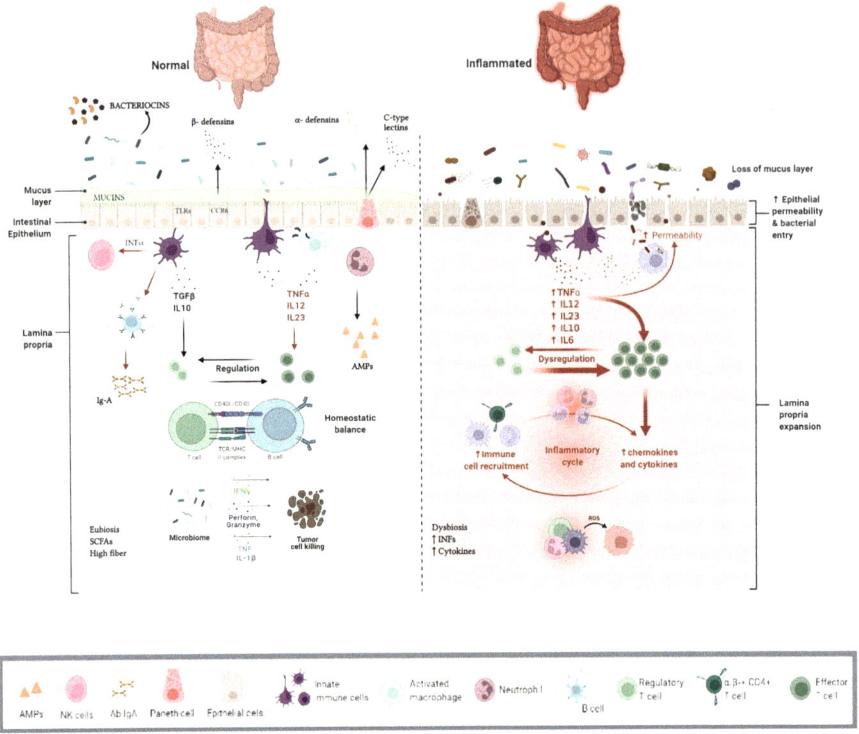

Fig. 8.1 Compares healthy and diseased guts in humans during COVID-19

both S1 and S2 subunits. The conformational changes in the S1 subunit expose the S2 cleavage site of the S2 subunit. During insufficient transmembrane protease, serine 2 (TMPRSS2), the virus-ACE2 complex triggers clathrin-mediated endocytosis, where cathepsins perform S2′ cleavage in endolysosomes. Cleavage of the S2′ site in the presence of TMPRSS2 exposes the fusion peptide, and the separation of S1 from S2 induces conformational changes in the S2 subunit and triggers fusion to the membrane. The fusion between the viral membrane and cellular membranes forms a fusion pore through which viral RNA is released into the host cell cytoplasm.

The other molecules that serve as receptors in SARS-CoV-2 infection are C-type lectins, DC-SIGN and L-SIGN, AXL and TIM1, phosphatidylserine receptors TIM and TAM, and CD147 (a transmembrane glycoprotein). Lectins bind to the surface (glycans) of the virion, promoting viral entry and intracellular adhesion. TIM and TAM bind to phosphatidylserine on the virion membrane to promote the entry of enveloped viruses. Increased viral entry was observed with higher levels of CD147 in SARS-CoV-2 and is a potential risk factor as it was upregulated in diabetic and obese patients. A remarkable modification of the nasopharyngeal microbiome is noticed in this study and assumed to have a proportional dysbiosis effect from the onset, treatment, and reduction of COVID-19 infection (Hoque et al. 2021) (Fig. 8.3).

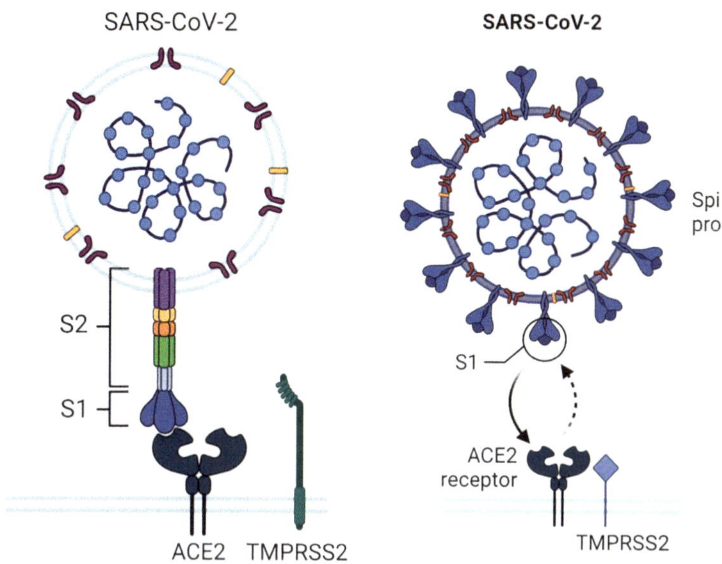

Fig. 8.2 The interaction of the S protein of SARS-COV-2 with the ACE2 receptor

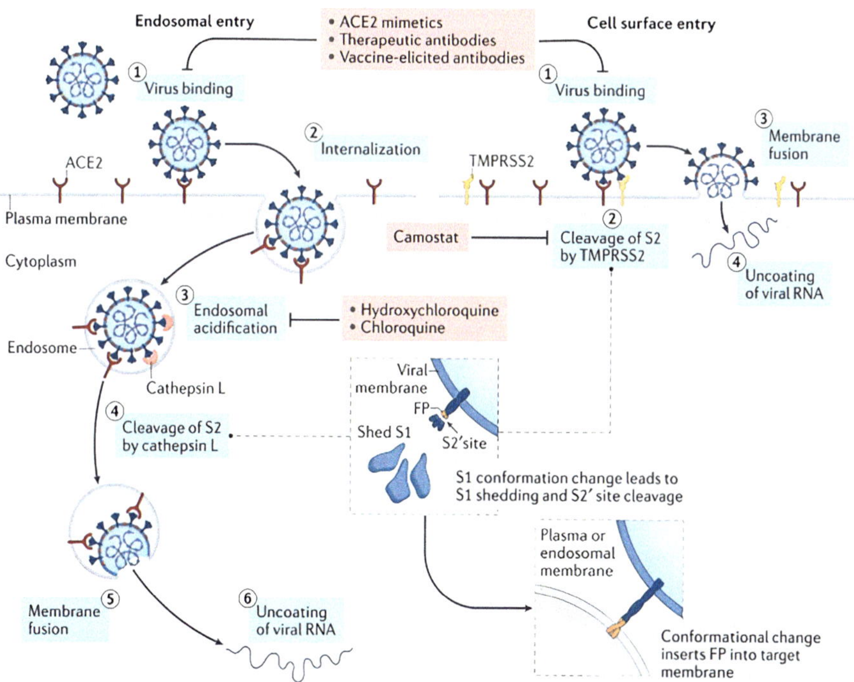

Fig. 8.3 Mechanism of SARS-COV-2 entry into cells (Jackson et al. 2022)

8.4 Variations in the Upper and Lower Respiratory Tract Microbiomes in the COVID-19 Patients

Upper vs. lower respiratory tract sampling, collection time, point or stage of infection, treatment with broad-spectrum antibiotics, invasive mechanical ventilation, and prolonged hospitalization are the vital factors that can determine the range and composition of the microbiome at specific sites. The respiratory tract microbiome of COVID-19 patients was analyzed in two experimental studies (using nasopharyngeal swabs and sputum) and detected using next-generation sequencing.

The oral and upper respiratory microbiomes include *Candida albicans* and *human alpha-herpesvirus*, which are the most common, and coinfection of COVID-19 with other viruses like *Influenza A/B*, enteroviruses, *Aspergillus* respiratory syncytial virus, and *Veillonella species* was found in some of the patients. Certain research studies identified microbiome makeup within COVID-19 throat samples, encompassing Haemophilus parainfluenzae, Neisseria cinerea, rhinovirus, Streptococcus mitis, Streptococcus bovis, Leptotrichia buccalis, and Rothia mucilaginosa. Additionally, bronchoalveolar samples exhibited Acinetobacter, Pseudomonas, Enterococcus, Lactobacillus, and Chryseobacterium. Some studies show that the diversity of pharyngeal microbiome (e.g., Bacteroidetes, Proteobacteria) decreased in older adult patients than in younger adult patients suffering from COVID-19 (Gaibani et al. 2021).

The lower respiratory microbiome in COVID-19 patients showed potentially pathogenic microorganisms like *Candida albicans*, human influenza virus, and *human alpha-herpesvirus* identified in the nasopharyngeal microbiome. Some studies described that the main microbiome composition in the throat of COVID-19 patients includes *Streptococcus bovis*, *Leptotrichia buccalis*, *Haemophilus parainfluenzae*, and *Neisseria cinerea*.

Whole-genome sequencing of BALF samples showed dysbiosis in oral and upper respiratory bacteria, which includes bacteria like *Acinetobacter*, *Sphingobium*, Enterobacterales, *Escherichia coli*, *Enterococcus*, *Rothia*, and *Lactobacillus*. They also identified fungi like *Cryptococcus*, *Cladosporium*, and *Alternaria* in the patients. A study based on 16S rRNA sequences indicates that *Acinetobacter* was the most common bacterial genus, followed by *Chryseobacterium*, *Burkholderia*, *Brevundimonas*, *Sphingobium*, and Enterobacterales in the lung tissues of deceased COVID-19 patients. *Cryptococcus* was identified as a prevalent fungus along with other species like *Issatchenkia*, *Wallemia*, *Cladosporium*, and *Alternaria*. Patients with both moderate and severe COVID-19 showed dysbiotic microbiomes and those treated with various antibiotics, under mechanical ventilation, and prolonged hospitalization.

The severity of SARS-CoV-2 infection-induced changes in the microbiome's diversity at the entry site of infection was high. Still, patients with mild illness did not show any significant changes in their diversity compared to healthy patients or patients with other viral respiratory tract infections. The patients admitted to the hospital had reduced microbiome diversity and increased dysbiosis of the oropharyngeal and nasopharyngeal microbiome. A detailed study of patients admitted to

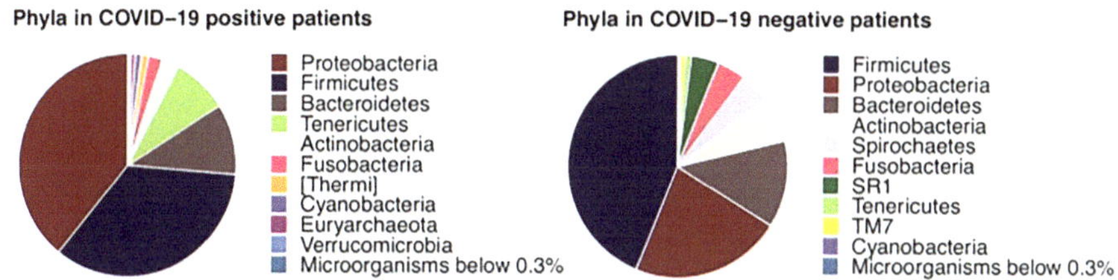

Fig. 8.4 Composition of the microbiome in COVID-19-positive and negative patients

an ICU for different medical conditions showed differences in the oral and gut microbiomes, which were found to have major microbial dysbiosis disturbing an individual's complete respiratory microbiome equilibrium (de Castilhos et al. 2022).

In healthy individuals, bacteria in the oral cavity include Actinobacteria (e.g., *Corynebacterium* spp., *Propionibacterium* spp.), Firmicutes (e.g., *Staphylococcus* spp.), and Proteobacteria. The nasopharyngeal system contains members of Firmicutes, *Staphylococcus*; Bacteroidetes, *Corynebacterium*; Proteobacteria, *Prevotella*; and commensal bacteria like *Streptococcus*, *Neisseria, and Haemophilus* spp. in the URT (upper respiratory tract) (Fig. 8.4).

Acinetobacter spp., *Clostridium hiranonis*, Enterobacteriaceae, *Pseudomonas alcaligenes*, and *Sphingobacterium* spp. were found in the lungs of critically ill COVID-19 patients. In contrast, *Haemophilus influenzae*, *Streptococcus* spp., and *Veillonella dispar* were found in the lungs of COVID-19-negative patients. The lung microbiota profile of critically ill patients with COVID-19 was dominated by the phyla Bacteroidetes (9%), Firmicutes (37%), and Proteobacteria (48%) (Hanada et al. 2018).

8.5 Impact of COVID-19 Infection on the Intestinal Microbiome

There are evident changes in the microbiome composition of the intestinal and respiratory microbiota during the infection. Some studies on COVID-19 infection have shown that the gut microbiota has more importance than the respiratory microbiota, as any change in the composition of the gut microbiota would adversely affect lung function and lead to gut dysbiosis. Another research study explains that any imbalance in the gut microbiome can result in illnesses of human health, indirectly affecting immunity (Jackson et al. 2022).

Macronutrient metabolism includes metabolites like short-chain fatty acids, alcohol, branched-chain fatty acids (acetate, propionate, and butyrate), amines, indoles, sulfur compounds, phenols, and glycerol and choline derivatives of the gut microbiome that influence human health. The commensals of gut microbiota like *Bacteroides* and Bifidobacteria secrete metabolites and immune signaling molecules that bind to receptors in innate cells such as dendritic cells and macrophages to modulate their functions. This helps regulate the development and function of the innate and adaptive immune systems, secreting antimicrobial peptides that can kill microbial pathogens directly or indirectly by modulating the host defense systems by competing for nutrients and the habitat site and maintaining the homeostasis of the body. The disruptions caused by the microbiota may alter the mechanisms of colonization resistance and affect the outcomes of infection. For example, dietary fiber can modify the microbiota's structure and function by producing SCFAs, such as acetate, butyrate, and propionate. They bind to the G-protein-coupled receptor (GPR43) and can stimulate AMPs, REGIIIγ and β-defensins, as shown in Fig. 8.5. SCFAs can diffuse through the membrane, acidify cytoplasm, and inhibit the growth of some pathogens. Alterations to the microbiota (due to antibiotics or a high-fat diet) can result in lowered B-cell-modulated production of IgA68 and increased permeability, leading to susceptibility to infection.

The studies showed that patients treated with antibiotics during hospitalization had a further depletion of bacterial species (symbionts) beneficial to host immunity. Figure 8.6 shows an increase in opportunistic bacterial and fungal pathogens: *Coprobacillus*, *Streptococcus*, *Actinomycetes*, *C. albicans*, *C. auris*, *Enterococcus*, *and Aspergillus niger*. The lack of microbiota and mycobiota, *Intestinibacter*, *Eubacterium*, *Fusicatenibacter*, *Ruminococcus*, *Clostridium ramosum* (Firmicutes phylum), Basidiomycota, Ascomycota, *and Penicillium citrinu*, can lead to the severity of the infection (Yamamoto et al. 2021).

An increase in the number of OPs revealed the levels of C-reactive protein, TNF-alpha, and IL-18, which are proinflammatory cytokines produced by the intestinal microbiota in the sera of COVID-19 patients compared with those in influenza patients and healthy individuals.

8.6 The Intestinal Dysbiosis Associated with COVID-19 Severity

The concept of dysbiosis has been broadly defined as the change in the composition of the resident commensal microbiota compared to those found in healthy individuals. This alteration causes the disruption of symbiosis between the host and microbes, with adverse consequences. The composition of a healthy human gut microbiome is still a question with a complicated answer that has yet to be addressed. Recent advancements by a working group of the International Life Sciences Institute North America show some of the challenges, like the high degree of intra- and inter-individual variation in the human microbiome, the lack of potential biomarkers to define and measure microbiome-host interactions, and the extreme consequences of dysbiosis on human physiology and disease.

Dysbiosis is described as an increase in potential pathogenic species and a decrease in beneficial organisms (e.g., *Bifidobacterium* and *Faecalibacterium* species) or a change in alpha diversity due to the production of harmful microbially derived compounds (e.g., hydrogen sulfide produced by sulfate-reducing bacteria). Higher alpha diversity is a marker of health in the GI tract but, conversely, a marker of dysbiosis in the vaginal microbiome.

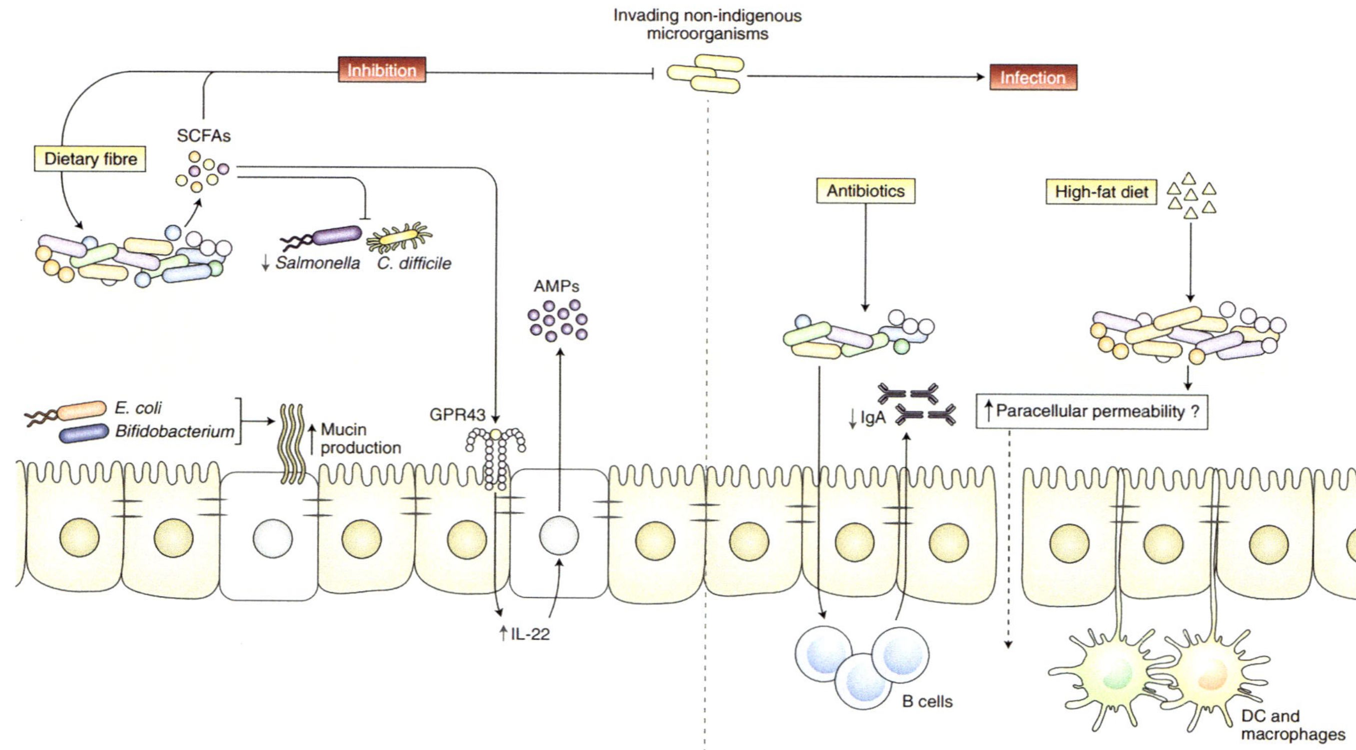

Fig. 8.5 Interference in the altered mechanisms of colonization resistance of microbiota and its effect on infection outcomes

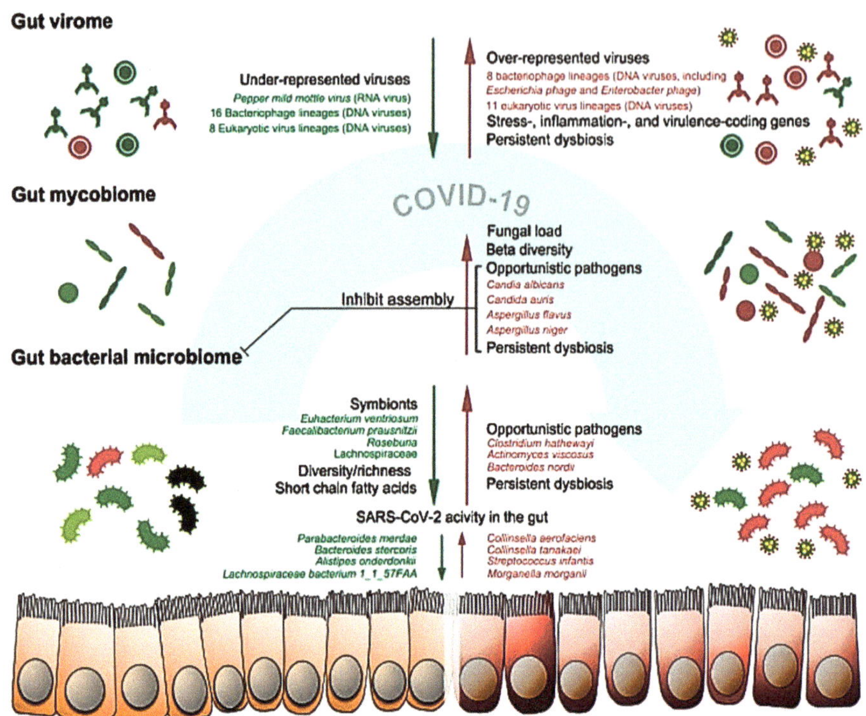

Fig. 8.6 *Modifying* gut virome, mycobiome, and the bacterial microbiome from healthy to COVID-19-infected patients (Zuo et al. 2021)

Reductions were observed in the endurance of bacteria and archaea such as Lactobacillus, Bifidobacterium, Clostridium butyricum, Lachnospiraceae, Prevotella, Roseburia, Ruminococcus bromii, Faecalibacterium, and Bacteroides. This shows a reduction in probiotic bacteria, neutrophil concentration, and IL-6 concentration compared to healthy individuals (Yamamoto et al. 2021).

8.7 Dysbiosis in the Fecal Microbiome of COVID-19 Patients

The microbiome and mycobiome dysbiosis in the gut of COVID-19 patients was observed to have an increase in the number of opportunistic pathogens (OPs) like *Actinomyces, Rothia*, *Enterococcus*, *Enterobacter*, *Streptococcus*, and *Klebsiella* species that pose a threat of a reduction in host immunity (Chakravorty et al. 2007). The intestinal instability, prolonged dysbiosis, and high viral transcription and replication began during hospitalization and lasted even after 2 weeks of recovery. Fecal samples with less SARS-CoV-2 infectivity show improved levels of bacteria like *Parabacteroides*, Lachnospiraceae, and *Bifidobacterium,* which produce SCFAs, GOS, and FOS vital for boosting host immunity.

Roseburia, *Faecalibacterium*, *Coprococcus*, and *Parabacteroides* showed lower abundance in COVID-19 patients than fungal pathogens like *Candida* and *Aspergillus* spp., which were found enriched in healthy individuals. Firmicutes, Bacteroidetes, Actinobacteria, and Proteobacteria phyla were observed to have enhanced gut and oral microbiota in both COVID-19 patients and healthy individuals.

Bacterial microbiota identified in the gut samples of COVID-19 patients are Bacteroidales, Enterobacterales, Clostridiales, Lactobacillales, and *Bifidobacteriales*, whereas the oral samples showed Lactobacillales, Micrococcales, Enterobacterales, and Selenomonadales (Libertucci and Young 2018).

8.8 Therapeutic Modulation of the Microbiome and Its Effects

Prebiotics and probiotics are used as therapies to prevent the colonization of pathogens, whereas fecal microbiome transplantation is used to clear the pathogens. This helps modulate the microbiota to prevent colonization and promote the clearance of pathogens. Some other treatments include the mechanisms of action that are different and unclear, and patients have yet to be studied (Harper et al. 2021).

8.8.1 Fecal Microbiota Transplantation in COVID-19 Patients

The transfer of a fecal suspension from a healthy donor to a recipient (a patient with dysbiosis) modulates the microbiota, which would indirectly help recover the health of a diseased individual by improving the functioning of the microbiota and decreasing the infection by replacing or restoring the functions in the body. It has proven highly effective against the decolonization of drug-resistant organisms to reduce or clear up the disease.

Studies have shown that using filtrates from healthy donor stools (fecal filtrate transfer), which includes transferring bacterial components and metabolites from donor to recipient, is more beneficial for treating infections. Evidence provided from studies has shown that outcomes of FMT were associated with alterations to the enteric virome and bacterial microbiota showing greater treatment success. When a diet high in resistant starch was eaten, *Ruminococcus bromii*, *Oscillibacter*, Firmicutes, Bacteroidetes, and *Eubacterium rectale* increased significantly.

Inulins found in various fruits, vegetables, and wheat have been shown to stimulate the growth of *Bifidobacterium* spp. and *Faecalibacterium by* increasing butyrate production. Fructooligosaccharide (FOS) and galactosaccharide (GOS) administration reduces the release of corticosterone, increases cecal acetate and propionate concentrations simultaneously, and reduces proinflammatory cytokines with an increase in interleukin (IL)-10, IL-8, and other anti-inflammatory cytokines (Cryan et al. 2019).

8.8.2 Prebiotics

Prebiotics stimulate the promotion of indigenous microbiota by participating in fermentation. They can inhibit or limit the development of the pathogen through the digestion of insoluble fiber sources to increase SCFA production, lactic acid, and peptidoglycan, which stimulate the innate immune system against pathogenic microorganisms. It has also been shown that pH alteration can affect the population of acid-sensitive species, regulate the virulence expression of pathogens, and inhibit the binding of pathogens to epithelial receptors.

The different types of prebiotics used for treating diseases or infections are oligosaccharide carbohydrates (OSCs), fructooligosaccharides, galactooligosaccharides (GOS), and resistant starch (RS). For example, fructooligosaccharide (FOS) increases the level of interleukin-4 (IL-4), a myeloid dendritic cell that improves the immune response in volunteers. Combining inulin and FOS can enhance antibody responses toward viral vaccines (diarrhea, measles).

The GOS improved the level of IL-8, IL-10, and C-reactive protein in the blood and the function of NK cells. SCFA increases the production of mucins and antimicrobial peptides, reduces pH, increases intestinal motility, and has anti-inflammatory properties significant for the health of the gut epithelium (Shin et al. 2022).

8.8.3 Probiotics

Probiotics are live microorganisms that help boost the health of the host when administered in an adequate amount. They are live microbial feed supplements that improve intestinal microbial balance by modulating the intestinal microbiota. The benefits of probiotic consumption include regulation of the intestinal microbiota, stimulation of the immune system, promoting the synthesis and bioavailability of nutrients, and reducing the symptoms of lactose intolerance and the risk of other diseases.

Probiotic products generally contain one or more microbial strains like *Lactobacillus, Enterococcus, Bifidobacterium, Lactococcus, Streptococcus, Saccharomyces cerevisiae, and Bacillus.* The FDA (Food and Drug Administration) must regulate the microorganisms used in these products for human consumption, and they must be GRAS (generally recognized as safe) organisms. The QPS (Qualified Presumption of Safety) criteria are used to maintain the safety assessment of bacterial supplements, safe usage, and the risk of acquired resistance to antibiotics (Shin et al. 2022). Probiotic microorganisms can produce enzymes like esterase, protease, and lipase and coenzymes A, Q, NAD, and NADP, which show the metabolism of antibiotics (bacitracin and lactacin), anticarcinogens, and immunomodulatory properties.

Most of these bacteria are probiotics, the good bacteria colonizing within your digestive tract that serve a beneficial purpose (produce vitamins, absorb nutrients from your food, and even help regulate your mood). Prebiotics are the natural dietary fiber (nonliving and nondigestible by humans) that nourishes our probiotic

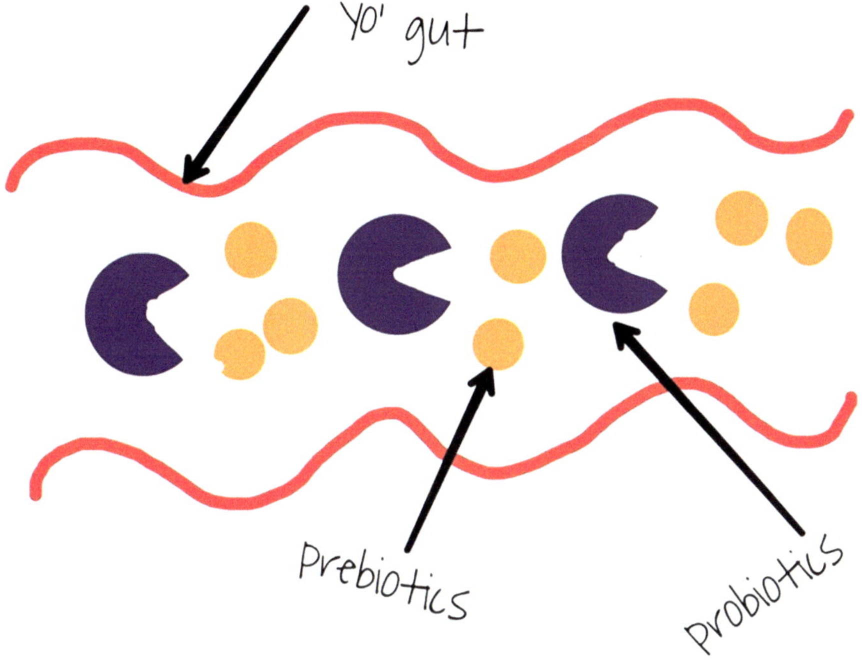

Fig. 8.7 Synbiotics

bacteria. The probiotics introduce good bacteria into the gut, and the prebiotics act as fertilizer for the good bacteria's growth, improving the ratio of good to bad bacteria. This directly correlates to your health and overall well-being (gut-brain axis) (Fig. 8.7).

Common medications have both a beneficial and harmful effect on gut microbes, according to new research. Antibiotics and gastric acid suppressants disrupt beneficial bacterial populations in the gut, while statins and ACE inhibitors are linked to improved bacterial composition and function (Shin et al. 2022) (Fig. 8.8).

Certain bacterial strains present in probiotics can also have an effect on central neuronal processes like neural communication, neurogenesis, the expression of neuropeptides, neurological inflammation, and even behavior. It helped with a wide range of conditions, from autism to melancholy and anxiety. "Psychobiotics" and "chobiotics," which aim to treat neurological and psychiatric diseases by altering the composition of the gut microbiota, emerged in response to the prevalence of these conditions. Also included are models based on flux balance analysis (FBA) used to foretell and comprehend how microorganisms will behave in a certain setting (Fig. 8.9).

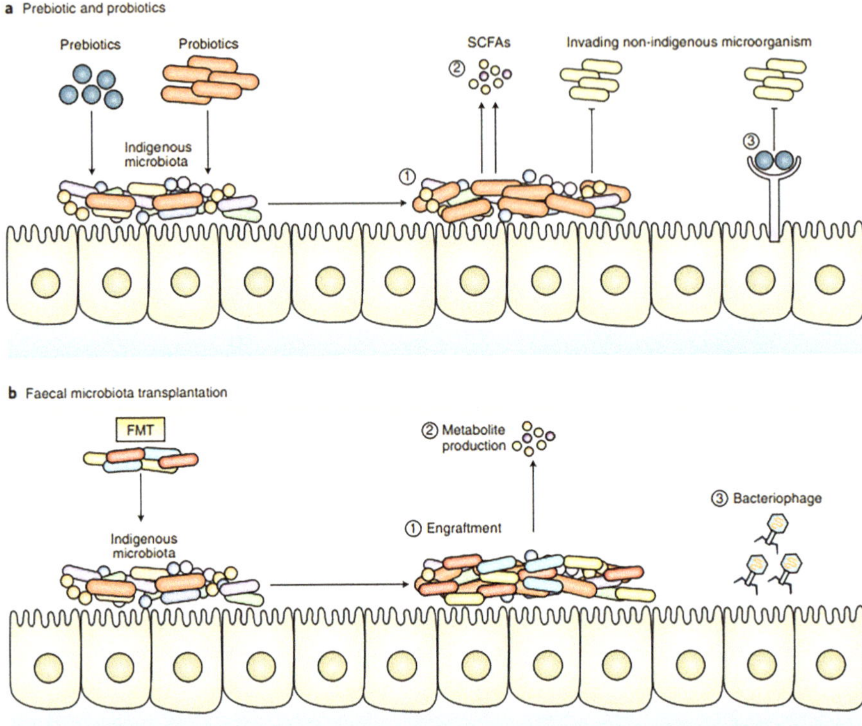

Fig. 8.8 The combination of these medications may provide the viable bacterial strain with a greater potential to fill a niche and restore its community structure and function by restoring the metabolome (Cryan et al. 2019)

8.8.4 The Gut-Associated Peptide Reg3g Connects the Microbiota of the Small Intestine to the Control of Energy Homeostasis, Blood Sugar Levels, and Gastrointestinal Activity

The composition of the gut microbiome and its adaptation to various environments in various metabolic diseases are key to understanding the necessary changes to maintain homeostasis. Fermentable fiber-rich inulin diets with gastric sleeve surgery (which Acyl homoserine lactones (Acyl-HSL) are acyl homoserine lactones that gram-negative bacteria produce, limiting amount of food you can eat) boost the production of the antibacterial peptide Reg3g in the intestines and the bloodstream. New approaches to therapy may take advantage of its role as a gut hormone that links the microbiota of the intestines to the remainder of the host's physiology (Lazar et al. 2018).

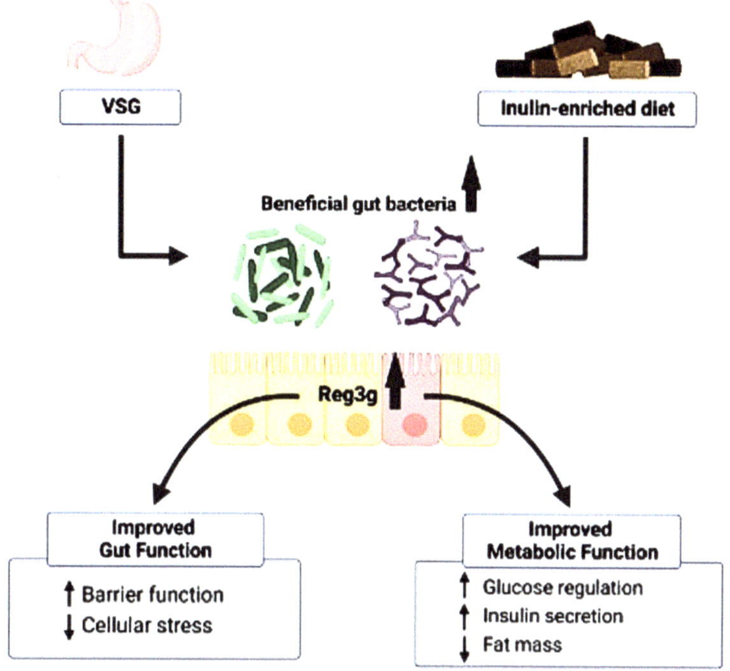

Fig. 8.9 Regulation of the small intestine with the gut peptide Reg3g

8.8.5 Microbe-Microbe Interactions

Bacteria use signaling molecules to share information and adapt their gene expression to their surroundings, which is especially important in highly competitive ecosystems with many coexisting species. The term for this phenomenon is "quorum sensing" (QS). Quantum signaling (QS) depends on the density of molecular language that controls cell phenotypic expression and behavior in response to external cues. This intercellular communication is divided into two categories. Interspecific communication among bacterial and eukaryotic/host cells is facilitated by interspecific interaction that utilizes a universal chemical language, the first form of cell-to-cell communication (Falcao et al. 2004).

Autoinducers (AIs) are tiny organic compounds that act like hormones and are part of this system. Acyl homoserine lactones (Acyl-HSL) are acyl homoserine lactones that gram-negative bacteria produce. Peptide compounds (AIP) are not diffusible in gram-positive bacteria. These cell-to-cell communication networks in bacteria were first reported as a way for bacteria to control the release of virulence genes, which are found in pathogens and play a crucial role in infection through cell density (Villapol 2020). This method allows commensal bacteria to control how much of the host they colonize (Fig. 8.10).

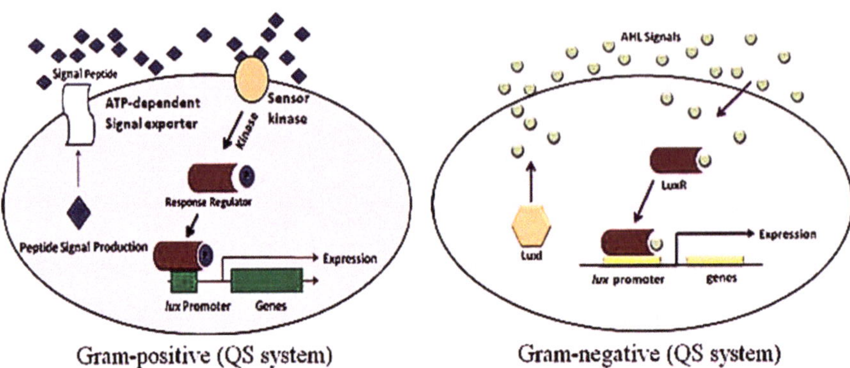

Fig. 8.10 Mechanism of the QS system in gram-positive and gram-negative bacteria

Cathelicidins, defensins, and AIs are all antimicrobial peptides made by eosinophils. They all work as signaling molecules within and between species.

Quorum sensing processes include measuring some of these processes, including bioluminescence, pathogenicity factor expression, biofilm formation, and conjugation. For instance, enteric pathogens use quorum sensing to control the expression of genes that encode virulence traits like mobility and type 3 secretion.

Thus, the use of quorum sensing ensures effective host colonization by detecting the presence of normal gut flora (Villapol 2020). This phenomenon sheds light on mitigating microbial infection against virulence factors and biofilm formation controlled by quorum sensing (Seibert et al. 2022) (Fig. 8.11).

8.9 Conclusion

8.9.1 The Microbiome Is an Extraordinary Helper: We Must Nurture Our Bodies' Microbes

- The makeup of the gut microbiome has an impact on host metabolism and general health.
- Take antibiotics exactly as prescribed. Antibiotic usage over an extended period is linked to a change in antibiotic-driven gut microbiome composition that upsets the body's normal microbial equilibrium.
- Therapeutic modulation of the microbiota is the future of diagnostics for infectious diseases.
- Identification of the imbalance (if any) in the changing microbiome of humans (testing for the microbiome).
- Developing noninvasive microbiome-based diagnostics (Peter et al. 2019).

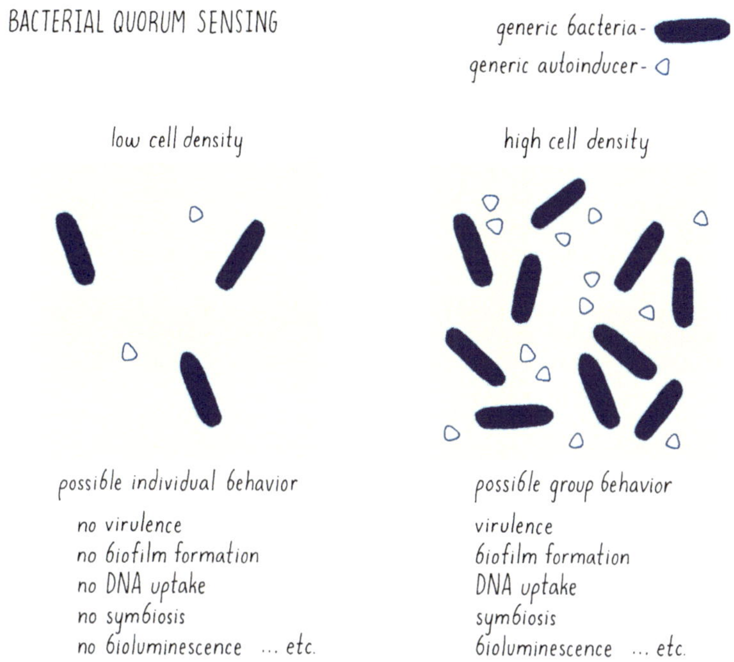

Fig. 8.11 Communication of bacteria via population density

- Increase the fiber in your diet. All plant foods, such as vegetables, fruits, and whole grains, contain fiber (Sudhakar and Padi 2022). Avoid these foods to keep your gut microorganisms happy (Atiatorme et al. 2022). These are foods that are heavy in sugar, fat, or processing.

Acknowledgments The authors would like to express their gratitude to the authorities of Andhra University for providing the necessary infrastructure and facilities for conducting this research. They would also like to acknowledge the support of the principal of Andhra University College of Science and Technology, Andhra University, Visakhapatnam, and the head of the Department of Biotechnology, AUCST, Andhra University, Visakhapatnam.

Conflict of Interest The authors declare that no financial or personal relationships with other people or organizations could inappropriately influence (bias) their work. Additionally, they have no competing interests related to this research.

References

Atiatorme E, Bramhachari PV, Kariali E, Sudhakar P (2022) A paradigm shift in the role of the microbiomes in environmental health and agriculture sustainability. In: Veera Bramhachari P (ed) Understanding the microbiome interactions in agriculture and the environment. Springer, Singapore, pp 83–101

Baghbani T, Nikzad H, Azadbakht J et al (2020) Dual and mutual interaction between microbiota and viral infections: a possible treatment for COVID-19. Microb Cell Factories 19:217

Chakravorty S, Helb D, Burday M, Connell N, Alland D (2007) A detailed analysis of 16S ribosomal RNA gene segments for the diagnosis of pathogenic bacteria. J Microbiol Methods 69(2):330–339

Cryan JF, Cowan M, Sandhu CS, Bastiaanssen KVS et al (2019) The microbiota-gut-brain axis. Physiol Rev 99(4):1877–2013

de Castilhos J, Zamir E, Hippchen T, Rohrbach R, Schmidt S, Hengler S et al (2022) Severe dysbiosis and specific Haemophilus and Neisseria signatures as hallmarks of the oropharyngeal microbiome in critically ill coronavirus disease 2019 (COVID-19) patients. Clin Infect Dis 75(1):e1063–e1071

Falcao JP, Sharp F, Sperandio V (2004) Cell-to-cell signaling in intestinal pathogens. Curr Issues Intest Microbiol 5:9–17

Gaibani P, Viciani E, Bartoletti M et al (2021) The lower respiratory tract microbiome of critically ill patients with COVID-19. Sci Rep 11:10103

Hanada S, Pirzadeh M, Carver KY, Deng JC (2018) Respiratory viral infection-induced microbiome alterations and secondary bacterial pneumonia. Front Immunol 9:2640

Harper A, Vijayakumar V, Ouwehand AC, ter Haar J, Obis D, Espadaler J, Binda S, Desiraju S, Day R (2021) Viral infections, the microbiome, and probiotics. Front Cell Infect Microbiol 10:596166. https://doi.org/10.3389/fcimb.2020.596166

Hoque MN, Sarkar MMH, Rahman MS, Akter S, Banu TA, Khan MS et al (2021) SARS-CoV-2 infection reduces the human nasopharyngeal commensal microbiome with the inclusion of pathobionts. Sci Rep 11(1):1–17

Jackson CB, Farzan M, Chen B et al (2022) Mechanisms of SARS-CoV-2 entry into cells. Nat Rev Mol Cell Biol 23:3–20

Shin J, Kramer NB, Shao Y, Abbott S et al (2022) The gut peptide Reg3g links the small intestine microbiome to the regulation of energy balance, glucose levels, and gut function. Cell Metab 34(11):1765–1778.e6

Konatala A, Parackel F, Sudhakar P (2021) Recent advancements in microbiome–immune homeostasis and their involvement in cancer immunotherapy. In: Microbiome in human health and disease. Springer, Singapore, pp 239–258

Lazar V, Ditu L-M, Pircalabioru GG, Gheorghe I, Curutiu C, Holban AM, Picu A, Petcu L, Chifiriuc MC (2018) Aspects of gut microbiota and immune system interactions in infectious diseases, immunopathology, and cancer. Front Immunol 9:1830

Libertucci J, Young VB (2018) The role of the microbiota in infectious diseases. Nat Microbiol 4(1):35–45

Liu TFD, Philippou E, Kolokotroni O, Siakallis G, Rahima K, Constantinou C (2022) Gut and airway microbiota and their role in COVID-19 infection and pathogenesis: a scoping review. Infection 50(4):815–847

Mohan SM, Sudhakar P (2022) Metagenomic approaches for studying plant–microbe interactions. In: Understanding the microbiome interactions in agriculture and the environment. Springer, Singapore, pp 243–254

Peter AE, Sudhakar P, Sandeep BV, Rao BG (2019) Antimicrobial and anti-quorum sensing activities of medicinal plants. In: Bramhachari P (ed) Implication of quorum sensing and biofilm formation in medicine, agriculture and food industry. Springer, Singapore, pp 189–213

Pola S, Padi DL (2021) Overview of human gut microbiome and its role in immunomodulation. In: Bramhachari PV (ed) Microbiome in human health and disease. Springer, Singapore, pp 69–82

Ronda C, Chen SP, Cabral V et al (2019) Metagenomic engineering of the mammalian gut microbiome in situ. Nat Methods 16:167–170

Sanapala P, Pola S (2021) Modulation of systemic immune responses through genital, skin, and oral microbiota: unveiling the fundamentals of human microbiomes. In: Bramhachari PV (ed) Microbiome in human health and disease. Springer, Singapore, pp 13–34

Seibert B, Cáceres CJ, Carnaccini S, Cardenas-Garcia S, Gay LC, Ortiz L et al (2022) Pathobiology and dysbiosis of the respiratory and intestinal microbiota in 14 months old golden syrian hamsters infected with SARS-CoV-2. PLoS Pathog 18(10):e1010734

Sudhakar P, Padi D (2022) Modifications in environmental microbiome and the evolution of viruses through genetic diversity. In: Veera Bramhachari P (ed) Understanding the microbiome interactions in agriculture and the environment. Springer, Singapore, pp 103–112

Villapol S (2020) Gastrointestinal symptoms associated with COVID-19: impact on the gut microbiome. Transl Res 226:57–69

Yamamoto S, Saito M, Tamura A, Prawisuda D, Mizutani T, Yotsuyanagi H (2021) The human microbiome and COVID-19: a systematic review. PLoS One 16(6):e0253293

Zuo T, Wu X, Wen W, Lan P (2021) Gut microbiome alterations in COVID-19: science direct. Genomics Proteomics Bioinformatics 19(5):679–688

Part II

Breast Milk, Skin and Urinary Microbiomes

The Human Breast Milk Microbiome: Establishment and Resilience of Microbiota over the Mother–Infant Relationship

9

Saqib Hassan, Ishfaq Hassan Mir, Meenatchi Ramu, Ayushi Rambia, Chinnasamy Thirunavukkarasu, George Seghal Kiran, Pallaval Veera Bramhachari, and Joseph Selvin

Abstract

Human milk provides a continuous supply of good bacteria to the infant's gut, which contributes to the maturation of the digestive and immunological systems in the developing infant. Nonetheless, the origin of bacterial populations in milk is unknown, and they have been suggested to come from maternal skin, the infant's mouth, and/or endogenously from the maternal digestive tract via a

S. Hassan
American Society for Microbiology, Washington, DC, USA

Department of Biotechnology, School of Bio and Chemical Engineering, Sathyabama Institute of Science and Technology, Chennai, Tamil Nadu, India

I. H. Mir · C. Thirunavukkarasu
Department of Biochemistry and Molecular Biology, School of Life Sciences, Pondicherry University, Puducherry, India

M. Ramu
Department of Biotechnology, Faculty of Science and Humanities, SRM Institute of Science and Technology, Chengalpattu, Tamil Nadu, India

A. Rambia · J. Selvin (✉)
Department of Microbiology, School of Life Sciences, Pondicherry University, Puducherry, India

G. S. Kiran
Department of Food Science and Technology, School of Life Sciences, Pondicherry University, Puducherry, India

P. V. Bramhachari
Department of Biosciences and Biotechnology, Krishna University, Machilipatnam, Andhra Pradesh, India

P. Veera Bramhachari (ed.), *Human Microbiome in Health, Disease, and Therapy*, https://doi.org/10.1007/978-981-99-5114-7_9

mechanism involving immune cells. Understanding the composition, roles, and assembly of the human milk microbiota has significant consequences not only for the development of the infant gut microbiota but also for breast health, as dysbiosis in milk bacteria can cause mastitis. Furthermore, host, microbial, medical, and environmental factors may influence the composition of the human milk microbiome, potentially affecting the mother–infant relationship.

Keywords

Breast milk microbiome · Human digestive tract · Mastitis · Gut microbiome

9.1 Introduction

Since the Human Microbiome Project came into being in 2007, it has been breaking our stereotypical understanding that the human body is more or less sterile and only a few sites harbor many microorganisms. As opposed to the belief that microbes only reside in the skin, upper respiratory tract, gut, and vaginal canal, microorganisms are also said to be found in human milk (McGuire and McGuire 2017). The earliest evidence of microbes in human milk was reported in the study on bacteria in the colostrum and milk of Guatemalan Indian Women by Wyatt and Mata. The presence of Enterobacteriaceae suggested poor environmental sanitation and personal hygiene among the studied population (McGuire and McGuire 2017; Wyatt and Mata 1969). This gave the impression that microbes in the milk contaminated it, mainly due to underlying causes. But now there is evidence that microorganisms are an innate part of human breast milk (HBM) (Stinson et al. 2021). It is also believed that the microbes in human milk contribute to the early colonization of the infant's gut and thus provide the necessary genes and antigens eminent for the growth and development of the newborn (McGuire and McGuire 2017).

Several genera of bacteria, viruses, archaea, and microeukaryotes have been reported to form the Human Milk Microbiome (HMM) (Stinson et al. 2021). Among these microbes, Firmicutes form the largest proportion, followed by Proteobacteria, Bacteroidetes, and Actinobacteria as the major classes (Kim and Yi 2020). A total of 329 genera have been detected with *Streptococcus, Staphylococcus, Bacteroides, Acinetobacter,* Enterobacteriaceae(f), Ruminococcaceae(f), *Bifidobacterium, Prevotella,* Clostridiales(o), *Corynebacterium, Akkermansia, Lactobacillus, Pseudomonas, Dialister, Stenotrophomonas, Blautia, Sphingomonas, Haemophilus, Neisseria,* Lachnospiraceae (f), *Rothia,* and *Faecalibacterium* in the order of their abundance (Kim and Yi 2020). The presence of these microbes in the HBM microbiome is mostly attributed to maternal skin and infant oral cavity during lactation, as well as the maternal gastrointestinal tract (Stinson et al. 2021). The variability and abundance of the microbes mentioned above depend on maternal age, health, lactation duration, mode of delivery, and geographical location (McGuire and McGuire 2017).

HBM microbiome is quintessential for infants because it is the best source of nutrition and certain immune components such as secretory antibodies, immune cells, antimicrobial proteins, cytokines, and human milk oligosaccharides (Kim and Yi 2020). All of these ensure proper growth and development of the infant and protection against several lifestyle disorders such as obesity, diabetes (both type 1 and 2), asthma, and cardiovascular diseases (Fitzstevens et al. 2017; Stinson et al. 2021). Therefore, a detailed study of the HBM microbiome can help improve formula-based infant food and replicate the microbiome from healthy mothers to ensure proper growth and development of neonates and infants.

9.2 Human Breast Milk (HBM) Microbiome and Its Importance

The HBM microbe is said to be composed of bacteria, viruses, archaea, and microeukaryotes. Their presence benefits the growing infant or may indicate signs of diseases and certain disorders (Fitzstevens et al. 2017; Stinson et al. 2021). HBM microbiome is responsible for the baby's gut colonization. Still, they may also have several other crucial roles, including affecting the maturation of the mucosal immune system, defending against infections, and assisting with digestion and nutritional absorption (Jeurink et al. 2013). HBM bacteria, in brief: bolster the gut immune system's homeostasis: Early microbial exposure is crucial to offer antigenic cues encouraging intestinal immune system maturation and improving intestinal homeostasis (Gensollen et al. 2016). By supporting a change from the predominate intrauterine T helper cell 2 immunological milieus to a TH1/TH2 balanced response and by inducing regulatory T cell development, the HBM microbiota specifically may enhance intestinal immune homeostasis. Additionally, a metagenomics analysis of the HBM microbiome revealed immunomodulatory DNA motifs that could aid in reducing excessive inflammatory reactions to bacterial colonization by enhancing intestinal functions. As HBM microbiome contains oligosaccharides indigestible by an infant's intestine, molecular analysis revealed that BM bacteria are metabolically active in producing short-chain fatty acids (SCFA). It has been proposed that this characteristic favors the proliferation of helpful bacteria against harmful taxa (Ward et al. 2013). Several BM isolates, including *Lactobacillus rhamnosus* and *crispatus*, were shown to inhibit the growth of pathogenic microorganisms in vitro studies. Other isolates had effects on an enteropathogenic *Salmonella enterica* strain that were both in vitro and in a mouse model that was bacteriostatic and/or bactericidal (Hirai et al. 2002). Breastfed infants also exhibit increased capacity for carbohydrate, amino acid, and nitrogen metabolism, cobalamin synthesis, membrane transport, oxidative stress, and human behavior and emotion (Lazar et al. 2019; Stinson et al. 2021; Valdes et al. 2018; Valles-Colomer et al. 2019) (Fig. 9.1).

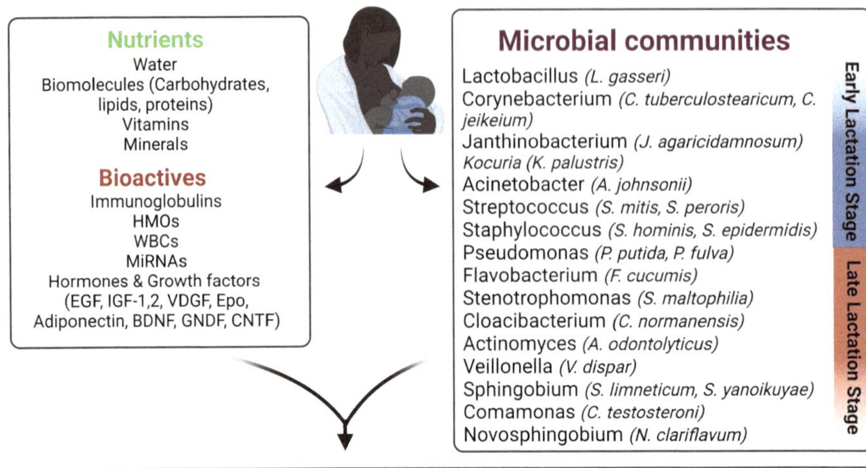

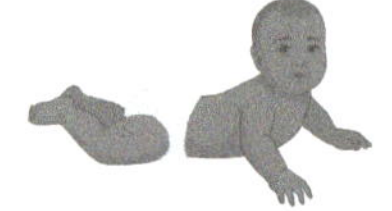

Fig. 9.1 Benefits of human milk microbiome for infants

9.3 Bacteria

Several genera of bacteria have been identified in the HBM through culture-dependent and culture-independent methods. The core genera primarily include *Staphylococcus, Streptococcus,* and *Pseudomonas.* These are found universally regardless of the lactating mothers' demographics, geographical location, or health status (Kim and Yi 2020; McGuire and McGuire 2017; Stinson et al. 2021). The most abundant genera found are *Staphylococcus, Streptococcus,* and *Propionibacterium; Bifidobacterium, Veillonella, Rothia,* and *Lactobacillus* are found in lower abundance (Stinson et al. 2021). *Corynebacterium, Ralstonia, Acinetobacter, Acidovorax, Pseudomonas, Bacteroides, Clostridium, Escherichia/Shigella, Gemella,* and *Enterococcus* are some of the other commonly found bacterial genera in the HBM (McGuire and McGuire 2017; Stinson et al. 2021).

Some of these microbes' early colonization of the infant's gut can help in short- and long-term health outcomes (Fitzstevens et al. 2017). Although found in low abundance, *Bifidobacterium* is the first to colonize the infant's gut and help utilize the glycans found in the HBM (Stinson et al. 2021). *Lactobacillus* in the gut provides resilience and reduction in the risk of diarrheal and other dysbiosis-related problems (Gomez-Gallego et al. 2016). Together these two bacteria help activate immunoglobulin A producing plasma cells in the neonatal gut (Khodayar-Pardo et al. 2014). Different bacteria also aid in decreasing the risk of respiratory tract infections, atopic dermatitis, asthma, obesity, type 1 and 2 diabetes, necrotizing enterocolitis, gastroenteritis, and inflammatory bowel disease (Fitzstevens et al. 2017; Gomez-Gallego et al. 2016).

9.4 Viruses

Members of certain viral families have been reported in infants up to 4 days of age. Bacteriophages are the most abundant, while certain eukaryotic viruses are the least. Members of Siphoviridae, Myoviridae, and Podoviridae families have also been reported in both HBM and breastfed infant stool (Stinson et al. 2021). Bacteriophages, in particular, help maintain the balance between different bacterial communities at different points in the early developmental age. Non-phage viruses from Papillomaviridae, Retroviridae, and Herpesviridae have also been reported (Stinson et al. 2021). Regarding the recent challenge of SARS-CoV-2 (COVID-19) in BM, there is no evidence of virus transmission during lactation (Zhu et al. 2021).

9.5 Archaea

Of the several studies on HBM samples, only two have reported the presence of archaeal DNA. The species identified were *Haloarcula marismortui, Halorhabdus utahensis, and Halomicrobium mukohataei* (Jiménez et al. 2015; Stinson et al. 2021). Although the presence of halophilic archaea is questionable in the HBM, a protective function has been attributed to their presence (Stinson et al. 2021).

9.6 Microeukaryotes

A few fungal species have also been reported in low diversity and biomass in the HBM, forming the HM mycobiome. These are generally members of *Malassezia, Davidiella, Ascomycota, Basidiomycota, Candida,* and *Saccharomyces* (Boix-Amorós et al. 2019; Jiménez et al. 2015; Stinson et al. 2021). Protozoal parasites, including *Giardia intestinalis* and *Toxoplasma gondii,* have also been reported (Jiménez et al. 2015; Stinson et al. 2021). However, the clinical significance of these fungal and protozoal species has not been reported yet.

9.7 Origin of Milk Microbiome

Now that there is a consensus on the presence of microbes in the HBM, the next step is to elucidate their source in the HBM. Several theories and evidence suggest two possible routes for the entry of microbes into the HBM. First is the entero-mammary pathway, which involves the translocation of gut microbiota to the mother's mammary glands (McGuire and McGuire 2017; Moossavi et al. 2019). The presence of certain gut microbiome anaerobes such as *Veillonella, Bacteroides,*

Parabacteroides, Clostridium, Collinsella, Faecalibacterium, Coprococcus, and *Blautia* in the HBM indicate their entry via the lymphatic system (Stinson et al. 2021). This has been supported by the presence of bacteria in the mesenteric lymph nodes of pregnant mice as opposed to nonpregnant mice, indicating an increase in bacterial translocation during gestation (Perez et al. 2007; Stinson et al. 2021). This has been further confirmed by the increase in lymphatics in the mammary tissues during the lactation period (Hitchcock et al. 2020; Schenkman et al. 1985; Stinson et al. 2021).

Second is the retrograde inoculation by the maternal skin and the infant's oral cavity (McGuire and McGuire 2017; Moossavi et al. 2019; Stinson et al. 2021). The presence of human skin commensals such as *S. epidermidis, S. hominis, S. haemolyticus, S. lugdunensis, Cutibacterium acnes,* and species of *Corynebacterium* in the HBM indicate their entry via the nipple into the mammary glands (Stinson et al. 2021). *S. epidermidis* has supported this in breastfed infants and its absence in formula-fed infants (Jiménez et al. 2008; Stinson et al. 2021). Reports have also suggested an exchange between the infant's oral and HBM microbiomes during the lactation period (Stinson et al. 2021). This has been supported by oral bacteria, such as *S. salivaris, S. mitis, Rothia mucilaginosa,* and *Gemella* spp., in the HBM of breastfed infants instead of bottle-fed infants (Biagi et al. 2018; Stinson et al. 2021).

The next question in our understanding of the seeding of the HBM is whether this occurs due to constant influx during the gestation and lactation period or as a result of a permanent mammary gland microbiome. The low abundance of bacteria in the HBM compared to other mucosal surfaces in the body suggests the constant influx approach. This is further supported by the fact that mammary epithelium tissues are not specialized for mucus secretion (Stinson et al. 2021). Another argument suggests that although the mammary epithelial cells do not secret mucus, they may act as a mucosal-like immune interface, thus allowing stratification and compartmentalization of resident microbes (Sakwinska and Bosco 2019; Stinson et al. 2021). This has been supported by the presence of *S. aureus* and lactic acid bacteria biofilms in the mammary epithelial cells in bovine models, favoring the permanent model of the mammary gland microbiome (Bouchard et al. 2015; Gomes et al. 2016; Stinson et al. 2021; Wallis et al. 2019).

9.8 Potential Factors Influencing Milk Microbiome

Several factors influence the HBM microbiome at all stages of gestation as well as lactation (Fig. 9.2). These include the following:-

1. Maternal factors.
2. Breastfeeding factors.
3. Early life factors.
4. Infant factors.
5. Milk environment.

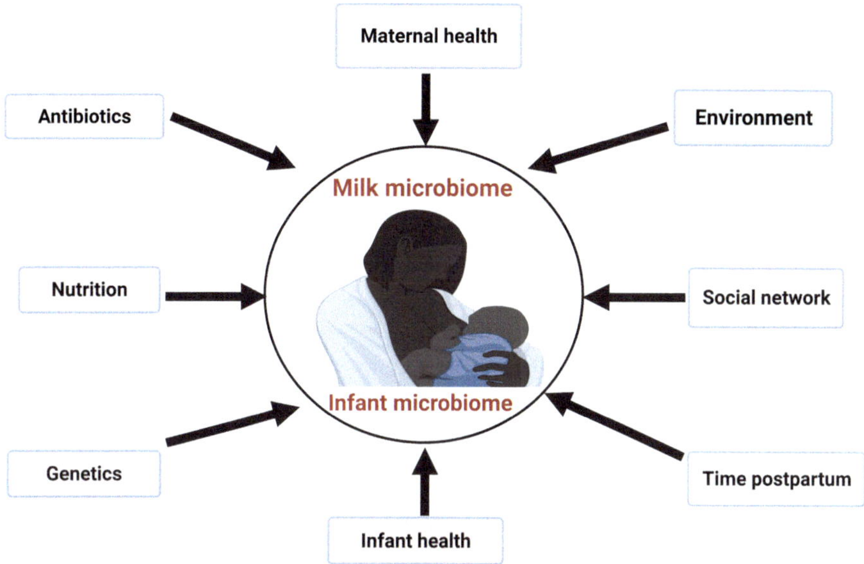

Fig. 9.2 Factors affecting human milk microbiome

Maternal factors such as body mass index (BMI), gestational age, ethnicity, and diet tend to influence the composition of the HBM microbiome (Cabrera-Rubio et al. 2012; Moossavi et al. 2019). Mothers with high BMI tend to have a more homogenous composition of bacteria than those with normal BMI (Cabrera-Rubio et al. 2012). HBM from obese mothers shows a higher abundance of *Staphylococcus* and *Lactobacillus* while a lower abundance of *Bifidobacterium* (Cabrera-Rubio et al. 2012). Lower levels of *Bifidobacterium* in the milk may lead to improper seeding of the infant's gut, thereby affecting the child's overall health. Gestational age tends to influence the *Bifidobacterium* concentration in the HBM. Preterm mothers have a higher *Bifidobacterium* count in the colostrum and during lactation. At the same time, there is no significant correlation between gestational age and any other bacterial genera (Khodayar-Pardo et al. 2014). Ethnicity, as well as diet, also influences the HBM microbiome. Intake of prebiotics and probiotics in the diet tends to reflect in the milk microbiome and influence the human milk's lipid profile (Gomez-Gallego et al. 2016; McGuire and McGuire 2017). A calorie-intensive and nutrient-rich diet is also associated with a higher abundance of Firmicutes, indicating that diet influences the HBM microbiome (McGuire and McGuire 2017; Williams et al. 2017). Women from different ethnicities show differences in the composition of breast milk metabolites (Gay et al. 2018).

Certain breastfeeding factors, such as whether the infant is breastfed or bottle-fed, or both and the lactation period influence the milk microbiome. Breastfed infants tend to have a more abundant microbiome than bottle-fed ones, especially that of *Bifidobacterium* (Stinson et al. 2021). It has also been found that colostrum has a more diverse microbiome than transition or mature milk (Cabrera-Rubio et al.

2012). The abundance of certain bacteria like *Bifidobacterium, Enterococcus* spp., *Staphylococcus, Streptococcus, Lactococcus, Veillonella, Leptotrichia,* and *Prevotella* has been reported to increase from colostrum to mature milk (Cabrera-Rubio et al. 2012; Khodayar-Pardo et al. 2014).

9.9 Mode of Delivery

Women who give birth vaginally or through emergency and elective caesarian show significant differences in their milk microbiome. For vaginal delivery, the HBM microbe shows an increase in *Bifidobacterium, Leuconostocaceae, Lactobacillus* spp., while the elective caesarian shows an increase in *Carnobacteriaceae* (Cabrera-Rubio et al. 2012; Gómez-Gallego et al. 2018; Khodayar-Pardo et al. 2014). The microbiome is more or less similar for vaginal and emergency caesarian procedures. This indicates that the difference does not arise due to the difference in procedure but the lack of physiological stress and hormonal signals which mediate the seeding of the milk microbiome (Boix-Amorós et al. 2016). Lactating mothers on antibiotics or chemotherapy also show a reduced bacterial diversity in the HBM microbiome (McGuire and McGuire 2017). Since the infant's oral cavity also seeds the HBM microbiome, it gets influenced by the infant's sex, as hormonal differences between the males and females alter the gut microbiome (Markle et al. 2013; Moossavi et al. 2019). All of the above-discussed factors tend to influence the composition of the HBM in terms of lipid profile, HMOs, cytokine levels, creatine, and riboflavin. The composition also tends to differ depending upon the demographic settings of the mother (Gay et al. 2018; Gomez-Gallego et al. 2016).

9.10 Global Variation in Dominant Bacterial Taxa

The HBM microbiome also varies depending on the geographic location of the lactating mothers. This global variation can be attributed to differences in diet, economic setting, and usual birth methods opted in the country. Several studies have shown that Caucasian Canadian women's milk microbiome predominantly contains *Staphylococcus, Pseudomonas, Streptococcus,* and *Lactobacillus,* while Streptococcus predominates in Mexican-American women (Davé et al. 2016; McGuire and McGuire 2017; Urbaniak et al. 2016). Another study showed that Spanish women's milk microbiome contained *Staphylococcus, Pseudomonas, Streptococcus,* and *Acinetobacter* (Boix-Amorós et al. 2019). The microbiome of women in the United States showed the presence of *Serratia* and *Corynebacteria,* while that of Finland showed the presence of *Leuconostoc, Weissella, Lactococcus,* and *Staphylococcus* (Cabrera-Rubio et al. 2012; McGuire and McGuire 2017). More studies regarding the Asian HBM microbiome are required to draw a better contrast.

9.11 Potential Factors to Modulate Milk Microbiota

Once we are clear with the source, difference, and importance of the HBM microbiome, it becomes evident to know how the milk microbiota can be modulated to mimic that of a healthy mother. Dietary interventions can prove helpful with the administration of probiotics and topical probiotics in improving the milk microbiome since the seeding route is both entero-mammary and retrograde (Stinson et al. 2021). The dairy industry predominantly uses phage therapy to modulate the bovine milk microbiome and protect bovine mothers from mastitis and other diseases. The same approach can also be used for the HBM microbiome (Dias et al. 2013; Porter et al. 2016; Stinson et al. 2021).

9.12 Role of Bacterial Extracellular Vesicles in HBM Microbiota (Their Involvement in the Vertical Transfer of Gut Microbiota)

The gastrointestinal tract of humans is housed in approximately 100 trillion microorganisms, which work in mutual harmony with the host to support the advancement of the host's defense mechanisms. The initial establishment of the gut microbiota throughout childhood has long-term consequences for human health. Several variables, including the type of delivery, antibiotic treatment, surroundings, and dietary exposure, are thought to trigger early infant gut colonization (Ojo-Okunola et al. 2018). Diet is vital in promoting the growth of the infant's immunity. The most significant driver of nourishment for newborns is human breast milk (HBM), which is recognized to include immunological factors such as immunoglobulins, cellular machinery of the immune system, antibacterial agents (like lactoferrin and lysozyme), cytokines, and other important factors (Kim and Yi 2020). Previously, microorganisms in HBM were considered a sign of harm or contamination, but multiple investigators have shown that HBM includes microbial species by employing culture-dependent strategies. It is understood that HBM includes many commensal microbes that may influence how the infant's gut colonizes (Murphy et al. 2017).

Extracellular vesicles (EVs) are membrane vesicles with a dimension of a few nanometers having a wide range of bioactive molecules within, including lipids, membrane proteins, cytoplasmic proteins, nucleic acids, and peptides found in the cytosol. Every microbe produces EVs, and this is a widely recognized fact. According to the properties of their membranes, microbes are subdivided into Gram-positive (G +) or Gram-negative (G-) bacteria. The EVs produced by G + bacteria are typically known as bacterial membrane vesicles, while those produced by G-bacteria are known as outer membrane vesicles (Brown et al. 2015). Microbe-derived EVs in bodily fluids like blood, urine, or feces suggest that these microbes can influence host cellular machinery by triggering host receptors, releasing different biologically active chemicals, or fostering EVs into the cells of hosts.

In an investigation by Kim and Yi, the top categories of bacteria found in mammalian breast milk they included Bacteroides, Acinetobacter, and Lactobacillus, whereas Streptococcus and Staphylococcus predominated in microbial samples (Kim and Yi 2020). Such findings suggest that the microorganisms producing EVs have a variance and do not always correspond with the microorganisms found in human breast milk. Additionally, this research indicates that EVs may aid in the vertical transmission of commensal microbes from women to their offspring. In addition, because such vesicles are abundant in microRNAs, they may also affect children's mucosal defenses and boost the range of their intestinal microbiomes (Macia et al. 2019).

Furthermore, Wang et al. (2019) examined the exosomes derived from human milk by contrasting the milk of term pregnancies to that of preterm pregnancies. Researchers observed that peptide contents of preterm and term milk exosomes differed significantly. These results suggest a possible variation in the bacterial EV makeup of human milk, which may impact a child's future health-related consequences. Human breast milk contains exosomal miRNAs (exomiRs) encased within the exosomes. These exomiRs are transported from the human milk to the newborn via the alimentary canal and may be extremely important for the immunological functioning of the child. The development of thymic regulatory T cells (Tregs), which prevents Th2-mediated atopic sensitization and atopic effector responses, is another benefit of milk-derived miRNAs. Exosomes generated from human and bovine milk exhibit exceptionally high concentrations of immune-modulatory miRNAs (miR-155, miR-146a, and miR-21) that have been demonstrated to be linked to thymic Treg development (Mirza et al. 2019).

It is crucial to separate EVs to examine their biochemical components and/or explicitly investigate their function to obtain familiarity with the roles played by EVs found in milk. The prevailing isolation procedures for milk EVs have been modified from procedures created for EV isolation from plasma or conditioned cultured media. Most of these techniques encompass size exclusion chromatography (SEC), density gradient centrifugation, commercial precipitation kits, and ultracentrifugation (Hu et al. 2021).

9.13 Conclusions and Future Perspectives

The "ideal" source of nutrition for infants is human milk. It is recommended to breastfeed, since it promotes healthy newborn development. Breastfeeding exposes the newborn to milk-associated bacteria, which may influence the newborn's microbiological, metabolic, and immunological health. Since this could alter the perception of the microbial BM ecosystem and have implications for infant health, more research is required to understand the connections between the microbial components of the HBM microbiome, including bacteria and fungi, archaea, and viruses. To improve our understanding of HBM microbiome regulation, it is essential to consider the complexity of BM and the variables that influence its composition.

Acknowledgments Saqib Hassan thanks the Indian Council of Medical Research (ICMR) for ICMR Research Associateship (Project ID: 2019-6981).

Ishfaq Hassan Mir gratefully acknowledges the Indian Council of Medical Research (ICMR), New Delhi, India, for financial assistance in the form of a Senior research fellowship [ICMR-SRF; S.No. 45/17/2022-/BIO/BMS].

References

Biagi E, Aceti A, Quercia S, Beghetti I, Rampelli S, Turroni S, Soverini M, Zambrini AV, Faldella G, Candela M, Corvaglia L, Brigidi P (2018) Microbial community dynamics in mother's milk and infant's mouth and gut in moderately preterm infants. Front Microbiol 9:2512. https://doi.org/10.3389/fmicb.2018.02512

Boix-Amorós A, Collado MC, Mira A (2016) Relationship between milk microbiota, bacterial load, macronutrients, and human cells during lactation. Front Microbiol 7:492. https://doi.org/10.3389/fmicb.2016.00492

Boix-Amorós A, Puente-Sánchez F, du Toit E, Linderborg KM, Zhang Y, Yang B, Salminen S, Isolauri E, Tamames J, Mira A, Collado MC (2019) Mycobiome profiles in breast milk from healthy women depend on mode of delivery, geographic location, and interaction with bacteria. Appl Environ Microbiol 85(9):e02994-18. https://doi.org/10.1128/aem.02994-18

Bouchard DS, Seridan B, Saraoui T, Rault L, Germon P, Gonzalez-Moreno C, Nader-Macias FME, Baud D, François P, Chuat V, Chain F, Langella P, Nicoli J, Le Loir Y, Even S (2015) Lactic acid bacteria isolated from bovine mammary microbiota: potential allies against bovine mastitis. PLoS One 10(12):e0144831. https://doi.org/10.1371/journal.pone.0144831

Brown L, Wolf JM, Prados-Rosales R, Casadevall A (2015) Through the wall: extracellular vesicles in gram-positive bacteria, mycobacteria and fungi. Nat Rev Microbiol 13(10):620–630

Cabrera-Rubio R, Collado MC, Laitinen K, Salminen S, Isolauri E, Mira A (2012) The human milk microbiome changes over lactation and is shaped by maternal weight and mode of delivery. Am J Clin Nutr 96(3):544–551. https://doi.org/10.3945/ajcn.112.037382

Davé V, Street K, Francis S, Bradman A, Riley L, Eskenazi B, Holland N (2016) Bacterial microbiome of breast milk and child saliva from low-income Mexican-American women and children. Pediatr Res 79(6):846–854. https://doi.org/10.1038/pr.2016.9

Dias R, Eller M, Duarte V, Pereira A, Silva C, Mantovani H, Oliveira L, Silva E, Depaula S (2013) Use of phages against antibiotic-resistant Staphylococcus aureus isolated from bovine mastitis. J Anim Sci 91:3930–3939. https://doi.org/10.2527/jas.2012-5884

Fitzstevens JL, Smith KC, Hagadorn JI, Caimano MJ, Matson AP, Brownell EA (2017) Systematic review of the human milk microbiota. Nutr Clin Pract 32(3):354–364. https://doi.org/10.1177/0884533616670150

Gay MCL, Koleva PT, Slupsky CM, du Toit E, Eggesbo M, Johnson CC, Wegienka G, Shimojo N, Campbell DE, Prescott SL, Munblit D, Geddes DT, Kozyrskyj AL, Dahl C, Haynes A, Hsu P, Mackay C, Penders J, Renz H et al (2018) Worldwide variation in human milk metabolome: indicators of breast physiology and maternal lifestyle? Nutrients 10(9):1151. https://doi.org/10.3390/nu10091151

Gensollen T, Iyer SS, Kasper DL, Blumberg RS (2016) How colonization by microbiota in early life shapes the immune system. Science 352(6285):539–544

Gomes F, Saavedra MJ, Henriques M (2016) Bovine mastitis disease/pathogenicity: evidence of the potential role of microbial biofilms. Pathog Dis 74(3):ftw006. https://doi.org/10.1093/femspd/ftw006

Gomez-Gallego C, Garcia-Mantrana I, Salminen S, Collado MC (2016) The human milk microbiome and factors influencing its composition and activity. Semin Fetal Neonatal Med 21(6):400–405. https://doi.org/10.1016/j.siny.2016.05.003

Gómez-Gallego C, Morales JM, Monleón D, du Toit E, Kumar H, Linderborg KM, Zhang Y, Yang B, Isolauri E, Salminen S, Collado MC (2018) Human breast milk NMR metabolomic profile

across specific geographical locations and its association with the milk microbiota. Nutrients 10(10):1355. https://doi.org/10.3390/nu10101355
Hirai C, Ichiba H, Saito M, Shintaku H, Yamano T, Kusuda S (2002) Trophic effect of multiple growth factors in amniotic fluid or human milk on cultured human fetal small intestinal cells. J Pediatr Gastroenterol Nutr 34(5):524–528
Hitchcock JR, Hughes K, Harris OB, Watson CJ (2020) Dynamic architectural interplay between leucocytes and mammary epithelial cells. FEBS J 287(2):250–266. https://doi.org/10.1111/febs.15126
Hu Y, Thaler J, Nieuwland R (2021) Extracellular vesicles in human milk. Pharmaceuticals (Basel) 14(10):1050. https://doi.org/10.3390/ph14101050
Jeurink PV, van Bergenhenegouwen J, Jimenez E, Knippels LM, Fernandez L, Garssen J et al (2013) Human milk: a source of more life than we imagine. Benef Microbes 4(1):17–30
Jiménez E, Delgado S, Maldonado A, Arroyo R, Albújar M, García N, Jariod M, Fernández L, Gómez A, Rodríguez JM (2008) Staphylococcus epidermidis: a differential trait of the fecal microbiota of breastfed infants. BMC Microbiol 8:143. https://doi.org/10.1186/1471-2180-8-143
Jiménez E, de Andrés J, Manrique M, Pareja-Tobes P, Tobes R, Martínez-Blanch JF, Codoñer FM, Ramón D, Fernández L, Rodríguez JM (2015) Metagenomic analysis of milk of healthy and mastitis-suffering women. J Hum Lact 31(3):406–415. https://doi.org/10.1177/0890334415585078
Khodayar-Pardo P, Mira-Pascual L, Collado MC, Martínez-Costa C (2014) Impact of lactation stage, gestational age and mode of delivery on breast milk microbiota. J Perinatol 34(8):599–605. https://doi.org/10.1038/jp.2014.47
Kim SY, Yi DY (2020) Analysis of the human breast milk microbiome and bacterial extracellular vesicles in healthy mothers. Exp Mol Med 52(8):1288–1297. https://doi.org/10.1038/s12276-020-0470-5
Lazar V, Ditu L-M, Pircalabioru G, Picu A, Petcu L, Cucu N, Chifiriuc M (2019) Gut microbiota, host organism, and diet trialogue in diabetes and obesity. Front Nutr 6:21. https://doi.org/10.3389/fnut.2019.00021
Macia L, Nanan R, Hosseini-Beheshti E, Grau GE (2019) Host- and microbiota-derived extracellular vesicles, immune function, and disease development. Int J Mol Sci 21(1):107
Markle J, Frank D, Mortin-Toth S, Robertson C, Feazel LM, Rolle-Kampczyk U, von Bergen M, McCoy K, Macpherson A, Danska J (2013) Sex differences in the gut microbiome drive hormone-dependent regulation of autoimmunity. Science 339:1084–1088
McGuire MK, McGuire MA (2017) Got bacteria? The astounding, yet not-so-surprising, microbiome of human milk. Curr Opin Biotechnol 44:63–68. https://doi.org/10.1016/j.copbio.2016.11.013
Mirza AH, Kaur S, Nielsen LB, Størling J, Yarani R, Roursgaard M, Mathiesen ER, Damm P, Svare J, Mortensen HB, Pociot F (2019) Breast milk-derived extracellular vesicles enriched in exosomes from mothers with type 1 diabetes contain aberrant levels of microRNAs. Front Immunol 10:2543
Moossavi S, Sepehri S, Robertson B, Bode L, Goruk S, Field CJ, Lix LM, de Souza RJ, Becker AB, Mandhane PJ, Turvey SE, Subbarao P, Moraes TJ, Lefebvre DL, Sears MR, Khafipour E, Azad MB (2019) Composition and variation of the human milk microbiota are influenced by maternal and early-life factors. Cell Host Microbe 25(2):324–335.e4. https://doi.org/10.1016/j.chom.2019.01.011
Murphy K, Curley D, O'Callaghan TF, O'Shea CA, Dempsey EM, O'Toole PW, Ross RP, Ryan CA, Stanton C (2017) The composition of human milk and infant faecal microbiota over the first three months of life: a pilot study. Sci Rep 7:40597
Ojo-Okunola A, Nicol M, du Toit E (2018) Human breast milk bacteriome in health and disease. Nutrients 10(11):1643
Perez PF, Doré J, Leclerc M, Levenez F, Benyacoub J, Serrant P, Segura-Roggero I, Schiffrin EJ, Donnet-Hughes A (2007) Bacterial imprinting of the neonatal immune system: lessons from maternal cells? Pediatrics 119(3):e724–e732. https://doi.org/10.1542/peds.2006-1649

Porter J, Anderson J, Carter L, Donjacour E, Paros M (2016) In vitro evaluation of a novel bacteriophage cocktail as a preventative for bovine coliform mastitis. J Dairy Sci 99(3):2053–2062. https://doi.org/10.3168/jds.2015-9748

Sakwinska O, Bosco N (2019) Host microbe interactions in the lactating mammary gland. Front Microbiol 10:1863. https://doi.org/10.3389/fmicb.2019.01863

Schenkman DI, Berman DT, Yandell BS (1985) Effect of stage of lactation on transport of colloidal carbon or Staphylococcus aureus from the mammary gland lumen to lymph nodes in Guinea pigs. J Dairy Res 52(4):491–500. https://doi.org/10.1017/s0022029900024432

Stinson LF, Sindi ASM, Cheema AS, Lai CT, Mühlhaüsler BS, Wlodek ME, Payne MS, Geddes DT (2021) The human milk microbiome: who, what, when, where, why, and how? Nutr Rev 79(5):529–543. https://doi.org/10.1093/nutrit/nuaa029

Urbaniak C, Angelini M, Gloor GB, Reid G (2016) Human milk microbiota profiles in relation to birthing method, gestation and infant gender. Microbiome 4:1. https://doi.org/10.1186/s40168-015-0145-y

Valdes AM, Walter J, Segal E, Spector TD (2018) Role of the gut microbiota in nutrition and health. BMJ 361:k2179. https://doi.org/10.1136/bmj.k2179

Valles-Colomer M, Falony G, Darzi Y, Tigchelaar EF, Wang J, Tito RY, Schiweck C, Kurilshikov A, Joossens M, Wijmenga C, Claes S, Van Oudenhove L, Zhernakova A, Vieira-Silva S, Raes J (2019) The neuroactive potential of the human gut microbiota in quality of life and depression. Nat Microbiol 4(4):623–632. https://doi.org/10.1038/s41564-018-0337-x

Wallis JK, Krömker V, Paduch J-H (2019) Biofilm challenge: lactic acid bacteria isolated from bovine udders versus Staphylococci. Foods 8(2):79. https://doi.org/10.3390/foods8020079

Wang X, Yan X, Zhang L, Cai J, Zhou Y, Liu H, Hu Y, Chen W, Xu S, Liu P, Chen T, Zhang J, Cao Y, Yu Z, Han S (2019) Identification and peptidomic profiling of exosomes in preterm human milk: insights into necrotizing enterocolitis prevention. Mol Nutr Food Res 63(13):e1801247

Ward TL, Hosid S, Ioshikhes I, Altosaar I (2013) Human milk metagenome: a functional capacity analysis. BMC Microbiol 13:1–12

Williams JE, Carrothers JM, Lackey KA, Beatty NF, York MA, Brooker SL, Shafii B, Price WJ, Settles ML, McGuire MA, McGuire MK (2017) Human milk microbial community structure is relatively stable and related to variations in macronutrient and micronutrient intakes in healthy lactating women. J Nutr 147(9):1739–1748

Wyatt RG, Mata LJ (1969) Bacteria in colostrum and milk of Guatemalan Indian women. J Trop Pediatr 15(4):159–162. https://doi.org/10.1093/tropej/15.4.159

Zhu F, Zozaya C, Zhou Q, De Castro C, Shah PS (2021) SARS-CoV-2 genome and antibodies in breastmilk: a systematic review and meta-analysis. Arch Dis Child Fetal Neonatal Ed 106(5):514–521

The Dynamics of Skin Microbiome: Association of Microbiota with Skin Disorders and Therapeutic Interventions

10

P. S. Seethalakshmi, Saqib Hassan, George Seghal Kiran, Pallaval Veera Bramhachari, and Joseph Selvin

Abstract

Skin, the body's largest organ, harbors a unique microbial community crucial for maintaining skin health. The cutaneous immune system and skin microbiome keep the pathogens at bay, and any anomaly generated in this tightly linked network culminates in skin abnormalities. Dysbiotic microbiome conditions are often observed in skin disorders suggesting their prominent role in protecting skin health. This chapter will discuss the components of the normal skin microbiome and its interactions with the immune system and the modern environment. In addition, the implications of cosmetics on skin microbiome, the association of microbiota with skin disorders, and therapeutic interventions have been discussed in detail.

P. S. Seethalakshmi · J. Selvin
Department of Biotechnology, School of Bio and Chemical Engineering, Sathyabama Institute of Science and Technology, Chennai, India

S. Hassan
Department of Biotechnology, School of Bio and Chemical Engineering, Sathyabama Institute of Science and Technology, Chennai, India

Future Leaders Mentoring Fellow, American Society for Microbiology, Washington, USA

G. S. Kiran
Department of Food Science and Technology, School of Life Sciences, Pondicherry University,
Puducherry, India

P. V. Bramhachari (✉)
Department of Biosciences & Biotechnology, Krishna University, Machilipatnam, Andhra Pradesh, India

P. Veera Bramhachari (ed.), *Human Microbiome in Health, Disease, and Therapy*, https://doi.org/10.1007/978-981-99-5114-7_10

Keywords

Dysbiosis · Rebiosis · Atopic dermatitis · Psoriasis · Acne vulgaris

10.1 Introduction

Before the advent of metagenomics, culture-based approaches were used to study the skin microbiota. The laboratory conditions do not permit the growth of most microorganisms, and many microorganisms outgrow the others (Staley and Konopka 1985), complicating microbiome studies. The advancement of DNA sequencing technologies introduces a new culture-free platform termed "metagenomics" to analyze the skin microbiome without bias (Handelsman 2004). Metagenomic analysis of microbiomes provides information regarding the identity of microbial populations and the relative abundance of each species in the population (Chen and Tsao 2013). This information becomes valuable in identifying the connection of skin microbiome to various diseases and the response of various therapies in the microbiota of the skin. Skin microbiota is classified into two types: resident and transient. Resident microbes are skin commensals that benefit the host. They reside on the skin surface for a long time and reestablish themselves in perturbation cases. Transient microbes, on the other hand, are mostly temporary residents of the skin that persist in the skin for hours or days and often skin surfaces from the surrounding environment (Kong and Segre 2012). The skin microbiome is dynamic and is under constant contact and influence of the external environment. The skin microbiome is susceptible to changes resulting from the everyday use of hygiene and cosmetic products. A symbiotic relationship between skin commensals and the host must be established to sustain healthy skin. Any interruptions that can upset this equilibrium pave way to infections and inflammatory responses in the skin. Many therapeutic interventions involving topical prebiotics, synbiotics, and microbial metabolites have been reported to influence skin microbiota positively. The upcoming sessions will deliberate on these aspects with future microbiome research prospectus.

10.2 The Components of the Skin Microbiome

The skin resides millions of bacteria, fungi, archaea, and viruses that make up the skin microbiome. The components of skin microbiota act as a physical barrier by preventing the invasion of pathogens. The human skin has diverse microenvironmental niches that are dry (palm and volar forearm), oily (chest, face, and back), and moist (groin, bend of the elbow, and knee) due to which the composition of microbial community varies from one physiological site to the other (Byrd et al. 2018). The microbiota can be unique to anatomic sites of the skin, such as follicles, sebaceous glands, and sweat glands (Kong and Segre 2012). The establishment of the skin microbiome starts during birth. Infants born vaginally acquire microbial members from the mother's vagina, whereas infants born via C-section receive microbes from the mother's skin (Dominguez-Bello et al. 2010; Mueller et al. 2015). The bacterial population of the buttocks, forehead, and arm changes as a function of age in infants, with a decline in the population of *Staphylococcus* and *Streptococcus* and expansion of other lesser predominant genera (Capone et al. 2011).

The microbial community changes during puberty due to hormonal fluctuations that influence sebum secretion and favor the colonization of lipophilic microorganisms such as *Malassezia* (Jo et al. 2016), *Cutibacterium*, and *Corynebacterium* (Oh et al. 2012). The adult skin microbiome comprises 19 bacterial phyla in which the phyla Actinobacteria, Firmicutes, Proteobacteria, and Bacteroidetes predominate (Grice et al. 2009). Four percent of skin microbial genes belong to archaea, where the phyla Thaumarchaeota predominates, followed by Euryarchaeota (Probst et al. 2013). Healthy skin hosts eukaryotic viruses belonging to Circoviridae, Papillomaviridae, and Polyomaviridae (Foulongne et al. 2012). The skin microbiota is influenced by several intrinsic factors such as metabolism, sleep, stress, gender, immunity, genetics, and external factors such as hygiene, climate, pollution, exposure to chemicals and sunlight, etc. (Skowron et al. 2021). Most skin microbiome studies have succeeded in characterizing the bacterial members of the microbiome, but the functional role of archaea and viromes is poorly understood and demands further investigation.

10.3 The Skin Microbiome-Immune System Interplay

The skin's immune response to wounds and infections could affect the colonization of the skin's microbiota. Pattern recognition receptor (PRR) of keratinocytes can get activated upon recognition of the pathogen-associated molecular pattern (PAMP) of a pathogen, which initiates the production of several antimicrobial peptides that kills pathogens like bacteria, viruses, and fungi. However, the immune system tolerates the skin commensals either by activating toll-like receptor (TLR) inhibitors or attenuating the TLR expression on cell surfaces (Fukao and Koyasu 2003; Strober 2004). To avoid unnecessary systemic inflammatory immune responses and to suppress adjuvant properties of skin commensals, the immune system has evolved to generate specific and compartmentalized responses. The skin commensals can amplify the innate immune response to pathogens by activating immune genes. *S. epidermidis* induces the expression of antimicrobial peptides (AMP) by keratinocytes via a TLR2-dependent mechanism (Lai et al. 2010). *S. epidermidis* produces phenol-soluble modulins that can act against the skin pathogen *S. aureus* (Cogen et al. 2010). Lipoteichoic acid, a ligand of TLR-2 receptor, derived from *S. epidermidis* has been associated with decreased TLR-3 inflammation mediated by keratinocytes (Lai et al. 2009) and has antiviral potential against vaccinia viruses by activating cathelicidin-producing mast cells (Wang et al. 2012). The metabolic products of skin commensals can also contribute to the elimination of potential pathogens. *Corynebacterium accolens*, a commensal seen around the nasal region, can metabolize the triacylglycerols into free fatty acids with potent antipneumococcal activity (Bomar et al. 2016). Short-chain fatty acids synthesized by *Cutibacterium acnes* through fermentation of glycerol, naturally found in the skin, can restrain the growth of *S. aureus* by bringing down the intracellular pH (Shu et al. 2013). AMPs from *S. epidermidis* and *Staphylococcus hominis* can synergize with human AMP LL-37 in defending *S. aureus* (Nakatsuji et al. 2017). During cutaneous infections, skin commensals have autonomous control in tuning local T- cell functioning and local inflammatory response (Naik et al. 2012). Disrupting the skin microbiome, immune system interaction can seriously affect the host's health. Mice treated with CR5 antagonist interrupted the complement pathway, resulting in an abnormal increase of Actinobacteria in the skin. The treatment also caused an overall reduction of immune filtration and expression of PRRs, chemokines, cytokines, and AMPs (Chehoud et al. 2013) (Fig. 10.1).

The skin microbiota also has a crucial role in host immune stimulation. The germ-free mice without bacterial colonization decreased immune genes' expression level in complement activation and binding (Chehoud et al. 2013). The immune system tolerates the skin commensals at the skin's surface. Penetration of bacteria to the dermis will induce the activation of commensal-reactive T cells in lymph nodes (Shen et al. 2014). The skin is the largest reservoir for memory T cells since it is home to approximately 20 billion effector lymphocytes (Clark et al. 2006). The cross talk of the skin microbiome immune system varies among innate and adaptive counterparts of the host immune system. A study by Scholz et al. (2014) discusses that in mice's adaptive immune system, MyD88-dependent innate responses and Langerhans cells have no control over the skin microbiome, but Rag1$^{-/-}$(a component of the adaptive immune system) altered the microbial consortia of mucosal surfaces in a steady-state manner. The alteration of the skin microbiome in immune-deficient individuals may change the host's immune response to pathogens. Patients

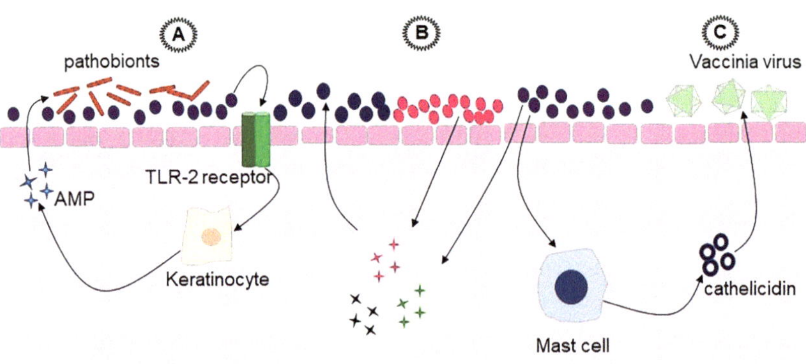

Fig. 10.1 The beneficial role of skin commensals in protecting the skin. A. *S. epidermidis* activates keratinocytes via TLR2 for expression of AMP that can kill pathogens of the skin B. *S. epidermidis* and *S. hominis* synthesize AMPs which can act synergistically with the AMP LL-37 produced by the host immune system against S. aureus. C. *S. epidermidis* activates mast cells to synthesize cathelicidin that acts against vaccinia virus

with *STAT1* and *STAT3* primary immunodeficiency inactivated primary leukocytes and subsequent cytokine response to *S. aureus* and *Candida albicans* associated with enrichment of *Acinetobacter* sp. (Smeekens et al. 2014). Skin microbiome plays an inevitable role in UV-induced immune suppression. The presence of skin microbiota nullifies the immunosuppressive response upon exposure to UV-B radiation and increases neutrophilic infiltration and epidermal hyperplasia. In contrast, its absence enhanced the number of macrophages and mast cells, suggesting that microbiota is involved in the modulation of cutaneous immune gene expression (Patra et al. 2019). Skin commensals can inhibit the colonization of other bacteria and may have a protective role in individuals where the adaptive immune system is disabled. *S. epidermidis* was found abundantly in the skin of mice with long-term immunodeficiency (Garcia-Garcerà et al. 2012). However, the extent to which commensals can fill the adaptive immune response gap in immune-compromised hosts will need further research.

10.4 Dysbiosis: The Common Anomaly in Skin Abnormalities

Dysbiosis results from the loss of microbial diversity and skin commensals with a subsequent increase of pathogens in the microbiota (Petersen and Round 2014). Growing evidence shows the obvious involvement of skin microbiota dysbiosis in skin disorders. Psoriasis vulgaris is a chronic inflammatory disease resulting from the production of cytokines tumor necrosis factor (TNF), interferon-γ, IL-17, and IL-22 and stimulation of Th-22, Th-1, and Th-17 cells (Lowes et al. 2008; Nograles et al. 2009). Changes in the microbial community of the skin at various physiological sites have been observed in psoriatic patients. Psoriatic lesions on the back were observed to be dominated by *Malassezia restricta*, *Brevibacterium*, and *Kocuria*, whereas *Malassezia sympodialis and Gordonia* dominated the microbiota of elbow psoriatic lesions. In addition, the coexistence of *Lactobacillus*, *Kocuria*, and *Streptococcus* with *Saccharomyces* has been found in psoriatic lesions from the elbow (Stehlikova et al. 2019). *Malassezia* sp. has been proven to promote the secretion of pro-inflammatory cytokines IL-1, IL-8, and IL-6 and TNF-α in keratinocytes (Watanabe et al. 2001), suggesting its involvement in predisposing to psoriatic conditions. An increase in the representation of *Corynebacterium kroppenstedtii*, *Corynebacterium simulans*, *Neisseria* sp., and *Finegoldia* sp. in psoriatic patients was reported by Olejniczak-Staruch et al. (2021).

S. aureus appears to be common in psoriatic skin with and without lesions, and colonization of *S. aureus* in mice models indicated the triggering of Th-17 polarization, confirming the involvement of the bacteria in psoriasis. Also, a significant fall in the population of skin commensals *C. acnes* and *S. epidermidis* in psoriatic patients concludes the role of dysbiosis in psoriasis (Chang et al. 2018). The elbow region in psoriatic patients has a uniform metagenome with a difference in metabolic functioning which entails amino acid and carbohydrate metabolism. It was revealed that protein export (ko03060) and bacterial secretion (ko0370) were scarce in psoriatic patients compared to healthy individuals. The altered microbiota of psoriatic patients can change the metabolic composition of plasma and enrich the branched-chain amino acid metabolism pathway of amino acids (isoleucine, valine, and leucine) and metabolism of lipids (α-linolenic acid, linoleic acid (Chen et al. 2021).

Another chronic inflammatory disorder linked to the dysbiotic conditions of the skin microbiome is atopic dermatitis (AD), characterized by pruritic skin lesions and eczema (Luger et al. 2021). AD patients exhibited an overall decrease in the diversity of skin microbiota and an increase in the *Staphylococcus* sp. (Pascolini et al. 2011; Kong et al. 2012). Dysbiotic states increase the skin's pH, favoring the colonization of pathobiont, such as *S. aureus*, to the skin (Rippke et al. 2004; Knor et al. 2011). The expression of host AMPs DEFB-2, DEFB-3, and cathelicidin is low in AD patients (Ong et al. 2002), prompting *S. aureus* to outgrow the skin microbiota. *S. aureus* colonization aggravates the severity of atopic dermatitis due to production of superantigens (Mallinckrodt 2000). The expression of Jun B in keratinocytes is a crucial factor determining the severity of AD. The loss of Jun B can increase the expression of MyD88 and IL-36 in keratinocytes favoring the colonization of *S. aureus* and dysbiosis of skin microbiota (Uluçkan et al. 2019).

On the contrary, skin microbiome studies in 12-month-old AD-affected infants did not show colonization of *S. aureus* in dysbiotic conditions (Kennedy et al. 2017). An overabundance of *S. epidermidis,* too, can complicate AD conditions due to the deleterious effect of cysteine protease produced by the bacteria on the skin (Cau et al. 2021). In the cases of acute AD, extensive colonization of *Corynebacterium bovis* has been reported by Kobayashi et al. (2015). Changes in the microbiota's metabolic profile have also been reported in AD subjects. The enzymes involved in citrulline and ammonia metabolism were significantly higher in the AD-associated microbiota (Chng et al. 2016). Filaggrin is an important protein involved in maintaining the integrity of the skin barrier, and mutations in the gene encoding it (FLG) have been frequently reported as a genetic factor in AD patients (Irvine et al. 2011). A recent finding suggests that *Staphylococcus caprae* was found more abundantly in AD patients with *FLG* mutation.

Acne vulgaris is a common chronic inflammatory skin disorder that affects 85% of adolescents worldwide (Bhate and Williams 2013). Comedone formation, follicular hyperkeratinization, excess sebum production, and inflammatory immune response generated by *C. acnes* are the cardinal factors that can contribute to the formation of acne (Bhambri et al. 2009; Bellew et al. 2011). The skin microbiome of acne-affected patients harbors bacteria belonging to Proteobacteria and Firmicutes, predominantly. In severe acnes, dysbiotic condition is more pronounced with abundance in the population of *Faecalibacterium*, *Klebsiella*, *Odoribacter*, and *Bacteroides* (Li et al. 2019). *C. acnes* produces various virulence factors like polyunsaturated fatty acid isomerase, lipase, dermatan sulfate-binding adhesins, and hemolysins (Christensen and Brüggemann 2014). The acne-associated phylotypes induce a pro-inflammatory response by stimulation of Th-1 and Th-17 resulting from enhanced expression of IFN-γ and IL-17 (Yu et al. 2016). *S. epidermidis*, a skin commensal, can alleviate the inflammatory response of *C. acnes* by activating miR-143 in keratinocytes responsible for reduced stability of TLR-2 mRNA (Xia

Table 10.1 An overview of dysbiotic states defined in some skin disorders

Skin condition	Abundant microbes in the skin under dysbiotic states	References
Aging	The abundance of species *Proteobacteria* and *Actinobacteria*	Kim et al. (2019)
Vitiligo	*Janibacter* and *Brevundimonas* (non-lesional) and *Enhydrobacter*, *Paracoccus* and *Staphylococcus* (lesional)	Ganju et al. (2016)
Dandruff	*Malassezia restricta* and *Staphylococcal* sp	Wang et al. (2015)
Melanoma	*Fusobacterium* and *Trueperella*	Mrázek et al. (2019)
Rosacea	*Campylobacter ureolyticus*, *Corynebacterium kroppenstedtii*	Rainer et al. (2020)

et al. 2016). The antibacterial components, polymorphic toxins, and epidermidin secreted by *S. epidermidis* control the population of *C. acnes* in the skin (Christensen et al. 2016). In contrast, some studies suggest the abundance of *C. acnes* in the follicles than *S. epidermidis* (Fitz-Gibbon et al. 2013; Barnard et al. 2016); higher counts of *S. epidermidis* were found in some follicular samples than *C. acnes* (Nakatsuji et al. 2013). *Staphylococcus capitis* has also been reported to produce phenol-soluble modulins that act specifically against *C. acnes* with less toxicity to keratinocytes and other commensals (O'Neill et al. 2020). Hence, defining the pathophysiology of acne vulgaris is difficult because of the disparities in different individuals' relative abundances of commensals.

Although the research in understanding the role of the skin microbiome in cancer is yet to progress, the involvement of skin microbiota has been reported in many studies. Mirvish et al. (2013) discuss the involvement *S. aureus*, *Chlamydophila pneumonia*, and *Borrelia burgdorferi* in cutaneous T-cell lymphoma. Vitiligo, an autoimmune disorder that causes loss of melanin resulting in discolored patches, has been linked to skin dysbiosis and lesions enriched with Proteobacteria, *Mycoplasma*, and *Streptococcus* (Bzioueche et al. 2021). Radiotherapy can cause adverse situations like radiotherapy-induced dermatitis and has been linked to the overrepresentation of *Stenotrophomonas*, *Staphylococcus*, and *Pseudomonas*, resulting in delayed healing. Hence, microbiome profiling can also be used as a prognosis for radiation therapies (Ramadan et al. 2021). Microbiome alterations have been linked to many other skin conditions, summarized in Table 10.1.

10.5 The Cross Talk of Skin Microbiome and Living Environment

The biodiversity of the human microbiome is a function of the biodiversity of its immediate environment and interactions with it (Von Hertzen et al. 2015). The skin microbiome changes more frequently over a while than the gut suggesting the ease with which exposure to environments can alter the microbial consortia of the skin

(Prescott et al. 2017). The skin microbiota varies between urban and rural populations due to the differences in exposure to soil, aquatic, and host allied microbial sources originating from the inhabitants with more indoor jobs in urban settings (Ying et al. 2015). People owning pets and have more exposure to outdoor environments possess diverse skin microbiota (Ross et al. 2017). Traditional environments and the microflora it comprises have been linked to exerting immune stimulation in humans. Exposure to green environments like agricultural lands was associated with less atopic sensitization (Ruokolainen et al. 2015). Children residing in rural environments have less risk of acquiring allergic conditions as they develop immune tolerance due to enhanced exposure to microbial antigens in farms (Riedler et al. 2000). The dust in rural houses harbors more diverse microbes and therefore carries a broad range of PAMPs than those in urban houses (Alenius et al. 2009). This also accounts for the decreased risk of acquiring allergic reactions in residents in rural settings.

Moreover, the skin microbiota of people residing in urban areas has also influenced the biodiversity in the vegetation of the adjacent environment (Hanski et al. 2012; Ruokolainen et al. 2015). Despite a similar richness in the skin microbiota of people from rural and urban places, the rural subjects possessed a larger intragroup variation than the urban subjects (Ying et al. 2015). Cohabiting couples shared a similar profile of overall skin microbiota, although the microbiota of thigh regions was more associated with biological sex than cohabiting partners (Ross et al. 2017). Even the altitude of living can significantly impact shaping the skin microbiota. Zeng et al. (2017) elucidate that the human population residing at high altitudes showed an abundance in the microbial population of Cellulomonadaceae, Xanthomonadaceae, *Paenibacillus*, *Arthrobacter*, and *Carnobacterium*. Moreover, the skin microbiome of humans and pigs residing in high altitudes was more similar, suggesting the possibility of convergent evolution of microbiota to adapt to new environments. The endogamous agriculturist Indian subpopulation showed unique microbial taxa in the skin microbiome, which included *Corynebacterium*, *Staphylococcus*, *Alloiococcus*, *Peptoniphilus*, *Streptococcus*, and *Anaerococcus*, indicating the role of environmental factors along with diet and host genetics shaping the skin microbiota (Chaudhari et al. 2020).

Westernization has resulted in the loss of microbial communities, and genes that were once part of the ancestral microbiome and have coevolved with the human body (Segata 2015). Urbanization can profoundly affect the microbial community of houses and increase the abundance of skin-associated microorganisms in urban dwellings. Frequent footwear use increased the abundance of *Staphylococci* sp. in the feet, and skin surfaces were enriched with skin-associated bacteria more than environmental bacteria in urban subjects (McCall et al. 2020). The high prevalence of human pathogens in urban environments increases people's susceptibility to infectious diseases and inflammatory diseases associated with pathogens. A study conducted by Cordain et al. (2002) found that the Western population was more prone to acne vulgaris, whereas indigenous tribes were unaffected by acne vulgaris (Steiner 1946). These studies imply that the environment and human microbiome are tightly interlinked, and any perturbations to the living environment can affect the microbiota and result in health disorders. Our knowledge regarding the link between environmental health, human health, and human microbiota is limited and has to progress further.

10.6 Implications of Cosmetic Products on Skin Microbiota

In vitro, ex vivo, and in vivo studies traditionally assess cosmetic regimens. In vitro studies are focused on assaying the antagonistic activity of cosmetic ingredients against skin pathogens, such as growth inhibition, biofilm inhibition, and expression of virulence factors. Ex vivo studies assess the same characteristics of the cosmetic ingredient in living tissues extracted from animals. In vivo studies screen the safety, toxicity, and efficacy of the cosmetic product containing the active ingredient. Assaying the effect of cosmetics on the skin is an innovative approach since it gives more clarity on the potential of the cosmetic substance to achieve rebiosis and improve skin health. This is studied in vitro with monoculture or coculture studies evaluating the growth of bacteria in the presence of an active ingredient (Peeters et al. 2008) or by qPCR tests to measure relative growth (Nakatsuji et al. 2013). 3D skin models are also currently employed to assess cosmetic ingredients as they provide favorable conditions for the growth of skin commensals. However, these models cannot provide the actual physiological response and represent the whole skin microbiome.

Using cosmetics reduced the population of *Staphylococcus*, *Cutibacterium*, and *Corynebacterium* in the cheeks; however, it increased the *Ralstonia* sp., capable of degrading cosmetic components (Lee et al. 2018). Synthetic ingredients in cosmetics were found to have less impact on enhancing skin microbial diversity (Wallen-Russell 2019). Routine skin care products can change the metabolome and microbiome of the skin, and these variations depend on the use of the product and the location where it is being applied to the body (Bouslimani et al. 2019). And antiperspirants changed the steroid levels in the armpits, contributing to a shift in microbiota toward microbes that can synthesize or biotransform steroids like *Enhydrobacter* and *Corynebacterium* (Bouslimani et al. 2019).

Pinto et al. (2021) studied the implications of 11 preservatives commonly found in cosmetic products on skin commensals such as *S. aureus*, *S. epidermidis*, and *C. acnes* using 3D skin models. Combinations of caprylyl glycol, hydroxy acetophenone, tocopherol, tetrasodium glutamate diacetate, and propanediol resulted in the rebiosis since these compounds selectively inhibited the growth of *C. acnes* and *S. aureus* without affecting the growth of *S. epidermidis*. The use of emollients in AD patients reduced the population of *Staphylococcus*. They enhanced microbial diversity indicating the importance of assaying cosmetics' ability to modulate the microbiome in the therapy of AD patients (Henley et al. 2014).

As evident from the above discussion, some studies have found skin care products to have harmful effects on the skin microbiome, whereas some have vouched for the positive influence of it on skin microbiota. Therefore, it's irrational to condemn the use of skin care products or even promote it. A reasonable solution to this dilemma is to encourage the evaluation of skin care products and their influence on the skin microbiome. Wallen-Russell and Wallen-Russell (2017) propose a benchmark value to determine whether skin care products can damage the skin microbiome. They say skin microbial diversity below 10,000 species after 150,000 sequences is unhealthy. Therefore, skin regimens need to be stringently evaluated for their effect on the skin microbiome and should be ascertained if they fall below the benchmark value of skin microbial diversity. The studies to evaluate skin care products for such analysis should be conducted over long durations to understand the length of time skin microbiota takes to readjust to a healthy state (Wallen-Russell 2019). Future studies must also focus on assessing the impact of toxic components in skin care products, such as paraben and sodium lauryl sulfate on skin microbial diversity.

10.7 Therapeutic Interventions Mediating Skin Rebiosis

Many studies have reported that reestablishing skin commensals can revert dysbiotic conditions and cure diseases (Sokol et al. 2008; Round and Mazmanian 2010; Ochoa-Reparaz et al. 2010). The reinstation of the microbial community to a healthy state was described as rebiosis (Petersen and Round 2014). Growing evidence suggesting the benefits of microbiota and its components in alleviating skin diseases, therapeutic strategies, and regimens targeted at establishing rebiosis have gained popularity. These include topical probiotics, prebiotics, synbiotics, microbial metabolites, skin postbiotics, and microbiota transplantation. Probiotics were redefined after FAO/WHO in 2001 as "live microorganisms that, when administered in adequate amounts, confer a health benefit on the host by the International Scientific Association of Probiotics and Prebiotics (ISAPP) in 2014 (Martín and Langella 2019)." Several topical probiotics were reported to be effective against AD. A topical probiotic cream consisting of lactic acid bacteria, *Streptococcus thermophilus*, administered for 2 weeks in AD patients, alleviated pruritis, scaling, and erythema (Di Marzio et al. 2003). Topical administration of *Lactobacillus johnsonii* for 3 weeks exhibited a significant drop in the load of *S. aureus* in AD lesions (Blanchet-Réthoré et al. 2017). Apart from lactic acid bacteria, bacteria originating from other environments have proven to have probiotic activities. Gram-negative bacteria *Vitreoscilla filiformis* isolated from hot springs improved seborrheic dermatitis (Gueniche et al. 2008). Application of *Lactobacillus plantarum* at 1×10^9 CFU daily to wounds suppressed inflammation due to burn wounds (Argenta et al. 2016), and collagen concentration was improved when the probiotic was used at 3×10^8 CFU daily (Satish et al. 2017). Topical application of probiotics comprising *L. plantarum* enhanced the production of IL-8 and decreased the number of polymorphonuclear cells infected chronic venous ulcers (Peral et al. 2010). A combination of probiotics and emollients has also been studied for microbiome-modulating potential. Emollient containing mannose, 4% niacinamide, 20% shea butter, and La Roche-Posay thermal spring water with *V. filiformis* decreased the severity of skin flares. It improved the skin microbiota by controlling the growth of *Staphylococcus* in AD subjects (Seité et al. 2017).

Prebiotics are defined as a selectively utilized substrate by host microorganisms conferring a health benefit (Gibson et al. 2017). Prebiotics in cosmetics can promote the growth of skin commensals (Al-Ghazzewi and Tester 2014). Topical administration of prebiotic 3′-sialyllactose enriched the *Bifidobacterium* population and activated differentiation of Treg cells in AD groups. In addition, the prebiotic suppressed the production of cytokines associated with AD, which includes IL-4, IL-5, IL-6, IL-13, IL-17, IFN-γ, TNF-α, and Tslp, indicating the therapeutic potential of the substrate in AD patients (Kang et al. 2020). Stettler et al. (2017) developed a new topical panthenol-containing emollient that increased the content of ceramide 3, cholesterol, and free fatty acids and fortified the growth of skin commensals. Balasubramaniam et al. (2020) reported the prebiotic potential of liquid coco-caprylate/caprate, a compound registered in the International Nomenclature of Cosmetic Ingredients, for the promotion of *S. epidermidis* colonization in the skin to act against UV-B. Oral administration of galactooligosaccharides was reported to improve wrinkles in the skin and reduce skin pigmentation (Jung et al. 2017; Suh et al. 2019). Topical application of cosmetic serum containing galactooligosaccharides, a prebiotic, increased skin's barrier function since its application improved water-holding capacity and reduced trans-epidermal water loss. The prebiotic also suppressed the growth of skin pathogens while promoting the growth of beneficial bacteria such as *Pediococcus* in the skin (Hong et al. 2020).

Synbiotics are "a mixture comprising live microorganisms and substrate(s) selectively utilized by host microorganisms that confers a health benefit on the host" (Swanson et al. 2020). A synbiotic formulation comprising of LAB strains *L. plantarum*, *Lactobacillus casei* ssp. *casei*, *Lactobacillus acidophilus*, *Lactobacillus gasseri*, and *Lactococcus lactis* ssp. *lactis* and konjac glucomannan hydrolysates showed antibacterial activity against *C. acnes* under in vitro conditions (Al-Ghazzewi and Tester 2014). *Lactobacillus rhamnosus* and selenium nanoparticles showed UV protective action with an SPF of 29.77 suggesting the potential of synbiotic regimens in routine skin care products (Kaur and Rath 2019). AD patients receiving daily synbiotic baths of probiotics *Bifidobacterium breve*, *Bifidobacterium animalis* subsp. *lactis*, *L. casei*, *L. gasseri*, *L. plantarum*, and *L. rhamnosus* and synbiotics maltodextrin, inulin, apple pectins were proven to significantly enhance colonization of probionts in the skin and reduce pruritis (Noll et al. 2021).

Microorganisms produce bioactive molecules that has anti-inflammatory and antimicrobial activity (Seethalakshmi et al. 2020). Lysates obtained from *V. filiformis* inhibited cutaneous inflammation by inactivating T effector cells and activating tolerogenic dendritic cells and regulatory Tr1 cells in AD mice (Volz et al. 2014). *Enterococcus faecalis* SL-5, a strain isolated from humans, produced enterocins whose topical application reduced *C. acnes* and pustule formation (Kang et al. 2009) hinting at its possible inclusion in treating acne. Bacteriocin AS-48 derived from *E. faecalis* also exhibited anti-*C. acnes* activity (Cebrián et al. 2018). *L. rhamnosus* G.G. lysate upregulated chemokine *CXCL2* and its receptor *CXCR2*, causing reepithelization in cutaneous wounds (Mohammedsaeed et al. 2015). Spermidine originating from *Streptococcus* improved lipid and collagen synthesis in aged cells, concluding the skin microbiome's role in aging (Kim et al. 2021). Postbiotics are the "preparation of inanimate microorganisms and/or their components that confers a health benefit on the host" (Salminen et al. 2021). Clinical studies in acne patients have proven that the postbiotic LactoSporin derived from *Bacillus coagulans* displayed superior activity to benzoyl peroxide in curing acne lesions (Majeed et al. 2020). Although skin microbiome transplantation studies are still in their infancy, transplantation of culturable gram-negative bacteria from healthy individuals to AD patients caused demoted growth of *S. aureus*, activation of the innate immune system, and improved barrier function (Myles et al. 2016). It can be expected that microbiome modulation with transplantation and biological regimens can replace the concurrent treatment options due to the strong association between skin microbiota and skin health.

10.8 Conclusions and Future Perspectives

Metagenomic investigations in the skin have changed our perspective of skin microbiome from a reservoir of pathogens and opportunistic pathogens to a diverse community of skin commensals and pathogens. This has led to new insights regarding the functional role of skin microbiota in maintaining skin health. Skin diseases are now being brought under microbiome and immune dysfunction. The interlink of skin dysbiosis and the immune system in skin diseases signifies the importance of maintaining microbiome-immune system homeostasis. Moreover, the negative implications of certain cosmetic regimens on skin microbiomes warrant reconsidering their use in everyday life. Western lifestyle and environment are proven to cause harmful effects on the skin microbiome, which could have severe health outcomes. Not much is known to us whether these alterations in skin microbiome caused by cosmetics and environmental changes are reversible. It would be worth investigating the interaction of skin microbiome with other microbiomes such as the lung, oral, and gut in various health disorders. Even though metagenomics has facilitated the identification of some crucial nonculturable microbial communities of the skin, the difficulty of culturing them in laboratory conditions remains the same. It can be anticipated that the multi-omics approach and synthetic biology may find new solutions to culture these microorganisms in the future.

References

Alenius H, Pakarinen J, Saris O, Andersson MA, Leino M, Sirola K et al (2009) Contrasting immunological effects of two disparate dusts–preliminary observations. Int Arch Allergy Immunol 149(1):81–90

Al-Ghazzewi FH, Tester RF (2014) Impact of prebiotics and probiotics on skin health. Benefic Microbes 5(2):99–107

Argenta A, Satish L, Gallo P, Liu F, Kathju S (2016) Local application of probiotic bacteria prophylaxes against sepsis and death resulting from burn wound infection. PLoS One 11(10):e0165294

Balasubramaniam A, Adi P, Keshari S, Sankar R, Chen CL, Huang CM (2020) Repurposing INCI-registered compounds as skin prebiotics for probiotic Staphylococcus epidermidis against UV-B. Sci Rep 10(1):1–10

Barnard E, Shi B, Kang D, Craft N, Li H (2016) The balance of metagenomic elements shapes the skin microbiome in acne and health. Sci Rep 6(1):1–12

Bellew S, Thiboutot D, Del Rosso JQ (2011) Pathogenesis of acne vulgaris: what's new, what's interesting and what may be clinically relevant. J Drugs Dermatol 10(6):582–585

Bhambri S, Del Rosso JQ, Bhambri A (2009) Pathogenesis of acne vulgaris: recent advances. J Drugs Dermatol 8(7):615–618

Bhate K, Williams HC (2013) Epidemiology of acne vulgaris. Br J Dermatol 168(3):474–485

Blanchet-Réthoré S, Bourdès V, Mercenier A, Haddar CH, Verhoeven PO, Andres P (2017) Effect of a lotion containing the heat-treated probiotic strain *lactobacillus johnsonii* NCC 533 on *Staphylococcus aureus* colonization in atopic dermatitis. Clin Cosmet Investig Dermatol 10:249

Bomar L, Brugger SD, Yost BH, Davies SS, Lemon KP (2016) Corynebacterium accolens releases antipneumococcal free fatty acids from human nostril and skin surface triacylglycerols. MBio 7(1):e01725–e01715

Bouslimani A, da Silva R, Kosciolek T, Janssen S, Callewaert C, Amir A et al (2019) The impact of skin care products on skin chemistry and microbiome dynamics. BMC Biol 17(1):1–20

Byrd AL, Belkaid Y, Segre JA (2018) The human skin microbiome. Nat Rev Microbiol 16(3):143

Bzioueche H, Sjödin KS, West CE, Khemis A, Rocchi S, Passeron T, Tulic MK (2021) Analysis of matched skin and gut microbiome of patients with vitiligo reveals deep skin dysbiosis: link with mitochondrial and immune changes. J Investig Dermatol 141(9):2280–2290

Capone KA, Dowd SE, Stamatas GN, Nikolovski J (2011) Diversity of the human skin microbiome early in life. J Investig Dermatol 131(10):2026–2032

Cau L, Williams MR, Butcher AM, Nakatsuji T, Kavanaugh JS, Cheng JY et al (2021) *Staphylococcus epidermidis* protease EcpA can be a deleterious component of the skin microbiome in atopic dermatitis. J Allergy Clin Immunol 147(3):955–966

Cebrián R, Arévalo S, Rubiño S, Arias-Santiago S, Rojo MD, Montalbán-López M et al (2018) Control of *Propionibacterium acnes* by natural antimicrobial substances: role of the bacteriocin AS-48 and lysozyme. Sci Rep 8(1):1–11

Chang HW, Yan D, Singh R, Liu J, Lu X, Ucmak D et al (2018) Alteration of the cutaneous microbiome in psoriasis and potential role in Th17 polarization. Microbiome 6(1):1–27

Chaudhari DS, Dhotre DP, Agarwal DM, Gaike AH, Bhalerao D, Jadhav P et al (2020) Gut, oral and skin microbiome of Indian patrilineal families reveal perceptible association with age. Sci Rep 10(1):1–13

Chehoud C, Rafail S, Tyldsley AS, Seykora JT, Lambris JD, Grice EA (2013) Complement modulates the cutaneous microbiome and inflammatory milieu. Proc Natl Acad Sci 110(37):15061–15066

Chen YE, Tsao H (2013) The skin microbiome: current perspectives and future challenges. J Am Acad Dermatol 69(1):143–155

Chen D, He J, Li J, Zou Q, Si J, Guo Y et al (2021) Microbiome and metabolome analyses reveal novel interplay between the skin microbiota and plasma metabolites in psoriasis. Front Microbiol 12:535

Chng KR, Tay ASL, Li C, Ng AHQ, Wang J, Suri BK et al (2016) Whole metagenome profiling reveals skin microbiome-dependent susceptibility to atopic dermatitis flare. Nat Microbiol 1(9):1–10

Christensen GJM, Brüggemann H (2014) Bacterial skin commensals and their role as host guardians. Benefic Microbes 5(2):201–215

Christensen GJ, Scholz CF, Enghild J, Rohde H, Kilian M, Thürmer A et al (2016) Antagonism between *Staphylococcus epidermidis* and *Propionibacterium acnes* and its genomic basis. BMC Genomics 17(1):1–14

Clark RA, Chong B, Mirchandani N, Brinster NK, Yamanaka KI, Dowgiert RK, Kupper TS (2006) The vast majority of CLA+ T cells are resident in normal skin. J Immunol 176(7):4431–4439

Cogen AL, Yamasaki K, Sanchez KM, Dorschner RA, Lai Y, MacLeod DT et al (2010) Selective antimicrobial action is provided by phenol-soluble modulins derived from Staphylococcus epidermidis, a normal resident of the skin. J Investig Dermatol 130(1):192–200

Cordain L, Lindeberg S, Hurtado M, Hill K, Eaton SB, Brand-Miller J (2002) Acne vulgaris: a disease of Western civilization. Arch Dermatol 138(12):1584–1590

Di Marzio L, Centi C, Cinque B, Masci S, Giuliani M, Arcieri A et al (2003) Effect of the lactic acid bacterium *Streptococcus thermophilus* on stratum corneum ceramide levels and signs and symptoms of atopic dermatitis patients. Exp Dermatol 12(5):615–620

Dominguez-Bello MG, Costello EK, Contreras M, Magris M, Hidalgo G, Fierer N, Knight R (2010) Delivery mode shapes the acquisition and structure of the initial microbiota across multiple body habitats in newborns. Proc Natl Acad Sci 107(26):11971–11975

Fitz-Gibbon S, Tomida S, Chiu BH, Nguyen L, Du C, Liu M et al (2013) *Propionibacterium acnes* strain populations in the human skin microbiome associated with acne. J Investig Dermatol 133(9):2152–2160

Foulongne V, Sauvage V, Hebert C, Dereure O, Cheval J, Gouilh MA et al (2012) Human skin microbiota: high diversity of DNA viruses identified on the human skin by high throughput sequencing. PLoS One 7(6):e38499

Fukao T, Koyasu S (2003) PI3K and negative regulation of TLR signaling. Trends Immunol 24(7):358–363

Ganju P, Nagpal S, Mohammed MH, Kumar PN, Pandey R, Natarajan VT et al (2016) Microbial community profiling shows dysbiosis in the lesional skin of vitiligo subjects. Sci Rep 6(1):1–10

Garcia-Garcerà M, Coscollà M, Garcia-Etxebarria K, Martín-Caballero J, González-Candelas F, Latorre A, Calafell F (2012) Staphylococcus prevails in the skin microbiota of long-term immunodeficient mice. Environ Microbiol 14(8):2087–2098

Gibson GR, Hutkins R, Sanders ME, Prescott SL, Reimer RA, Salminen SJ et al (2017) Expert consensus document: the international scientific association for probiotics and prebiotics (ISAPP) consensus statement on the definition and scope of prebiotics. Nat Rev Gastroenterol Hepatol 14(8):491

Grice EA, Kong HH, Conlan S, Deming CB, Davis J, Young AC et al (2009) Topographical and temporal diversity of the human skin microbiome. Science 324(5931):1190–1192

Gueniche A, Knaudt B, Schuck E, Volz T, Bastien P, Martin R et al (2008) Effects of nonpathogenic gram-negative bacterium *Vitreoscillafiliformis* lysate on atopic dermatitis: a prospective, randomized, double-blind, placebo-controlled clinical study. Br J Dermatol 159(6):1357–1363

Handelsman J (2004) Metagenomics: application of genomics to uncultured microorganisms. Microbiol Mol Biol Rev 68(4):669–685

Hanski I, von Hertzen L, Fyhrquist N, Koskinen K, Torppa K, Laatikainen T et al (2012) Environmental biodiversity, human microbiota, and allergy are interrelated. Proc Natl Acad Sci 109(21):8334–8339

Henley JB, Ing RMM, Hana Zelenkova MD (2014) Microbiome of affected and unaffected skin of patients with atopic dermatitis before and after emollient treatment. J Drugs Dermatol 13(11):1365–1372

Hong KB, Hong YH, Jung EY, Jo K, Suh HJ (2020) Changes in the diversity of human skin microbiota to cosmetic serum containing prebiotics: results from a randomized controlled trial. J Pers Med 10(3):91

Irvine AD, McLean WH, Leung DY (2011) Filaggrin mutations associated with skin and allergic diseases. N Engl J Med 365(14):1315–1327

Jo JH, Deming C, Kennedy EA, Conlan S, Polley EC, Ng WI et al (2016) Diverse human skin fungal communities in children converge in adulthood. J Investig Dermatol 136(12):2356–2363

Jung EY, Kwon JI, Hong YH, Suh HJ (2017) Evaluation of anti-wrinkle effects of duoligo, composed of lactulose and Galactooligosaccharides. Prev Nutr Food Sci 22(4):381

Kang BS, Seo JG, Lee GS, Kim JH, Kim SY, Han YW et al (2009) Antimicrobial activity of enterocins from *enterococcus faecalis* SL-5 against *Propionibacterium acnes*, the causative agent in acne vulgaris, and its therapeutic effect. J Microbiol 47(1):101–109

Kang LJ, Oh E, Cho C, Kwon H, Lee CG, Jeon J et al (2020) 3′-Sialyllactose prebiotics prevents skin inflammation via regulatory T cell differentiation in atopic dermatitis mouse models. Sci Rep 10(1):1–13

Kaur K, Rath G (2019) Formulation and evaluation of U.V. protective synbiotic skin care topical formulation. J Cosmet Laser Ther 21(6):332–342

Kennedy EA, Connolly J, Hourihane JOB, Fallon PG, McLean WI, Murray D et al (2017) Skin microbiome before development of atopic dermatitis: early colonization with commensal staphylococci at 2 months is associated with a lower risk of atopic dermatitis at 1 year. J Allergy Clin Immunol 139(1):166–172

Kim HJ, Kim JJ, Myeong NR, Kim T, Kim D, An S et al (2019) Segregation of age-related skin microbiome characteristics by functionality. Sci Rep 9(1):1–11

Kim G, Kim M, Kim M, Park C, Yoon Y, Lim DH et al (2021) Spermidine-induced recovery of human dermal structure and barrier function by skin microbiome. Commun Biol 4(1):1–11

Knor T, Meholjić-Fetahović A, Mehmedagić A (2011) Stratum corneum hydration and skin surface pH in patients with atopic dermatitis. Acta Dermatovenerol Croat 19(4):242–247

Kobayashi T, Glatz M, Horiuchi K, Kawasaki H, Akiyama H, Kaplan DH et al (2015) Dysbiosis and Staphylococcus aureus colonization drives inflammation in atopic dermatitis. Immunity 42(4):756–766

Kong HH, Segre JA (2012) Skin microbiome: looking back to move forward. J Investig Dermatol 132(3):933–939

Kong HH, Oh J, Deming C, Conlan S, Grice EA, Beatson MA et al (2012) Temporal shifts in the skin microbiome associated with disease flares and treatment in children with atopic dermatitis. Genome Res 22(5):850–859

Lai Y, Di Nardo A, Nakatsuji T, Leichtle A, Yang Y, Cogen AL et al (2009) Commensal bacteria regulate toll-like receptor 3–dependent inflammation after skin injury. Nat Med 15(12):1377

Lai Y, Cogen AL, Radek KA, Park HJ, MacLeod DT, Leichtle A et al (2010) Activation of TLR2 by a small molecule produced by *Staphylococcus epidermidis* increases antimicrobial defense against bacterial skin infections. J Investig Dermatol 130(9):2211–2221

Lee HJ, Jeong SE, Lee S, Kim S, Han H, Jeon CO (2018) Effects of cosmetics on the skin microbiome of facial cheeks with different hydration levels. Microbiology 7(2):e00557

Li CX, You ZX, Lin YX, Liu HY, Su J (2019) Skin microbiome differences relate to the grade of acne vulgaris. J Dermatol 46(9):787–790

Lowes MA, Kikuchi T, Fuentes-Duculan J, Cardinale I, Zaba LC, Haider AS et al (2008) Psoriasis vulgaris lesions contain discrete populations of Th1 and Th17 T cells. J Investig Dermatol 128(5):1207–1211

Luger T, Amagai M, Dreno B, Dagnelie MA, Liao W, Kabashima K et al (2021) Atopic dermatitis: role of the skin barrier, environment, microbiome, and therapeutic agents. J Dermatol Sci 102(3):142–157

Majeed M, Majeed S, Nagabhushanam K, Mundkur L, Rajalakshmi HR, Shah K, Beede K (2020) Novel topical application of a postbiotic, LactoSporin®, in mild to moderate acne: a randomized, comparative clinical study to evaluate its efficacy, Tolerability and Safety. Cosmetics 7(3):70

Mallinckrodt V (2000) Colonization with superantigen-producing *Staphylococcus aureus* is associated with increased severity of atopic dermatitis. Clin Exp Allergy 30(7):994–1000

Martín R, Langella P (2019) Emerging health concepts in the probiotics field: streamlining the definitions. Front Microbiol 10:1047

McCall LI, Callewaert C, Zhu Q, Song SJ, Bouslimani A, Minich JJ et al (2020) Home chemical and microbial transitions across urbanization. Nat Microbiol 5(1):108–115

Mirvish JJ, Pomerantz RG, Falo LD Jr, Geskin LJ (2013) Role of infectious agents in cutaneous T-cell lymphoma: facts and controversies. Clin Dermatol 31(4):423–431

Mohammedsaeed W, Cruickshank S, McBain AJ, O'Neill CA (2015) Lactobacillus rhamnosus G.G. lysate increases re-epithelialization of keratinocyte scratch assays by promoting migration. Sci Rep 5(1):1–11

Mrázek J, Mekadim C, Kučerová P, Švejstil R, Salmonová H, Vlasáková J et al (2019) Melanoma-related changes in skin microbiome. Folia Microbiol 64(3):435–442
Mueller NT, Bakacs E, Combellick J, Grigoryan Z, Dominguez-Bello MG (2015) The infant microbiome development: mom matters. Trends Mol Med 21(2):109–117
Myles IA, Williams KW, Reckhow JD, Jammeh ML, Pincus NB, Sastalla I et al (2016) Transplantation of human skin microbiota in models of atopic dermatitis. JCI Insight 1(10):e86955
Naik S, Bouladoux N, Wilhelm C, Molloy MJ, Salcedo R, Kastenmuller W et al (2012) Compartmentalized control of skin immunity by resident commensals. Science 337(6098):1115–1119
Nakatsuji T, Chiang HI, Jiang SB, Nagarajan H, Zengler K, Gallo RL (2013) The microbiome extends to subepidermal compartments of normal skin. Nat Commun 4(1):1–8
Nakatsuji T, Chen TH, Narala S, Chun KA, Two AM, Yun T et al (2017) Antimicrobials from human skin commensal bacteria protect against Staphylococcus aureus and are deficient in atopic dermatitis. Sci Transl Med 9(378):eaah4680
Nograles KE, Zaba LC, Shemer A, Fuentes-Duculan J, Cardinale I, Kikuchi T et al (2009) IL-22–producing “T22” T cells account for upregulated IL-22 in atopic dermatitis despite reduced IL-17–producing TH17 T cells. J Allergy Clin Immunol 123(6):1244–1252
Noll M, Jäger M, Lux L, Buettner C, Axt-Gadermann M (2021) Improvement of atopic dermatitis by synbiotic baths. Microorganisms 9(3):527
O’Neill AM, Nakatsuji T, Hayachi A, Williams MR, Mills RH, Gonzalez DJ, Gallo RL (2020) Identification of a human skin commensal bacterium that selectively kills cutibacterium acnes. J Investig Dermatol 140(8):1619–1628
Ochoa-Reparaz J, Mielcarz DW, Wang Y, Begum-Haque S, Dasgupta S, Kasper DL, Kasper LH (2010) A polysaccharide from the human commensal *Bacteroides fragilis* protects against CNS demyelinating disease. Mucosal Immunol 3(5):487–495
Oh J, Conlan S, Polley EC, Segre JA, Kong HH (2012) Shifts in human skin and nares microbiota of healthy children and adults. Genome Med 4(10):1–11
Olejniczak-Staruch I, Ciążyńska M, Sobolewska-Sztychny D, Narbutt J, Skibińska M, Lesiak A (2021) Alterations of the skin and gut microbiome in psoriasis and psoriatic arthritis. Int J Mol Sci 22(8):3998
Ong PY, Ohtake T, Brandt C, Strickland I, Boguniewicz M, Ganz T et al (2002) Endogenous antimicrobial peptides and skin infections in atopic dermatitis. N Engl J Med 347(15):1151–1160
Pascolini C, Sinagra J, Pecetta S, Bordignon V, De Santis A, Cilli L et al (2011) Molecular and immunological characterization of Staphylococcus aureus in pediatric atopic dermatitis: implications for prophylaxis and clinical management. Clin Dev Immunol 2011:718708
Patra V, Wagner K, Arulampalam V, Wolf P (2019) Skin microbiome modulates the effect of ultraviolet radiation on cellular response and immune function. iScience 15:211–222
Peeters E, Nelis HJ, Coenye T (2008) Comparison of multiple methods for quantification of microbial biofilms grown in microtiter plates. J Microbiol Methods 72(2):157–165
Peral MC, Rachid MM, Gobbato NM, Martinez MH, Valdez JC (2010) Interleukin-8 production by polymorphonuclear leukocytes from patients with chronic infected leg ulcers treated with lactobacillus plantarum. Clin Microbiol Infect 16(3):281–286
Petersen C, Round JL (2014) Defining dysbiosis and its influence on host immunity and disease. Cell Microbiol 16(7):1024–1033
Pinto D, Ciardiello T, Franzoni M, Pasini F, Giuliani G, Rinaldi F (2021) Effect of commonly used cosmetic preservatives on skin resident microflora dynamics. Sci Rep 11(1):1–7
Prescott SL, Larcombe DL, Logan AC, West C, Burks W, Caraballo L et al (2017) The skin microbiome: impact of modern environments on skin ecology, barrier integrity, and systemic immune programming. World Allergy Organ J 10(1):1–16
Probst AJ, Auerbach AK, Moissl-Eichinger C (2013) Archaea on human skin. PloS One 8(6):e65388

Rainer BM, Thompson KG, Antonescu C, Florea L, Mongodin EF, Bui J et al (2020) Characterization and analysis of the skin microbiota in rosacea: a case–control study. Am J Clin Dermatol 21(1):139–147

Ramadan M, Hetta HF, Saleh MM, Ali ME, Ahmed AA, Salah M (2021) Alterations in skin microbiome mediated by radiotherapy and their potential roles in the prognosis of radiotherapy-induced dermatitis: a pilot study. Sci Rep 11(1):1–11

Riedler J, Eder W, Oberfeld G, Schreuer M (2000) Austrian children living on a farm have less hay fever, asthma and allergic sensitization. Clin Exp Allergy 30(2):194–200

Rippke F, Schreiner V, Doering T, Maibach HI (2004) Stratum corneum pH in atopic dermatitis. Am J Clin Dermatol 5(4):217–223

Ross AA, Doxey AC, Neufeld JD (2017) The skin microbiome of cohabiting couples. mSystems 2(4):17

Round JL, Mazmanian SK (2010) Inducible Foxp3+ regulatory T-cell development by a commensal bacterium of the intestinal microbiota. Proc Natl Acad Sci 107(27):12204–12209

Ruokolainen L, Von Hertzen L, Fyhrquist N, Laatikainen T, Lehtomäki J, Auvinen P et al (2015) Green areas around homes reduce atopic sensitization in children. Allergy 70(2):195–202

Salminen S, Collado MC, Endo A, Hill C, Lebeer S, Quigley EM et al (2021) The international scientific association of probiotics and prebiotics (ISAPP) consensus statement on the definition and scope of postbiotics. Nat Rev Gastroenterol Hepatol 18:1–19

Satish L, Gallo PH, Johnson S, Yates CC, Kathju S (2017) Local probiotic therapy with lactobacillus plantarum mitigates scar formation in rabbits after burn injury and infection. Surg Infect 18(2):119–127

Scholz F, Badgley BD, Sadowsky MJ, Kaplan DH (2014) Immune mediated shaping of microflora community composition depends on barrier site. PLoS One 9(1):e84019

Seethalakshmi PS, Rajeev R, Kiran GS, Selvin J (2020) Promising treatment strategies to combat Staphylococcus aureus biofilm infections: an updated review. Biofouling 36:1–23

Segata N (2015) Gut microbiome: westernization and the disappearance of intestinal diversity. Curr Biol 25(14):R611–R613

Seité S, Zelenkova H, Martin R (2017) Clinical efficacy of emollients in atopic dermatitis patients - relationship with the skin microbiota modification. Clin Cosmet Investig Dermatol 10:25–33

Shen W, Li W, Hixon JA, Bouladoux N, Belkaid Y, Dzutzev A, Durum SK (2014) Adaptive immunity to murine skin commensals. Proc Natl Acad Sci 111(29):E2977–E2986

Shu M, Wang Y, Yu J, Kuo S, Coda A, Jiang Y et al (2013) Fermentation of *Propionibacterium acnes*, a commensal bacterium in the human skin microbiome, as skin probiotics against methicillin-resistant *Staphylococcus aureus*. PLoS One 8(2):e55380

Skowron K, Bauza-Kaszewska J, Kraszewska Z, Wiktorczyk-Kapischke N, Grudlewska-Buda K, Kwiecińska-Piróg J et al (2021) Human skin microbiome: impact of intrinsic and extrinsic factors on skin microbiota. Microorganisms 9(3):543

Smeekens SP, Huttenhower C, Riza A, van de Veerdonk FL, Zeeuwen PL, Schalkwijk J, van der Meer JW, Xavier RJ, Netea MG, Gevers D (2014) Skin microbiome imbalance in patients with STAT1/STAT3 defects impairs innate host defense responses. J Innate Immun 6(3):253–262

Sokol H, Pigneur B, Watterlot L, Lakhdari O, Bermúdez-Humarán LG, Gratadoux JJ et al (2008) *Faecalibacterium prausnitzii* is an anti-inflammatory commensal bacterium identified by gut microbiota analysis of crohn disease patients. Proc Natl Acad Sci 105(43):16731–16736

Staley JT, Konopka A (1985) Measurement of in situ activities of nonphotosynthetic microorganisms in aquatic and terrestrial habitats. Annu Rev Microbiol 39(1):321–346

Stehlikova Z, Kostovcik M, Kostovcikova K, Kverka M, Juzlova K, Rob F et al (2019) Dysbiosis of skin microbiota in psoriatic patients: co-occurrence of fungal and bacterial communities. Front Microbiol 10:438

Steiner PE (1946) Necropsies on okinawans. Anatomic and pathologic observations. Arch Pathol 42(4):359–380

Stettler H, Kurka P, Lunau N, Manger C, Böhling A, Bielfeldt S et al (2017) A new topical panthenol-containing emollient: results from two randomized controlled studies assessing

its skin moisturization and barrier restoration potential, and the effect on skin microflora. J Dermatol Treat 28(2):173–180
Strober W (2004) Epithelial cells pay a toll for protection. Nat Med 10(9):898–900
Suh MG, Hong YH, Jung EY, Suh HJ (2019) Inhibitory effect of galactooligosaccharide on skin pigmentation. Prev Nutr Food Sci 24(3):321
Swanson KS, Gibson GR, Hutkins R, Reimer RA, Reid G, Verbeke K et al (2020) The international scientific association for probiotics and prebiotics (ISAPP) consensus statement on the definition and scope of synbiotics. Nat Rev Gastroenterol Hepatol 17(11):687–701
Uluçkan Ö, Jiménez M, Roediger B, Schnabl J, Díez-Córdova LT, Troulé K et al (2019) Cutaneous immune cell-microbiota interactions are controlled by epidermal JunB/AP-1. Cell Rep 29(4):844–859
Volz T, Skabytska Y, Guenova E, Chen KM, Frick JS, Kirschning CJ et al (2014) Nonpathogenic bacteria alleviating atopic dermatitis inflammation induce IL-10-producing dendritic cells and regulatory Tr1 cells. J Investig Dermatol 134(1):96–104
Von Hertzen L, Beutler B, Bienenstock J, Blaser M, Cani PD, Eriksson J et al (2015) Helsinki alert of biodiversity and health. Ann Med 47(3):218–225
Wallen-Russell C (2019) The role of everyday cosmetics in altering the skin microbiome: a study using biodiversity. Cosmetics 6(1):2
Wallen-Russell C, Wallen-Russell S (2017) Meta analysis of skin microbiome: new link between skin microbiota diversity and skin health with proposal to use this as a future mechanism to determine whether cosmetic products damage the skin. Cosmetics 4(2):14
Wang Z, MacLeod DT, Di Nardo A (2012) Commensal bacteria lipoteichoic acid increases skin mast cell antimicrobial activity against vaccinia viruses. J Immunol 189(4):1551–1558
Wang L, Clavaud C, Bar-Hen A, Cui M, Gao J, Liu Y et al (2015) Characterization of the major bacterial–fungal populations colonizing dandruff scalps in Shanghai, China, shows microbial disequilibrium. Exp Dermatol 24(5):398–400
Watanabe S, Kano R, Sato H, Nakamura Y, Hasegawa A (2001) The effects of *Malassezia* yeasts on cytokine production by human keratinocytes. J Investig Dermatol 116(5):769–773
Xia X, Li Z, Liu K, Wu Y, Jiang D, Lai Y (2016) Staphylococcal LTA-induced miR-143 inhibits propionibacterium acnes-mediated inflammatory response in skin. J Investig Dermatol 136(3):621–630
Ying S, Zeng DN, Chi L, Tan Y, Galzote C, Cardona C et al (2015) The influence of age and gender on skin-associated microbial communities in urban and rural human populations. PLoS One 10(10):e0141842
Yu Y, Champer J, Agak GW, Kao S, Modlin RL, Kim J (2016) Different propionibacterium acnes phylotypes induce distinct immune responses and express unique surface and secreted proteomes. J Investig Dermatol 136(11):2221–2228
Zeng B, Zhao J, Guo W, Zhang S, Hua Y, Tang J et al (2017) High-altitude living shapes the skin microbiome in humans and pigs. Front Microbiol 8:1929

Advances in Human Urinary Microbiome: A Role Beyond Infections

11

Kishore Kumar Godisela and Pallaval Veera Bramhachari

Abstract

The Human Microbiome Project heralded a new age within the study of microbial communities. In recent years, using advanced culturing techniques and molecular approaches, the urogenital system may contain a significant number of bacteria in both healthy and asymptomatic people. Environmental factors affecting microbial colonization in the urinary tract include pH, food availability, oxygen stress, adhesion sites, osmolarity, and immune interaction. The urinary microbiome varies in keeping with age and sex. Urinary incontinence, bladder disease, kidney stone development, and urolithiasis are all associated with changes within the urinary microbiome. Research shows that commensal species, as in urinary tract and urogenital tract microbiomes, like *Lactobacillus crispatus*, can protect against uropathogenic invasion. Antibiotics can also induce changes within the urobiome, which may cause many complications. Probiotics have been used to alter the intestinal microbiome. This chapter focuses on the function of the urobiome in some urinary diseases, as well as attempts to quantify, restore, and/or conserve the urinary microbiome's native, safe ecology.

Keywords
Microbiome · Probiotics · Urinary incontinence · Metagenomics · Urinary tract infection · Urinary incontinence

K. K. Godisela (✉)
Department of Biochemistry, SRR Govt. Arts & Science College, Karimnagar, Telangana, India

P. V. Bramhachari
Department of Biosciences & Biotechnology, Krishna University, Machilipatnam, Andhra Pradesh, India

P. Veera Bramhachari (ed.), *Human Microbiome in Health, Disease, and Therapy*, https://doi.org/10.1007/978-981-99-5114-7_11

11.1 Introduction

The Human Microbiome Project (HMP) instigated in 2008 ushered in a new age of research into our microbial communities. Strikingly, the microorganisms outnumbered our cells tenfold before then, and the vast majority of them were unknown because they had yet to be cultured and studied (Cho and Blaser 2012; Grice and Segre 2012; Human Microbiome Project Consortium 2012). The HMP has given several results and a glimpse into the complex and significant roles these microbes play in health, from metabolism and phenotype modulation (such as obesity) to the development of the innate immune system and disease etiology (due to microbial ecosystem imbalance). Discoveries challenge our flow of thinking and approaches, and they brief change in perspective toward elective arrangements in disease diagnosis, prevention, and treatments. The HMP zeroed in on five destinations: gastrointestinal, oral, skin, nasal, and urogenital. The urinary framework was not a piece of the examination because the urine was viewed as sterile. In recent years, using sophisticated culturing methods and molecular approaches, we discovered that the urinary system represents a large number of bacteria in both healthy and asymptomatic people (Hilt et al. 2014; Lewis et al. 2013).

This section gives an overview of the compositions and associations of the urinary microbiome with different ages and genders. Examples of illnesses and disorders resulting from changes in microbial communities illustrate the effect of these microbes on health.

11.2 The Human Urinary Microbiome: An Unexpected Niche Becomes the Center of Interest

Given our current knowledge about the human microbiome, also referred to as the urobiome, microbial species in every location that communicates with the surface world remains rational.

11.2.1 The Urinary Tract's Environmental Niche

The human urinary tract (UT) is often broadly divided into upper and lower compartments. The ureters and kidneys are located in the upper partition. In contrast, the urethra and bladder are located in the lower compartment (Ingersoll and Albert 2013) (Fig. 11.1). Urothelium is a transitional epithelium that forms a thin layer of the GAG in the luminous surface, bladder, and proximal urethra of the lower urinary tract (Ingersoll and Albert 2013; Anand et al. 2012). The apical, separated cells of the urothelium, known as umbrella cells, serve as an important boundary among urinary waste products from internal body tissues.

Environmental factors influencing microbial colonization include pH, nutrient availability, oxygen tension, adhesion sites, osmolarity, and immune interactivity (81). The pH of the urine is normally acidic for each individual person, although the

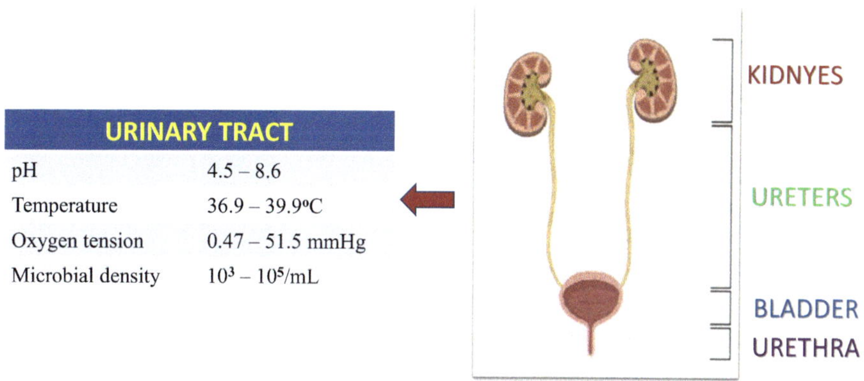

Fig. 11.1 Microbial niches in the human body's urinary tract and their environmental features

pH of the healthy urine is between 5 and 8 (Reitzer and Zimmern 2019). Since many microorganisms need either aerobic or anaerobic metabolism, oxygen in the UT affects the ecology and spatial organization of the UT microbiota (Shannon et al. 2019). Human urine is composed of various soluble components, as well as amino acids, osmolytes, electrolytes, and carbohydrates. A catalog of over 2600 composites has been found in the urine (Reitzer and Zimmern 2019). There are, however, other nutrient sources in the UT. The urothelium is covered with a thin GAG layer to protect and lubricate the tissue underneath (Ingersoll and Albert 2013). It is not fully understood what the human UT GAG layer is composed of. However, a bed of excessive viscous material, specifically amino acids, mucins, more than a few GAGs, and other complex carbohydrates, are present in the epithelium vaginally but carefully regarding the mucosa. Interestingly, many human commensals known to colonize the UT, such as *Lactobacillus*, *Bifidobacterium*, and *Streptococcus*, have enzymes that degrade extracellular GAGs into smaller palatable carbohydrates (Zuniga et al. 2018).

11.2.2 The Urinary Microbiome Differs Between Populations

11.2.2.1 Gender

According to the study, men and women have different urinary microbiomes, most likely due to anatomical differences and hormone levels. *Lactobacillus* species predominate in the urine microbiome of women of childbearing age (Kok and Erasmus 2016). Other microbiomes reported by molecular methods (16S rRNA gene sequencing) in women isolates included members of the Actinobacteria (e.g., *Actinomyces*, *Arthobacter*) and Bacteroidetes (e.g., *Bacteroides*) phyla's that were missing in male samples (Shannon et al. 2019). Regarding the key microbiome in varying abundance in the urinary tract throughout life, men and women have mainly three genera: *Lactobacillus*, *Corynebacterium*, *and Streptococcus* (Fouts et al. 2012). The difference in urinary microbiome composition between males and females may also be due to the different compositions of their urine. Females excrete more citrate but less calcium and oxalate, whereas males excrete more creatinine (Ipe et al. 2016). The various constituents may favor certain microbes' survival in this niche, and these microbes, in turn, contribute to the health of their hosts. Women, for example, are more prone to urinary tract infections (UTIs) due to anatomical reasons.

11.2.2.2 Age

Children and adults contain distinct microbiome populations, most likely due to urinary metabolites from lifestyle changes, hygiene practices, and voiding behaviors. Adults have different bacterial genera depending on their age. Hormone fluctuations may affect the microbiome. Several studies (Fouts et al. 2012; Tang 2017) found that *Lactobacillus* species increased throughout puberty and declined after menopause. Women who become pregnant may change their microbiota, while those who do not tend to have more stable microbial communities (Brubaker and Wolfe 2017). Constipation and urinary incontinence can affect the survival and sustainability of microbes as people age. Changes within the urinary microbiome over time are linked to various diseases. One example is the decline in vaginal *Lactobacillus spp.* in women after menopause, which causes uropathogens to colonize, resulting in a rise in urinary tract diseases. Finally, the *Oxalobacter formigenes* have been linked to the development of urolithiasis (kidney stones). Its invasion in the gut appears to be age based, as it is present in children up to the age of 8 years old and instead begins to decrease by the age of 12 into adulthood (Dwyer et al. 2012).

11.3 How Do We Sample the Microbiota of Human Urine?

Without a UTI, the human urinary microbiome has poor biomass, with <100 to <10^5 CFU per milliliter of urine (Rowe and Juthani-Mehta 2013). As a result, attempts to classify a native microbial community should be considered by taking minimal starting materials in urine samples (Karstens et al. 2018). The sampling procedure is an important factor for any research evaluating the microbiome. Over the last 10 years, three key sampling techniques have been used to attain urine samples suggested to denote the urinary microbiota:

(a) Suprapubic aspiration (SPA)
(b) Gathering of urine by a transurethral catheter (TUC)
(c) Collection of midstream clean-catch (CC) urine

Each of these strategies has benefits and drawbacks. Many sampling techniques disrupt the local microbial populations. When using any sampling process, remember that sampling ordering can lead to contamination of serial sampling sites (Table 11.1).

Table 11.1 Methods for studying the urinary microbiome

Topics	Methods
Study design	Longitudinal & cross sectional
Sample collection	CC, TUC & SPA
Data and metadata acquiring	Culture based, 16S rRNA sequencing Whole-genome metagenomics
Data analysis	Taxonomic profiling Functional analysis & meta-analysis

11.3.1 Urine Microbial Culturing

The capacity to observe and affirm feasible microbial residents, achieved by trials obtained with such a TUC from a women's urinary tract, is a vital advantage of urine culturing techniques. Standard urine culture is used to diagnose UTIs. Urine is plated onto agar plates containing 5% sheep blood agar and MacConkey agar plates and incubated aerobically at 35 °C for 24 h to obtain quantitative colony counts (Hilt et al. 2014). This process employs a variety of urine volumes, specific intervals of time, and culture media, as well as anaerobic, aerobic, and CO_2-augmented conditions. By contrast, the normal urine culture was intended to detect the most important uropathogenic *E. coli* and the related growth criteria for other bacterial species. Price et al. discovered in a 2016 analysis that traditional urine culture lost 77% with wholly uropathogens and 88% with non-*E.coli* uropathogens spotted via expanded-spectrum quantitative urine culture (EQUC). The standardized EQUC procedure detected 84% of harmful uropathogens sampled through the extended EQUC practice, while standardized urine culture spotted 33% of possible uropathogens (Price et al. 2016).

11.3.2 Urinary Microbiome Metagenomic Sequencing

Researchers used NGS-based metagenomic sequencing approaches to determine bacterial communities without cultural bias. There are two main methods for metagenomic sequencing: whole-genome shotgun metagenomic sequencing (WGMS) and 16S rRNA amplicon sequencing (Quince et al. 2017; Moustafa et al. 2018). Both methods focus on next-generation sequencing (NGS) technologies. WGMS experiments are frequently contaminated. Human genomic contamination, for example, accounted for 1.3–99.9% of sequencing reads attained by clinical urine trials. Samples must be sequenced to an adequate read depth to control the host contamination and correctly assay the microbial community. Sample preparation and DNA extraction techniques can be streamlined to enrich microbial DNA. With access to the entire metagenome, researchers may establish the population structure and taxonomic history of the native microbiota and systematically identify their genetic potential.

11.3.3 The Human UT Microbiome's Taxonomic Profile

Numerous taxonomic profiling studies have been conducted to classify the human urinary microbiome. Many studies have assessed taxonomic richness among healthy and disease states using 16S rRNA amplicon sequencing and specialized culture techniques such as EQUC (Siddiqui et al. 2011). The main taxa found in the stable urinary microbiome are organisms that are known to be fastidious, slow-growing microbes. Firmicutes, Bacteroidetes, Actinobacteria, *Fusobacteria*, and Proteobacteria are the five main species and often include the genera *Lactobacillus*, *Corynebacterium*, *Prevotella*, *Staphylococcus*, *and Streptococcus* (Brubaker and Wolfe 2017). The vaginal microbiome of healthy adolescent girls is controlled primarily by *Lactobacillus* organisms. *Lactobacillus crispatus*, *Lactobacillus iners*, *Lactobacillus gasseri*, and *Lactobacillus jensenii* are the utmost common *Lactobacilli* reported there within the vaginal microbiome (Ceccarani et al. 2019). The vaginal microbiome is considered significant in vaginal pH management and the protection of various urogenital infections (Ceccarani et al. 2019).

11.4 The Urinary Microbiome in the Disease

Several studies have reported a connection between certain conditions and changes in the cutaneous, stomach, colon and intestinal microbiome in relation to HMP (Cho and Blaser 2012). In the stable urinary tract, various bacterial species were identified. The changes in this microbiota were related to urinary incontinence (UI), neurogenic bladder dysfunction (NBD), urologic cancer, STI, interstitial cystitis (IC), and chronic prostatitis/chronic pelvic pain (CP/CPPS).

11.4.1 Changes in the Urinary Microbiome in UI

Urgency UI (UUI), stress UI (SUI), and mixed UI (MUI) are some of the most common UI complaints around the world. Six experiments have established a potential part for the urinary microbiome in UUI and SUI. According to Pearce et al., the *Gardnerella* and lower *Lactobacillus* load were higher than those of the non-UUI microbiomes for patients with UUI (Pearce et al. 2014). The overwhelming diversity of 14 microbial operational taxonomic units in UUI and non-UI patients has been identified by Karstens et al. (2016) as the most acute symptom of UUI in persons with a limited diversity of bacteria. The UM plays a defined role in UUI and the response to UUI care, according to the studies published (Pearce et al. 2015). Future UUI patient research will be very helpful for evaluating whether variations in UM between the gender and age contribute to UUI sensitivity and for efforts to improve the therapy for this disorder.

11.4.2 The Role of the Urinary Microbiome in Urologic Cancer

The function of the UM in some UT cancers remains unknown. The most significant risks in urothelial bladder cancer, common urologic cancer, were tobacco smoke and exposure to aromatic amines and polycyclic aromatic hydrocarbons in the air (Burger et al. 2013). A preliminary UM study of a few urothelial cell carcinoma patients was carried out by Xu et al. (UCC). They found variations between the healthy and the microbiome population of UCC patients. *Streptococcus* was abundant in the urine of these patients; however, larger sample sizes are required to validate these results (Xu et al. 2014; Bakare et al. 2018). Beneficial associations have also been found in the vaccine for Bacillus Calmette-Guérin (BCG) between the bladder and attenuated *Mycobacterium tuberculosis*. For nearly four decades, this vaccine has been the most common therapy for intermediate- and high-risk non-muscle-invasive bladder tumors (Lenis et al. 2020). According to studies, bacteria administered in the bladder with BCG vaccine have been reported to be inflammatory and result in an immune antitumor, which plays a key role in the antitumor effect. More research is needed to determine the link between cancer treatments causing dysbiosis in the UT and the risk of some urologic disorders. At baseline, healthy women possess an elevated prevalence of *Mycobacteria* and additional *Actinomycetes* (Lee and Stern 2019), which are thought to inhibit cancer progression or development. While the research is still in its early stages, some data suggests correlations between some urinary microbe profiles and the likelihood of recurrence, progression, and treatment (Wu et al. 2018).

11.4.3 Modifications of UMs in Other Urinary Diseases

Numerous studies show that UM can be directly or indirectly associated with certain urinary disorders. These findings pose new concerns regarding potential cause-and-effect relationships. IC urine samples showed a substantial increase in *Lactobacillus* genus richness and a decline in all-inclusive abundance and ecological variety. There has been evidence in some investigations that specified *Lactobacillus* species like *L. gasseri*, and *L. delbrueckii* may be related to UUI and UTI (Maillet et al. 2019). *Lactobacillus* and *Corynebacterium* genera were significantly enriched in urine samples from healthy control bladders, while other microbial types, such as *Enterococcus*, *Escherichia*, and *Klebsiella*, became prominent in NBD samples (Magri et al. 2019). Bacterial species in normal culture that do not grow, including *Sneathia*, *Gemella*, *Aerococcus*, *Anaerococcus*, *Veillonella*, and *Prevotella*, clearly dominate the STI men's UM (Nelson et al. 2010).

11.4.4 The Microbiome and the Production of Calcium Oxalate Stones

Multiple issues influence the progress of calcium oxalate in the kidneys and urinary oxalate excretion. Enteric colonization with *Oxalobacter formigenes* could minimize intestinal oxalate, oxalate intake, and urinary excretion, possibly lowering the danger of calcium oxalate stone formation. According to the hypothesis, Kaufman and others have discovered that a 70% reduction in the risk of kidney stone recurrence is associated with gastrointestinal invasion with *O. formigenes*. According to a global survey, 38–77% of the ordinary population and 17% of stone formers are populated by *O. formigenes* (Kaufman et al. 2008). Based on these preliminary results, *O. formigenes* colonization was thought to be a successful tool for reducing the risk of calcium oxalate stones. However, since *O. formigenes* is an anaerobe with fastidious growth requirements, preparing it as a good prophylactic probiotic poses challenges. The in vitro properties needed by a good probiotic strain were examined by Ellis et al. (2016) to improve *O. formaigene's* clinical applicability. The results show that a specific *O. formigenes* strain (OxCC13) can thrive in the lack of oxalate, is aerotolerant, and can be freeze-dried or blended with yogurt for extended periods.

11.4.5 The Effect of Antibiotics on UM

Given the large variety of antibiotics on the market, they affect pathogens and healthy microbiota. While recent treatment recommendations for uncomplicated UTIs prohibit fluoroquinolones due to collateral damage to commensal microbiota, such drugs are still widely used. The urinary microbiome of kidney transplants, where there was decreased microbial diversity and increased prevalence of potentially pathogenic species, was significantly different from that of healthy controls (Milam et al. 2017). Enhanced genes aimed at enzymes like dihydrofolate synthase not embarrassed by the trimethoprim-sulfamethoxazole are linked to the choice of antibiotic-resistant microorganisms in the prophylactic regime, as in urinary microbiota. Based on these results, assessing and producing appropriate prophylactic regimens that do not encourage antibiotic resistance is effective.

11.5 Role of Prebiotics, Probiotics, and Diet in Urologic Diseases

The medicinal use of probiotic bacteria as a therapy for various ailments is debatable (Waigankar and Patel 2011). The use of probiotics to alter the intestinal microbiome has been used. Several clinical studies have investigated the function of helpful strains in urogenital diseases, bladder cancer, and the growth of kidney stones (Table 11.2). Probiotics have recently arisen as a potentially effective or adjuvant medication for the therapy and preclusion of urinary tract infections (UTIs).

Table 11.2 Probiotics, prebiotics, and diet modifications used in urologic disorders

Prebiotics/probiotics/diet modification	Administration	Urinary diseases treated
Lactobacillus drinks and berry juice	Oral	Urinary tract infection
Bacillus Calmette-Gue´rin immunotherapy	Intravesical	Bladder cancer
Lactobacillus casei	Oral	Bladder cancer
Oxalabacter formigenes	Oral	Urolithiasis
L. casei strain Shirota	Oral	Bladder cancer
Supplemental calcium	Oral	Urolithiasis
Diet low in sodium and animal protein	Oral	Urolithiasis
Lactobacillus rhamnosus GR-1	Vaginal	Urinary tract infection

The efficient treatment of UTI has been shown for *Lactobacillus* strains such as *L. rhamnosis* GR1, *L. fermentum* RC-14, and *L. reuteri* B-54 (Reid and Bruce 2006). In two experimental trials, 138 patients with superficial transitional cell carcinoma of the bladder were given an oral *Lactobacillus casei* treatment for prevention. The results indicated that Shirota strain of *L. casei* could help prevent and treat non-muscle-invasive bladder tumors (Aso and Akazan 1992). Finally, dietary patterns that alter the microbiome may be essential to developing urologic pathologies. As a result, dietary changes may be the first step in eliminating such urinary disorders.

11.6 Future Prospectives

A basic understanding of the urinary microbiome is expanding, and most remains to be learned. More longitudinal and interventional trials must be used to investigate causality beyond associations in observational epidemiological studies. To date, the mainstream of urinary microbiome research often uses cross-sectional cohort formats to equate diseased patients to healthy controls. 16S rRNA amplicon screening has been the predominant method of mapping the human urinary microbiome in the past decade. Yet, no main microbiome in the microbiome niches of a bladder, UT, or UGT has been established. Instead of recognizing a core taxonomic enhancement, it could be insightful to create a core genetic enrichment requiring the application of WGMS. The human gut microbiome has a catalog of microbial genes that is 150 times larger than the human genome and has various functional possibilities. Modeling the group dynamics of UT-resident microbial populations throughout infection is a significant research opportunity in the community. Frequently used sequence depths during UTI cannot be used to outline low plentiful UT microbial populations. Extensive metagenome sequencing and complete density calculations can aid in determining the future of presumed commensal populations throughout a UTI (Karstens et al. 2018).

The availability of human metagenomic data sets to the public is critical for translational development because it enables large-scale meta-analyses and objective scrutiny of published findings (Langille et al. 2018). Development and democratic data can be the basis for consensus microbial group models of human UT and UGT in health and disease. A 2010 study discovered that urinary tract infections have a major effect on morale and that there is a strong correlation between UTI occurrence and depression in older adults (Eriksson et al. 2010). Inorder to support a live biotherapeutic pipeline that could benefit people with drug-resistant UI symptoms, the useful consequences of few *Lactobacillus* species and some additional urinary commensals must be further investigated.

Nevertheless, the findings reported to date open the door to more studies on novel diagnostic, prognostic, and predictional biomarkers grounded on microbiomes used in research into clinical urology. Moreover, there are growing signs that may be helpful to monitor UM in the very first phase, to minimize the risk of urinary disease, or cure it, amid previous controversies over the use of probiotics, prebiotics, and nutritional changes for urological care.

Acknowledgments The authors gratefully acknowledge SRR Govt. Arts & Science College, Karimnagar, Telangana and Krishna University, Machilipatnam, for the support extended.

Conflict of Interest The author declares that they have no competing interests.

References

Anand M, Wang C, French J, Isaacson-Schmid M, Wall LL, Mysorekar IU (2012) Estrogen affects the glycosaminoglycan layer of the murine bladder. Female Pelvic Med Reconstr Surg 18:148–152

Aso Y, Akazan H (1992) Prophylactic effect of a lactobacillus casei preparation on the recurrence of superficial bladder cancer. BLP study group. Urol Int 49:125–129

Bakare SO, Adebayo AS, Awobode HO, Onile OS, Agunloye AM, Isokpehi RD, Anumudu CI (2018) Arsenicosis in bladder pathology and schistosomiasis in Eggua, Nigeria. Trans R Soc Trop Med Hyg 112:230–237

Brubaker L, Wolfe AJ (2017) The female urinary microbiota, urinary health and common urinary disorders. Ann Transl Med 5:34

Burger M, Catto JW, Dalbagni G, Grossman HB, Herr H, Karakiewicz P, Kassouf W, Kiemeney LA, La Vecchia C, Shariat S, Lotan Y (2013) Epidemiology and risk factors of urothelial bladder cancer. Eur Urol 63:234–241

Ceccarani C, Foschi C, Parolin C, D'Antuono A, Gaspari V, Consolandi C, Laghi L, Camboni T, Vitali B, Severgnini M, Marangoni A (2019) Diversity of vaginal microbiome and metabolome during genital infections. Sci Rep 9:14095

Cho I, Blaser MJ (2012) The human microbiome: at the interface of health and disease. Nat Rev Genet 13:260–270

Kok DJ, Erasmus MC, Relations to urinary tract infection, hormonal status, gender and age., 17th international EAUN meeting, Munich, Germany. (2016)

Dwyer ME, Krambeck AE, Bergstralh EJ, Milliner DS, Lieske JC, Rule AD (2012) Temporal trends in incidence of kidney stones among children: a 25-year population based study. J Urol 188:247–252

Ellis ML, Dowell AE, Li X, Knight J (2016) Probiotic properties of Oxalobacter formigenes: an in vitro examination. Arch Microbiol 198:1019–1026

Eriksson I, Gustafson Y, Fagerstrom L, Olofsson B (2010) Do urinary tract infections affect morale among very old women? Health Qual Life Outcomes 8:73

Fouts DE, Pieper R, Szpakowski S, Pohl H, Knoblach S, Suh MJ, Huang ST, Ljungberg I, Sprague BM, Lucas SK, Torralba M, Nelson KE, Groah SL (2012) Integrated next-generation sequencing of 16S rDNA and metaproteomics differentiate the healthy urine microbiome from asymptomatic bacteriuria in neuropathic bladder associated with spinal cord injury. J Transl Med 10:174

Grice EA, Segre JA (2012) The human microbiome: our second genome. Annu Rev Genomics Hum Genet 13:151–170

Hilt EE, McKinley K, Pearce MM, Rosenfeld AB, Zilliox MJ, Mueller ER, Brubaker L, Gai X, Wolfe AJ, Schreckenberger PC (2014) Urine is not sterile: use of enhanced urine culture techniques to detect resident bacterial flora in the adult female bladder. J Clin Microbiol 52:871–876

Human Microbiome Project Consortium (2012) Structure, function and diversity of the healthy human microbiome. Nature 486:207–214

Ingersoll MA, Albert ML (2013) From infection to immunotherapy: host immune responses to bacteria at the bladder mucosa. Mucosal Immunol 6:1041–1053

Ipe DS, Horton E, Ulett GC (2016) The basics of bacteriuria: strategies of microbes for persistence in urine. Front Cell Infect Microbiol 6:14

Karstens L, Asquith M, Davin S, Stauffer P, Fair D, Gregory WT, Rosenbaum JT, McWeeney SK, Nardos R (2016) Does the urinary microbiome play a role in urgency urinary incontinence and its severity? Front Cell Infect Microbiol 6:78

Karstens L, Asquith M, Caruso V, Rosenbaum JT, Fair DA, Braun J, Gregory WT, Nardos R, McWeeney SK (2018) Community profiling of the urinary microbiota: considerations for low-biomass samples. Nat Rev Urol 15:735–749

Kaufman DW, Kelly JP, Curhan GC, Anderson TE, Dretler SP, Preminger GM, Cave DR (2008) Oxalobacter formigenes may reduce the risk of calcium oxalate kidney stones. J Am Soc Nephrol 19:1197–1203

Langille MGI, Ravel J, Fricke WF (2018) "available upon request": not good enough for microbiome data! Microbiome 6:8

Lee JA, Stern JM (2019) Understanding the link between gut microbiome and urinary stone disease. Curr Urol Rep 20:19

Lenis AT, Lec PM, Chamie K, Mshs MD (2020) Bladder cancer: a review. JAMA 324:1980–1991

Lewis DA, Brown R, Williams J, White P, Jacobson SK, Marchesi JR, Drake MJ (2013) The human urinary microbiome; bacterial DNA in voided urine of asymptomatic adults. Front Cell Infect Microbiol 3:41

Magri V, Boltri M, Cai T, Colombo R, Cuzzocrea S, De Visschere P, Giuberti R, Granatieri CM, Latino MA, Largana G, Leli C, Maierna G, Marchese V, Massa E, Matteelli A, Montanari E, Morgia G, Naber KG, Papadouli V, Perletti G, Rekleiti N, Russo GI, Sensini A, Stamatiou K, Trinchieri A, Wagenlehner FME (2019) Multidisciplinary approach to prostatitis. Arch Ital Urol Androl 90:227–248

Maillet F, Passeron A, Podglajen I, Ranque B, Pouchot J (2019) Lactobacillus delbrueckii urinary tract infection in a male patient. Med Mal Infect 49:226–228

Modena BD, Milam R, Harrison F, Cheeseman JA, Abecassis MM, Friedewald JJ, Kirk AD, Salomon DR (2017) Changes in urinary microbiome populations correlate in kidney transplants with interstitial fibrosis and tubular atrophy documented in early surveillance biopsies. Am J Transplant 17:712–723

Moustafa A, Li W, Singh H, Moncera KJ, Torralba MG, Yu Y, Manuel O, Biggs W, Venter JC, Nelson KE, Pieper R, Telenti A (2018) Microbial metagenome of urinary tract infection. Sci Rep 8:4333

Nelson DE, Van Der Pol B, Dong Q, Revanna KV, Fan B, Easwaran S, Sodergren E, Weinstock GM, Diao L, Fortenberry JD (2010) Characteristic male urine microbiomes associate with asymptomatic sexually transmitted infection. PLoS One 5:e14116

Pearce MM, Hilt EE, Rosenfeld AB, Zilliox MJ, Thomas-White K, Fok C, Kliethermes S, Schreckenberger PC, Brubaker L, Gai X, Wolfe AJ (2014) The female urinary microbiome: a comparison of women with and without urgency urinary incontinence. MBio 5:e01283–e01214

Pearce MM, Zilliox MJ, Rosenfeld AB, Thomas-White KJ, Richter HE, Nager CW, Visco AG, Nygaard IE, Barber MD, Schaffer J, Moalli P, Sung VW, Smith AL, Rogers R, Nolen TL, Wallace D, Meikle SF, Gai X, Wolfe AJ, Brubaker L, Pelvic Floor Disorders N (2015) The female urinary microbiome in urgency urinary incontinence. Am J Obstet Gynecol 213:347.e1–347.11

Price TK, Dune T, Hilt EE, Thomas-White KJ, Kliethermes S, Brincat C, Brubaker L, Wolfe AJ, Mueller ER, Schreckenberger PC (2016) The clinical urine culture: enhanced techniques improve detection of clinically relevant microorganisms. J Clin Microbiol 54:1216–1222

Quince C, Walker AW, Simpson JT, Loman NJ, Segata N (2017) Shotgun metagenomics, from sampling to analysis. Nat Biotechnol 35:833–844

Reid G, Bruce AW (2006) Probiotics to prevent urinary tract infections: the rationale and evidence. World J Urol 24:28–32

Reitzer L, Zimmern P (2019) Rapid growth and metabolism of Uropathogenic Escherichia coli in relation to urine composition. Clin Microbiol Rev 33:1

Rowe TA, Juthani-Mehta M (2013) Urinary tract infection in older adults. Aging Health 9:10.2217/ahe.13.38

Shannon MB, Limeira R, Johansen D, Gao X, Lin H, Dong Q, Wolfe AJ, Mueller ER (2019) Bladder urinary oxygen tension is correlated with urinary microbiota composition. Int Urogynecol J 30:1261–1267

Siddiqui H, Nederbragt AJ, Lagesen K, Jeansson SL, Jakobsen KS (2011) Assessing diversity of the female urine microbiota by high throughput sequencing of 16S rDNA amplicons. BMC Microbiol 11:244

Tang J (2017) Microbiome in the urinary system-a review. AIMS Microbiol 3:143–154

Waigankar SS, Patel V (2011) Role of probiotics in urogenital healthcare. J Midlife Health 2:5–10

Wu P, Zhang G, Zhao J, Chen J, Chen Y, Huang W, Zhong J, Zeng J (2018) Profiling the urinary microbiota in male patients with bladder cancer in China. Front Cell Infect Microbiol 8:167

Xu W, Yang L, Lee P, Huang WC, Nossa C, Ma Y, Deng FM, Zhou M, Melamed J, Pei Z (2014) Mini-review: perspective of the microbiome in the pathogenesis of urothelial carcinoma. Am J Clin Exp Urol 2:57–61

Zuniga M, Monedero V, Yebra MJ (2018) Utilization of host-derived glycans by intestinal lactobacillus and Bifidobacterium species. Front Microbiol 9:1917

Part III

Microbiome Diagnostics, Therapeutics and Bioinformatics

Role of Microbiomes in Defining the Metabolic and Regulatory Networks that Distinguishes Between Good Health and a Continuum of Disease States

12

Satyanagalakshmi Karri, Manohar Babu Vadela, and Vijay A. K. B. Gundi

Abstract

In this world, animals are teeming with different microbes comprising bacteria, fungi, and viruses. Animal-microbe interactions have become an interesting research area because of their beneficial and significant role in human life. The human microbiome project reported the significance of gut microbes and also explained complex diversity in a way to equilibrium maintained. For instance, gut microbes help control the colonization of exogenous pathogens. The beneficial role of microbes in humans extended the knowledge from individual taxa to a level of an ecosystem. However, rapid-growing technology provided a great understanding of individual microbes. Little is known about the microbiota association with animals and humans and the significance of microbial consortia in the ecosystem. The knowledge of microbes and host metabolism and their influence on modifying the microbiota ecosystem is important to understand microbiota's beneficial and pathogenic efficacy. This chapter outlined the current knowledge of microbiota and microbiome in ecosystems and their significant role in human and animal life.

Keywords

Pathogens · Dysbiosis · Communal bacteria · Hologenome theory · Animal-microbe interaction

S. Karri · M. B. Vadela · Vijay A. K. B. Gundi (✉)
Department of Biotechnology, Vikrama Simhapuri University, Nellore, Andhra Pradesh, India

P. Veera Bramhachari (ed.), *Human Microbiome in Health, Disease, and Therapy*, https://doi.org/10.1007/978-981-99-5114-7_12

12.1 Introduction

The deep evolutionary history among organisms was revealed by drawing genetic relatedness and phylogenetic relationship by the advanced game-changing technology, the next-generation DNA sequencing method. It allowed us to recognize the microorganisms' true diversity and functional activity and also helped us understand the interdependence between multicellular organisms and their microbiota. The interaction between diverse taxonomic groups and complex multicellular organisms is difficult to comprise in a single chapter. Instead, we focused on a domain of bacteria and multicellular animal species to understand the interaction and significance of bacteria in animal life. However, the role of microbes in animal origin does not prevent the perspectives of the evolution of complex multicellularity but applies a functional and ecological dimension to these considerations (Grosberg and Strathmann 2007).

It is important to understand comprehensively that humans act as hosts for microbiomes, and interaction with diverse extracellular microorganisms begins with birth (Favier et al. 2002). It was observed that the gut microbiota during the first year of life showed unique patterns and the initial gut colonization helped establish an adult's gut microbiota, as Ley et al. (2005) proved. The mice microbial population was very similar to their mother gut microbiota, and the relationship between them as a factor determined the gut microbial population (Ley et al. 2005). In the human gut, more than 50 bacterial phyla have been detected; among these, two phyla were dominated, phylum Bacteroidetes and the Firmicutes, and the individuals harbor more than 1000 microbial species alone (Claesson et al. 2009; Xu and Gordon 2003).

The population of bacteria throughout the GIT distributed unevenly per gram of the stomach and duodenal contents contains 10–10^3 bacterial population per gram, in the small intestine between 10^4 and 10^7 bacteria cells, and rising to between 10^{11} and 10^{12} bacteria per gram in the large intestine (Sekirov et al. 2010). Moreover, the species belonging to phyla Firmicutes were predominant in the small intestines, and phylum Bacteroidetes were more abundant in the colon. When the transverse section of GIT was studied, a vast diverse microbiota with different environments was observed across the intestine. The genera, including *Bacteroides*, *Bifidobacterium*, *Streptococcus*, *Enterococcus*, *Clostridium*, *Lactobacillus*, and *Ruminococcus*, were the common luminal community, whereas *Clostridium*, *Lactobacillus*, and *Enterococcus* were detected in the small intestine (Swidsinski et al. 2005). As a population, 10–100 trillion microorganisms inhabit the adult intestine, benefiting us in different ways. For example, microbiota in the gut allows one to get calories from indigestible polysaccharides in the diet. The microbiota produces a group of enzymes called glycoside hydrolases and polysaccharide lyases to digest these complex polysaccharides, which cannot be encoded and synthesized by the human genome (Sonnenburg et al. 2005). Fasting-induced adipocyte factor (Fiaf) is a mouse angiopoietin-like family member protein, and microbial suppression promotes leanness (Bäckhed et al. 2004).

12.2 Microorganisms in the Evolution of Animals

About 3.8 billion years ago, the evolution of cellular life began, and the biosphere has been dominated by bacteria, archaea, and eukaryotic microbes (Pace et al. 2012). Then, eukaryotic cells evolved about 1.5 billion years ago. From the literature, it is known that eukaryotic cells arose as a result of an endosymbiotic process by which an alpha-proteobacterium was converted into the mitochondrion. Consequently, this evolutionary incidence led that all eukaryotes have mitochondria derived through the endosymbiotic process. The exploration and duplication of genome size during the endosymbiotic process increased the metabolic efficiency in eukaryotic cells (Lane 2014). Indeed, biologists had demonstrated 100 years ago the role of microbiota in the health and diseases of higher organisms. Currently, culture-free molecular techniques allow us to understand that the symbiotic microbial genetic information far exceeds that of their hosts. The microbiota and host interaction have received growing attention, such as the mode of interaction, which leads to diversity and abundance of the symbiont, advantage or harm to the host, and how the genes are responsible for their interaction (Zilber-Rosenberg and Rosenberg 2008). The symbiotic relation of microbiota and animals left with a gap "if microbiota plays a vital role in eukaryotic life, what is the effect of microbes in the evolution of these higher organisms." It has been applied that the "Hologenome Theory" (Box 1) of evolution addresses the question raised in the evolution of animal life (Zilber-Rosenberg and Rosenberg 2008). Rosenberg and others developed the hologenome theory of evolution in coral microbiology (Rosenberg et al. 2007). According to them, the well-established empirical key points are a fundamental base for evolution. The animals started symbiotic relations with microbiota and are transmitting to generations. The association of holobiont affects each other in the environment.

12.2.1 Role of Gut Microbiota

The unicellular eukaryotes and bacteria are the inhabitants of the human body. Diverse bacterial communities dominate the human gut; among these phyla, Bacteroidetes, Firmicutes, and Actinobacteria are predominant organisms (Neish 2009). The human body, such as the skin surface, gastrointestinal tract, genitourinary tract, oral cavity, respiratory tract, and ear, is colonized with diverse bacterial species (Chiller et al. 2001; Neish 2009; Verstraelen 2008). Among all these organs, the gastrointestinal tract contains the most bacterial populations, and the colon alone is estimated to have 70% of the microbes of the human body (Ley et al. 2006a; Whitman et al. 1998). Gut microbiota plays a vital role in human health, including immunity and metabolic activities (Fig 12.1).

Moreover, gut microbes encode several genes than a host, and they actively metabolize food that is not able to execute by a host. Gut microbes produce vitamins and essential and nonessential amino acids required by humans. In addition, gut microbes involve in biochemical pathways of carbohydrate oxidation and digest

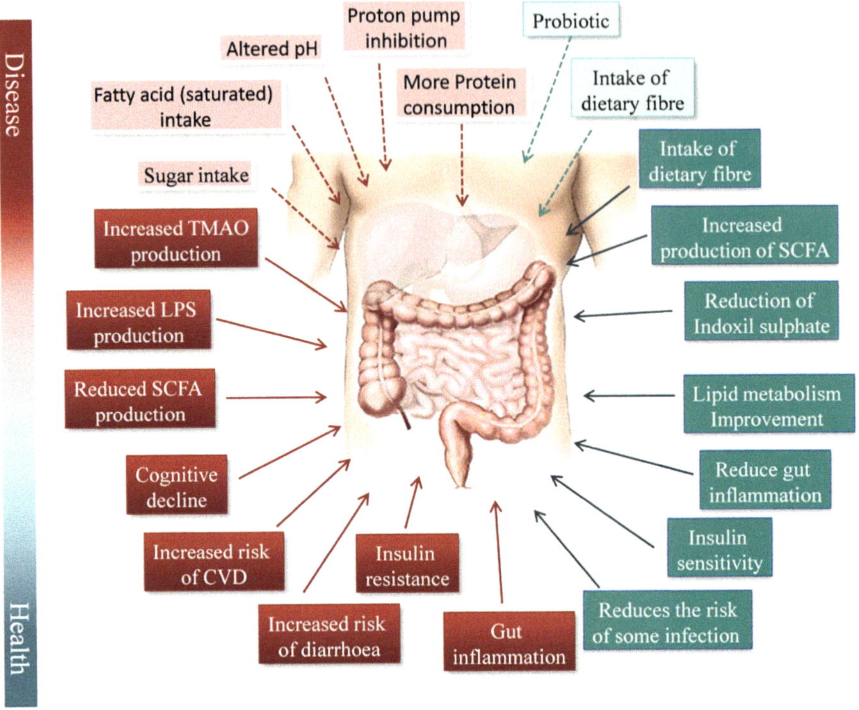

Fig. 12.1 Schematic representation of the role of the gut microbiota in health and disease

unabsorbable carbohydrates like starch, pectin, and cellulose. Therefore, it helps to recover energy and substrates to the host cell and can also be used for the growth and survival of the bacteria in the gut (Vyas and Ranganathan 2012). The GIT comprises the many molecules that can be used as nutrients by microbes majorly colonized in the colon or large intestine, harboring harmful and beneficial bacteria. Advanced technologies like DNA sequencing and computational methods helped us understand microbiota's impact on human health and how the microbiota relates to extending the influence from physiological to psychological (Fig. 12.1).

12.2.2 Gut Microbiomes in Metabolism

It was proven that the metagenome of gut microbes is about 150-fold higher than the human genome (Qin et al. 2010). Metabolic pathways are linked with major functions related to host maintenance, such as nervous system development and intestine development and regulation. In general, the digestion of dietary foods such as carbohydrates, proteins, and fat and fermentation occur to provide energy through these metabolic pathways. Human beings cannot degrade complex polysaccharides to convert them into energy. Consequently, these undigested materials enter the colon and are processed by microbiota inhabited in the colon. Normally, 10–60 g of carbohydrates, mostly insoluble starch, oligosaccharides, raffinose, and lactose, are converted into useful compounds by the action of microbiota (Ouwehand et al. 2005). By a sequence of fermentation processes, the complex carbohydrates convert into glucose by producing different hydrolytic enzymes from microorganisms. The microbial metabolic degradation process is continued depending on the type of functional groups of the dietary compounds (Rossi et al. 2005). The initial degradation starts with microbial species of *Bacteroides* sp. and *Ruminococcus* sp. (Flint et al. 2008). Gases including H_2, CO_2, methane, short-chain fatty acids, butyrate, and propionate are major by-products formed during fermentation. In this process, a short-chain fatty acid, butyrate, is produced by *Clostridium* cluster XIVa that mediates using *CoA* transferase, alternately using a rare pathway butyrate-kinase activity (Louis et al. 2014).

In contrast, propionate is synthesized by three major pathways called succinate pathway (*Bacteroidetes*), propanediol pathway (*Roseburia inulinivorans*), and acrylate pathway (*Megasphaera elsdenii*) (Reichardt et al. 2014). It is a continuous and chain-linked process in which a product of one microorganism acts as a substrate for another organism. It was also observed that metabolites formed from fermentation are not found in feces, indicating the effective degradation of carbohydrates (Samuel and Gordon 2006).

The commonly found bacteria in proteolysis are *Bacteroides*, *Clostridium*, *Propionibacterium*, *Fusobacterium*, *Lactobacillus*, and *Streptococcus*. Initially, complex proteins are cleaved into free amino acids and short peptides with the action of bacterial peptidases and proteases. Then, these components undergo fermentation and produce metabolites such as branched-chain fatty acids, gases, and trace amounts of ammonia and phenolic compounds (Macfarlane and Macfarlane 2006). Propionate short-chain fatty acid is synthesized from amino acids such as aspartate, alanine, threonine, and methionine. In contrast, butyrate is produced from glutamate, lysine, and histidine, respectively, cysteine, serine, and methionine. For the fermentation of complex proteins, exfoliated epithelial cells and pancreatic enzymes are also involved in the colon (Macfarlane and Macfarlane 2006). From the proteins, by-products like amines and ammonia act as toxins, increasing colon cancer chances (Bingham et al. 1996). Aromatic amino acids, including tyrosine, phenylacetic acid, and tryptophan, are fermented by bacterial species belonging to *Clostridium*, *Bacteroides*, *Bifidobacterium*, *Lactobacillus*, and *Peptostreptococcus*. Tryptophan is an essential amino acid that converts into indole propionic acid. Kynurenine is a major product of tryptophan fermentation by *Clostridium sporogenes* and *Lactobacillus* spp. and forms quinolinic acid and kynurenic acid, which act as an agonist and antagonist of glutamate receptors, respectively (Kim and Park 2013). A minor part of tryptophan is used to synthesize serotonin, which is used by the brain, and serotonin is regulated by tryptamine, which is formed by the decarboxylation of tryptophan (Sandyk 1992). The sulfur amino acids such as cysteine and methionine are very important in the cellular functions of a host. These amino acids are produced by amino acid fermentation and bacterial sulfate reduction, in which 200 bacterial species were identified to reduce sulfate in the gastrointestinal tract (Leloup et al. 2009). Lipids digest in the liver and convert into bile acids. Gut microbes degrade bile acid and its conjugate amino acid using bile salt hydrolase enzyme, which helps to detoxify the bile acids. The bacterial species belonging to *Bacteroides*, *Bifidobacterium*, *Clostridium*, *Lactobacillus*, and *Listeria* are known to involve in this process (Jones et al. 2008). Approximately, 30 types of bile acids maintained by the gut microbiota only are circulating in human body (García-Cañaveras et al. 2012).

12.2.3 Symbiotic and Dysbiosis Relationship with Microbiomes

The gut microbiota helps develop cells and tissues and secretes butyrate that helps regulate cell growth and differentiation. These microbiota help the cells degenerate from a neoplastic to a nonneoplastic phenotype. Studying the germ-free mice colonized with *B. thetaiotaomicron* explained the importance of microbes in the structural and morphological development of the gut (Stappenbeck et al. 2002).

Most of the gut bacterial lineages produce antimicrobial components to compete with other microbes for nutrients, thereby helping to control the pathogen entities and influencing the host immune system's development. The development of B cells, T cells, and regulatory T helper cells depends on gut microbial signals. It was evidenced that butyrate synthesized by microbes in the gut inhibits NF-kB and employs immunomodulatory effects (Maslowski et al. 2009). An innate immune system triggers toll-like receptors that bind to specific microbial compounds such as lipopolysaccharide, flagellin, and peptidoglycan in the mucosa. Further, this innate immune system activates nuclear factor-kB pathways, mitogen-activated protein kinase, and caspase-dependent signaling cascade pathways to produce cytokines, chemokines, and phagocytes, respectively, resulting in protection to commensal bacteria and helps to develop homeostasis of the immune system and inflammatory response to foreign bacteria (Fig 12.2) (Kinross et al. 2011).

In general, changes in the intestine ecosystem lead to host illness. The alteration or imbalance of normal gut microbiota triggers the development of dysbiosis. Intestinal enzymes and IgA establish the first line of defense against pathogens. Dysbiosis may cause diseases such as inflammatory bowel disease, cancer, hypercholesterolemia, obesity, diabetes, and infections. The lung is one of the important organs that is exposed to microbes through inhalation. Over the last decade, culture-independent technology had changed our understanding of the interaction between the microbiota and the lung, revealing that the lung is not sterile (Dickson et al. 2016). The pathogens associated with lung disease and pneumonia include *Streptococcus pneumoniae*, *Staphylococcus aureus*, *Klebsiella pneumoniae*, *Haemophilus influenzae*, *Chlamydophila psittaci*, *Mycoplasma pneumoniae*, and *Legionella pneumophila*. Most bacteria on the skin play an important role as either commensal or mutualistic. Microbiota on the skin is important in the maturation and homeostasis of cutaneous immunity, which helps to produce antimicrobial peptides from keratinocytes and interleukins. It was found that the major microbiota on the skin is *Corynebacterium*. Diseases associated with microorganisms are given in Table 12.1 (Fig. 12.2).

12.2.4 Microbiomes in Cancer

Microbial dysbiosis in the gut can induce carcinoma by activating the pathways involved in carcinogenesis. It also encourages the pro-inflammatory effects by activating the toll-like receptors (TLRs), resultantly increasing the production of pro-inflammatory factors from mucosal cells, thereby increasing carcinogenesis (Lam et al. 2017). Alteration in the gut microbiota initiates and develops cancer (gastric cancer, colorectal cancer (CRC), hepatocellular carcinoma (HCC), pancreatic cancer, breast cancer, and melanoma) in the host. In a study comparing fecal microbiota between cancer and normal patients, the fecal microbiota from cancer patent was induced into a carcinogen received germ-free mice and observed that these mice developed carcinogenic properties (Wong et al. 2017). It was also found that long-term antibiotic exposure altered the microbiota and increased the risk of colorectal cancer in mice (Bullman et al. 2017).

Microorganisms such as *Human papillomavirus*, *Helicobacter pylori*, and *Hepatitis B virus* are considered etiological factors and contribute to 20% of cancer worldwide. Most of the bacteria are found to trigger carcinogenic pathways by producing metabolites in the host. It was noticed that *H. pylori* produces a virulence factor, CagA, and activates chronic gastritis and gastric carcinogenesis. The CagA activates NF-κB, ERK/MAPK, Wnt/β-catenin, Ras, and STAT3 pathways, resultantly increase inflammatory cytokine production (IFN-γ, TNF-α, IL-1, IL-6, and IL-10) and activate immune responses, thereby promote cell scattering and proliferation. The details of bacterial pathogens associated with cancer, mode of action, and triggered and suppressed pathway are given in Table 12.2.

12.2.5 Microbiomes in Obesity

Gut microbes play an important role in the host immune system. Microbes can contact the epithelium due to dysbiosis, resulting in an inflammatory response mediated by toll-like receptors (TLR). The TLRs play an important role in recognizing bacterial cells through lipopolysaccharides present in bacterial cell walls. TLR4 and TLR5 receptors bind to bacterial flagellin (Ringel 2017). These reactions trigger the production of cytokines and chemokines, which are pro-inflammatory compounds (Ringel 2017), and then activate the tumor necrosis factor-alpha (TNF-α) (O'Neill et al. 2013). The TLR4 involved in the inflammatory reaction is linked with insulin resistance in mice (Cani et al. 2007). An animal model experiment observed increased plasma LPS, elevated glucose levels, hyperinsulinemia, and weight gain in mice fed a high-fat diet (Vijay-Kumar et al. 2010). In another study, mice with high lipopolysaccharides lead to a decreased *Bifidobacterium* spp. Humans with diabetes and obesity showed high levels of lipopolysaccharides and inflammation,

Table 12.1 A partial list of microbial-associated diseases in human

Diseases	Related microbiome community	Description of relation	Reference
Autism	Gut-associated *Clostridium* and *Sutterella* species	Toxin production and aberrations in the fermentation of products by altered gut microbes	Ding et al. (2017)
Cardiovascular disease	Gut microflora	The gut microbiome is linked to trimethylamine (TMA) and TAM-N-oxide formation	Tang and Hazen (2014)
Atherosclerosis	Gut microflora	Gut microbiome metabolism of cholesterol and lipid can affect the atherosclerotic plaque formation	Jonsson and Bäckhed (2017)
Chronic skin wounds	*Staphylococcus aureus, Pseudomonas aeruginosa, Proteus mirabilis*	Play a role in the development of chronicity and delaying wound healing	Bessa et al. (2015)
Atopic dermatitis	*Staphylococcus aureus, Corynebacterium bovis*	Eczema formation and T-helper two-cell responses trigger the dysbiosis of skin microbial flora	Kobayashi et al. (2015)
Colorectal cancer	Gut-associated pathogens (*S. bovis, Bacteroides fragilis, Fusobacterium nucleatum*, etc.)	Microorganisms possibly cause DNA damage in epithelial cells and initiate to develop the cancer	Gao et al. (2017)
Cystic fibrosis	*Pseudomonas aeruginosa*	Colonization of hypermutable strains in respiratory airways and evaluation of antibiotic resistance and virulence potentially linked to the disease	Oliver et al. (2000) and Lynch and Bruce (2013)
Type 1 diabetes	Gut microbiota	A significant decrease in microbiome alpha diversity is linked to the development of clinical symptoms	Kostic et al. (2015)
Type 2 diabetes	Gut microflora	Increased virulence factors and antibiotic resistance genes of the gut microbiome may induce insulin resistance	(Wang et al. 2017)
Irritable bowel syndrome	Mucosal and fecal gut microbiota	Changes in gut microbiota activate the mucosal innate immune response and nociceptive sensory pathways, leading to dysregulation of the enteric nervous system	Simrén et al. (2013)
Parkinson's disease	Gut microbiota	Alterations of gut bacteria lead to failure to regulate motor deficits, microglia activation, and αSyn protein aggregation, representing disease risk factors	Sampson et al. (2016)

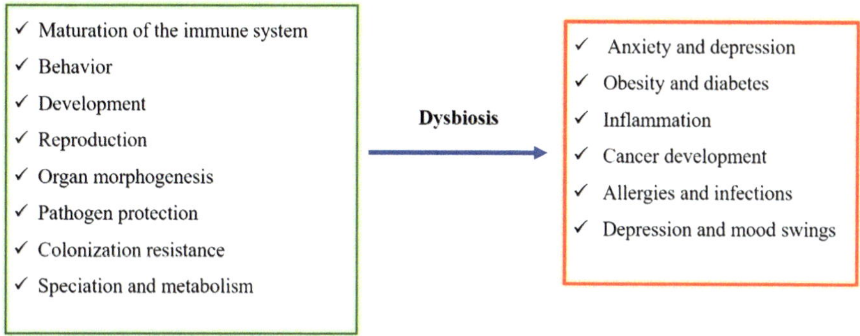

Fig. 12.2 Effect of dysbiosis on human health occurred by altered gut microbiome

and it can be concluded that LPS can increase insulin sensitivity (van der Crabben et al. 2009). The fecal content of the obese mice contained increased levels of Firmicutes and decreased levels of Bacteroidetes compared with control mice (Turnbaugh et al. 2009) and also noted increased levels of butyrate and acetate (Ley et al. 2006b). Similar results were observed in human with obese microbiota (Fernandes et al. 2014). An increased level of Actinobacteria, decreased levels of Bacteroidetes, and no significant difference in Firmicutes were observed (Schwiertz et al. 2010). Another study found high levels of Bacteroidetes and lower levels of Firmicutes than their lean counterparts (Allin et al. 2018).

Recent studies found that the major microbial clusters are Firmicutes and Bacteroidetes. Still, decreased diversity is linked to increased BMI in healthy humans. In contrast, Proteobacteria was higher, and an abundance of *Bifidobacterium*, *Faecalibacterium*, and butyrate-producing bacteria was decreased in obese than overweight or healthy humans. Insulin signaling pathways trigger enzymes to be downregulated, resulting in changes in short-chain fatty acid production, and insulin regulation leads to obesity development (Jia et al. 2017).

12.2.6 Microbiomes in Diabetes

Diabetes is a metabolic disorder mediated by an imbalance in glucose metabolism. Type 1 represents 10%, and type 2 diabetes represents 90% (Mohammad and Ahmad 2016). Fewer butyrate-producing microbes and increased *Lactobacillus* sp. were observed in type 2 diabetes human patients. Likewise, a positive correlation between Bacteroidetes to Firmicutes ratio and the glucose levels in plasma as well as a group of bacterial clusters such as *Bacteroides-Prevotella* and *Clostridium coccoides-Eubacterium rectal* were found in humans with type 2 diabetes. It was also observed elevated levels of Betaproteobacteria in type 2 diabetes. This might be possible because of high levels of endotoxin production, wherein the LPS play a key role in enhancing the pathogenesis of type 2 diabetes (Larsen et al. 2010). In patients with metabolic syndrome, the antibiotic treatment with vancomycin reduced the

Table 12.2 A partial list of gut pathogens and their associated gastrointestinal cancers with possible mechanisms

Bacterial species	Cancers	Virulence factor	Mechanisms	Reference
Helicobacter pylori	Gastric cancer	VacA	1. Activate ERK/MAPK, VEGF, and Wnt/β-catenin pathway 2. Induce cell autophagy and vacuolation 3. Methylate CpG islands of E-cadherin, TFF2, and FOXD3 4. Inhibit PI3K/Akt and GSK3 pathway	Mashima et al. (2008) and Ricci (2016)
		CagA	1. Activate NF-κB, ERK/MAPK, Wnt/β-catenin, Ras, and STAT3 pathway 2. Promote cell scattering and proliferation 3. Increase inflammatory cytokine production (IFN-γ, TNF-α, IL-1, IL-6, and IL-10) 4. Activate immune responses	Odenbreit et al. (2000), Wang et al. (2014) and Yong et al. (2015)
Fusobacterium nucleatum, *Peptostreptococcus stomatis*	Gastric cancer	NR	NR	Coker et al. (2018) and Dias-Jácome et al. (2016)
Enteropathogenic *Escherichia coli*	Colorectal cancer	CIF	1. Modify cytoskeleton 2. Induce G2/M arrest 3. Induce DNA damage	Collins et al. (2011) and Khan (2015)
		CNF	1. Promote cell proliferation 2. Inhibit the process of program cell death 3. Stimulate rho GTPases 4. Trigger G1–S transition and induce DNA replication	Collins et al. (2011) and Nougayrède et al. (2005)
		CDT	1. Induce DNA damage 2. Induce pro-inflammatory molecule production (TNF-a, COX-2, NF-κB, and IL-6)	Khan (2015) and Nesić et al. (2004)
		Intimin	1 Downregulate DNA mismatch repair	Maddocks et al. (2009)
		Colibactin	1. Induce DNA double-strand breaks 2. Induce chromosomal instability 3. Induce senescence-associated secretory phenotype	Bonnet et al. (2014), Cougnoux et al. (2014), and Cuevas-Ramos et al. (2010)

(continued)

Table 12.2 (continued)

Bacterial species	Cancers	Virulence factor	Mechanisms	Reference
Fusobacterium nucleatum	Colorectal cancer	Fap2	1. Induce microRNA-21 and NF-κB expression 2. Activate TLR4 and Wnt/β-catenin pathway 3. Combine with gal-GalNAc	Abed et al. (2016) and Yang et al. (2017)
		FadA	1. Upregulate NF-κB signaling 2. Activate Wnt/β-catenin pathway 3. Form bacterial aggregation	Allen-Vercoe and Jobin (2014), Hold (2016), and Rubinstein et al. (2013)
		NR	1. Activate autophagy pathway 2. Mediate CRC chemoresistance	Yu et al. (2017)
Enterotoxigenic *Bacteroides fragilis*	Colorectal cancer	BFT	1. Activate Wnt/β--catenin, STAT3, and NF-κB pathway 2. Induce E-cadherin cleavage 3. Induce Th17 T cell responses 4. Promote cell proliferation 5. Increase the expression of c-Myc and cyclin-D1 6. Promote the release of pro-inflammatory factors	Chung et al. (2018), Goodwin et al. (2011), Wu et al. (2009), and Zhou and Fang (2018)
Streptococcus bovis	Colorectal cancer	WEA	1. Increase the production of COX-2 and IL-8 2. Promote cell proliferation	Alazmi et al. (2006) and Biarc et al. (2004)
Helicobacter pylori	Colorectal cancer	BabA CagA, VacA,	1. Produce ROS and NO 2. Promote cell proliferation 3. Decrease cell apoptosis 4. Increase the production of COX-2 and IL-8	Collins et al. (2011) and Papastergiou et al. (2016)
Peptostreptococcus anaerobius	Colorectal cancer	NR	1. Activate SREBP-2 and increase cholesterol synthesis 2. Regulate TLR and AMPK pathway 3. Increase the level of ROS 4. Promote cell proliferation	Tsoi et al. (2017)

(continued)

Table 12.2 (continued)

Bacterial species	Cancers	Virulence factor	Mechanisms	Reference
Streptococcus gallolyticus	Colorectal cancer	NR	1. Promote cell proliferation 2. Increase the level of c-Myc, β-catenin, and PCNA	Kumar et al. (2017)
Enterococcus faecalis	Colorectal cancer	NR	1. Elicit high levels of ROS 2. Induce genetic instability 3. Activate Wnt/β-catenin pathway	de Almeida et al. (2018), Fearon (2011), and Wang et al. (2012)
Desulfovibrio sp.	Colorectal cancer	H_2S	1. Increase the production of sulfur radicals and ROS 2. Induce direct DNA damage 3. Transform CA to DCA 4. Promote cell proliferation	Balamurugan et al. (2008), Deplancke and Gaskins (2003), Ridlon et al. (2005), and Scanlan et al. (2009)
Helicobacter hepaticus	Hepatocellular carcinoma	CDT	1. Mediate γ-H_2AX foci formation 2. Remodel cytoskeleton 3. Induce nuclear translocation of NF-κB 4. Increase the expression of p21 and Ki-67	Péré-Védrenne et al. (2016) and Péré-Védrenne et al. (2017)
		NR	1. Induce oxidative DNA injury 2. Impair metabolic detoxification 3. Activate immune response 4. Promote cell proliferation 5. Activate Wnt/β-catenin pathway	Alyamani et al. (2007), Fox et al. (2010), and García et al. (2011)
Helicobacter pylori	Hepatocellular carcinoma	VacA, CagA, LPS	1. Activate NF-κB pathway 2. Increase the production of IL-8 and AP-1	Fox et al. (2010), Queiroz et al. (2006), and Sakr et al. (2013)

(continued)

Table 12.2 (continued)

Bacterial species	Cancers	Virulence factor	Mechanisms	Reference
Porphyromonas gingivalis	Pancreatic cancer	LPS	1. Trigger innate and adaptive immunity 2. Activate NF-κB and STAT3 pathway 3. Stimulate TLR2 and TLR4	Michaud (2013)
Helicobacter pylori		LPS, ammonia	1. Increase K-ras gene mutation 2. Activate NF-κB, STAT3, and MAPK pathways 3. Increase the production of IL-8 and AP-1 4. Activate immune responses 5. Increase the production of antiapoptotic and pro-proliferative proteins	Daniluk et al. (2012), di Magliano and Logsdon (2013), and Fukuda et al. (2011)

AP-1 activator protein-1, BabA blood group antigen-binding adhesin, BFT *Bacteroides fragilis* toxin, CA cholic acid, CagA cytotoxin-associated gene A, CDT cytolethal distending toxin, CIF cycle-inhibiting factor, CNF cytotoxic necrotizing factor, COX-2 cyclo-oxygenase 2, CRC colorectal cancer, DCA deoxycholic acid, ERK extracellular signal-regulated protein kinases, FadA fusobacterium adhesion A, FOXD3 forkhead box D3, H2S hydrogen sulfide, IFN-γ interferon-γ, IL interleukin, LPS lipopolysaccharide, MAPK mitogen-activated protein kinase, NF-κB nuclear factor-κB, NO nitric oxide, NR not reported, γ-H2AX phosphorylated form of H2A histone family member X, PCNA proliferating cell nuclear antigen, ROS reactive oxygen species, SRB sulfate-reducing bacteria, SREBP sterol regulatory element binding proteins, STAT3 signal transducers and activators of transcription 3, TFF2 trefoil factor 2, TLR toll-like receptor, TNF-α tumor necrosis factor-α, VacA vacuolating cytotoxin A, VEGF vascular endothelial growth factor, WEA wall-extracted antigen

abundance of gram-positive bacteria, which is involved in the production of butyrate. This condition was correlated with impaired insulin sensitivity. It was well understood that the occurrence of disease pathogenesis might be due to lowered levels of butyrate-producing gut microorganisms in type 2 diabetes patients (Vrieze et al. 2012). In exploring gut microbiota in 345 type 2 diabetes patients and control subjects, it was found that through a study, Qin et al. studied 345 type 2 diabetes and control subjects' gut microbial data using a metagenome-wide association study and found that *Bacteroides caccae*, *Clostridium hathewayi*, *Clostridium ramosum*, *Clostridium symbiosum*, *Eggerthella lenta*, *E. coli Clostridiales sp. SS3/4*, *E. rectale*, *Faecalibacterium prausnitzii*, *Roseburia intestinalis*, and *Roseburia inulinivorans* are well documented opportunistic pathogens that causes the infection (Qin et al. 2012). The gut microbes can be a biomarker for predicting type 2 diabetes. However, little is known about type 1 diabetes and microorganisms. In a clinical study with 81 patients with type 1 diabetes and 40 control subjects, it was shown that intestinal permeability was significantly increased in type 1 diabetes, which means that intestinal barriers are significantly not functioning (Bosi et al. 2006).

12.2.7 Microbiomes in Psychotic Disorders

Autism spectrum disorder (ASD) is a neurobehavioral disorder characterized by impaired social interaction and communication, resulting in restricted and repetitive behavior. Autism is considered the most important disorder in ASD (Fig 12.2). The gut microbiota was studied in an affected person and compared with normal humans. The bacterial population belonging to the *Clostridium* genus was found to be ten times higher in affected human fecal samples (Favier et al. 2002). Another study also observed an imbalance of Bacteroidetes and Firmicutes phyla. Other bacterial communities belonging to genera such as *Bifidobacterium*, *Lactobacillus*, *Sutterella*, *Prevotella*, and *Ruminococcus* and family Alcaligenaceae were also increased compared with control fecal samples (Finegold et al. 2010).

Changes in the microbial ecosystem alter the production of potent pro-inflammatory endotoxin lipopolysaccharides. These lipopolysaccharides greatly impact the modulation of the central nervous system and increase the emotional control such as the amygdala, leading to alteration in psychological brain activities and neuropeptide synthesis (Kastin and Pan 2010).

12.2.8 Clinical Applications

In a study, *Lactobacillus reuteri* was able to repair the damaged intestinal mucosa. Antibiotics, stress, food, and infection with pathogens are general factors causing inflammation in the intestine. Microbiota plays an important role in maintaining the mucosal barrier and invasion of pathogens. Tumor necrosis factor (TNF) is an important key factor in regulating autoimmune diseases such as arthritis, inflammatory bowel disease, and acute inflammation states, including sepsis, necrotizing enterocolitis, and intestinal hypoxia. By administrating with *Lactobacillus* in the intestinal epithelia of mice, it has been observed that damaged mucosa was repaired and intestinal organoids grew well (Wu et al. 2020).

Prebiotics are considered nondigestible food components that promote the communal microbiota's growth. Prebiotics with dietary supplements help to improve health. In addition to prebiotics, probiotics also give tremendous health benefits. Probiotics are defined as living microorganisms introduced into the gut for potential beneficial effects. It was proven that probiotics shift microbial communities, help grow beneficial microbes, reduce inflammation, and reduce obesity and diabetes (Casacchia et al. 2019).

12.3 Conclusions and Future Perspectives

At present, the understanding of bacterial evolution and its association with animals reveals the interaction of microbes with animals and their biological functions. New technologies such as whole genome sequencing, genome array, and culture-independent technologies have changed our perspective of the microbial world and revealed that microorganisms are the centrality to the biological process of animals and can be considered a macrobiological world of the microbiota. From this chapter, we can understand that resident microbes play decisive roles in determining the metabolic and regulatory networks that define good health and a spectrum of disease states.

Acknowledgments The authors thank the Department of Biotechnology, Vikrama Simhapuri University, Nellore, 524 320, Andhra Pradesh, India, for its constant support.

Conflict of Interest The authors declare that there is no conflict of interest.

References

Abed J, Emgård JE, Zamir G, Faroja M, Almogy G, Grenov A, Sol A, Naor R, Pikarsky E, Atlan KA, Mellul A, Chaushu S, Manson AL, Earl AM, Ou N, Brennan CA, Garrett WS, Bachrach G (2016) Fap2 mediates Fusobacterium nucleatum colorectal adenocarcinoma enrichment by binding to tumor-expressed gal-GalNAc. Cell Host Microbe 20(2):215–225. https://doi.org/10.1016/j.chom.2016.07.006

Alazmi W, Bustamante M, O'Loughlin C, Gonzalez J, Raskin JB (2006) The association of Streptococcus bovis bacteremia and gastrointestinal diseases: a retrospective analysis. Dig Dis Sci 51(4):732–736. https://doi.org/10.1007/s10620-006-3199-7

Allen-Vercoe E, Jobin C (2014) Fusobacterium and enterobacteriaceae: important players for CRC? Immunol Lett 162(2 Pt A):54–61. https://doi.org/10.1016/j.imlet.2014.05.014

Allin KH, Tremaroli V, Caesar R, Jensen BAH, Damgaard MTF, Bahl MI, Licht TR, Hansen TH, Nielsen T, Dantoft TM, Linneberg A, Jørgensen T, Vestergaard H, Kristiansen K, Franks PW, Hansen T, Bäckhed F, Pedersen O (2018) Aberrant intestinal microbiota in individuals with prediabetes. Diabetologia 61(4):810–820. https://doi.org/10.1007/s00125-018-4550-1

Alyamani EJ, Brandt P, Pena JA, Major AM, Fox JG, Suerbaum S, Versalovic J (2007) Helicobacter hepaticus catalase shares surface-predicted epitopes with mammalian catalases. Microbiology 153(Pt 4):1006–1016. https://doi.org/10.1099/mic.0.29184-0

Bäckhed F, Ding H, Wang T, Hooper LV, Koh GY, Nagy A, Semenkovich CF, Gordon JI (2004) The gut microbiota as an environmental factor that regulates fat storage. Proc Natl Acad Sci U S A 101(44):15718–15723. https://doi.org/10.1073/pnas.0407076101

Balamurugan R, Rajendiran E, George S, Samuel GV, Ramakrishna BS (2008) Real-time polymerase chain reaction quantification of specific butyrate-producing bacteria, Desulfovibrio and enterococcus faecalis in the feces of patients with colorectal cancer. J Gastroenterol Hepatol 23(8 Pt 1):1298–1303. https://doi.org/10.1111/j.1440-1746.2008.05490.x

Bessa LJ, Fazii P, Di Giulio M, Cellini L (2015) Bacterial isolates from infected wounds and their antibiotic susceptibility pattern: some remarks about wound infection. Int Wound J 12(1):47–52. https://doi.org/10.1111/iwj.12049

Biarc J, Nguyen IS, Pini A, Gossé F, Richert S, Thiersé D, Van Dorsselaer A, Leize-Wagner E, Raul F, Klein JP, Schöller-Guinard M (2004) Carcinogenic properties of proteins with pro-

inflammatory activity from Streptococcus infantarius (formerly S.bovis). Carcinogenesis 25(8):1477–1484. https://doi.org/10.1093/carcin/bgh091
Bingham SA, Pignatelli B, Pollock JR, Ellul A, Malaveille C, Gross G, Runswick S, Cummings JH, O'Neill IK (1996) Does increased endogenous formation of N-nitroso compounds in the human colon explain the association between red meat and colon cancer? Carcinogenesis 17(3):515–523. https://doi.org/10.1093/carcin/17.3.515
Bonnet M, Buc E, Sauvanet P, Darcha C, Dubois D, Pereira B, Déchelotte P, Bonnet R, Pezet D, Darfeuille-Michaud A (2014) Colonization of the human gut by E. coli and colorectal cancer risk. Clin Cancer Res 20(4):859–867. https://doi.org/10.1158/1078-0432.ccr-13-1343
Bosi E, Molteni L, Radaelli MG, Folini L, Fermo I, Bazzigaluppi E, Piemonti L, Pastore MR, Paroni R (2006) Increased intestinal permeability precedes clinical onset of type 1 diabetes. Diabetologia 49(12):2824–2827. https://doi.org/10.1007/s00125-006-0465-3
Bullman S, Pedamallu CS, Sicinska E, Clancy TE, Zhang X, Cai D, Neuberg D, Huang K, Guevara F, Nelson T, Chipashvili O, Hagan T, Walker M, Ramachandran A, Diosdado B, Serna G, Mulet N, Landolfi S, Ramon Y, Cajal S, Fasani R, Aguirre AJ, Ng K, Élez E, Ogino S, Tabernero J, Fuchs CS, Hahn WC, Nuciforo P, Meyerson M (2017) Analysis of Fusobacterium persistence and antibiotic response in colorectal cancer. Science 358(6369):1443–1448. https://doi.org/10.1126/science.aal5240
Cani PD, Amar J, Iglesias MA, Poggi M, Knauf C, Bastelica D, Neyrinck AM, Fava F, Tuohy KM, Chabo C, Waget A, Delmée E, Cousin B, Sulpice T, Chamontin B, Ferrières J, Tanti JF, Gibson GR, Casteilla L, Delzenne NM, Alessi MC, Burcelin R (2007) Metabolic endotoxemia initiates obesity and insulin resistance. Diabetes 56(7):1761–1772. https://doi.org/10.2337/db06-1491
Casacchia T, Scavello F, Rocca C, Granieri MC, Beretta G, Amelio D, Gelmini F, Spena A, Mazza R, Toma CC, Angelone T, Statti G, Pasqua T (2019) Leopoldia comosa prevents metabolic disorders in rats with high-fat diet-induced obesity. Eur J Nutr 58(3):965–979. https://doi.org/10.1007/s00394-018-1609-1
Chiller K, Selkin BA, Murakawa GJ (2001) Skin microflora and bacterial infections of the skin. J Investig Dermatol Symp Proc 6(3):170–174. https://doi.org/10.1046/j.0022-202x.2001.00043.x
Chung L, Thiele Orberg E, Geis AL, Chan JL, Fu K, DeStefano Shields CE, Dejea CM, Fathi P, Chen J, Finard BB, Tam AJ, McAllister F, Fan H, Wu X, Ganguly S, Lebid A, Metz P, Van Meerbeke SW, Huso DL, Wick EC, Pardoll DM, Wan F, Wu S, Sears CL, Housseau F (2018) Bacteroides fragilis toxin coordinates a pro-carcinogenic inflammatory cascade via targeting of colonic epithelial cells. Cell Host Microbe 23(2):203–214.e5. https://doi.org/10.1016/j.chom.2018.01.007
Claesson MJ, O'Sullivan O, Wang Q, Nikkilä J, Marchesi JR, Smidt H, de Vos WM, Ross RP, O'Toole PW (2009) Comparative analysis of pyrosequencing and a phylogenetic microarray for exploring microbial community structures in the human distal intestine. PLoS One 4(8):e6669. https://doi.org/10.1371/journal.pone.0006669
Coker OO, Dai Z, Nie Y, Zhao G, Cao L, Nakatsu G, Wu WK, Wong SH, Chen Z, Sung JJY, Yu J (2018) Mucosal microbiome dysbiosis in gastric carcinogenesis. Gut 67(6):1024–1032. https://doi.org/10.1136/gutjnl-2017-314281
Collins D, Hogan AM, Winter DC (2011) Microbial and viral pathogens in colorectal cancer. Lancet Oncol 12(5):504–512. https://doi.org/10.1016/s1470-2045(10)70186-8
Cougnoux A, Dalmasso G, Martinez R, Buc E, Delmas J, Gibold L, Sauvanet P, Darcha C, Déchelotte P, Bonnet M, Pezet D, Wodrich H, Darfeuille-Michaud A, Bonnet R (2014) Bacterial genotoxin colibactin promotes colon tumour growth by inducing a senescence-associated secretory phenotype. Gut 63(12):1932–1942. https://doi.org/10.1136/gutjnl-2013-305257
Cuevas-Ramos G, Petit CR, Marcq I, Boury M, Oswald E, Nougayrède JP (2010) Escherichia coli induces DNA damage in vivo and triggers genomic instability in mammalian cells. Proc Natl Acad Sci U S A 107(25):11537–11542. https://doi.org/10.1073/pnas.1001261107
Daniluk J, Liu Y, Deng D, Chu J, Huang H, Gaiser S, Cruz-Monserrate Z, Wang H, Ji B, Logsdon CD (2012) An NF-κB pathway-mediated positive feedback loop amplifies Ras activity to pathological levels in mice. J Clin Invest 122(4):1519–1528. https://doi.org/10.1172/jci59743

de Almeida CV, Taddei A, Amedei A (2018) The controversial role of enterococcus faecalis in colorectal cancer. Ther Adv Gastroenterol 11:1756284818783606. https://doi.org/10.1177/1756284818783606

Deplancke B, Gaskins HR (2003) Hydrogen sulfide induces serum-independent cell cycle entry in nontransformed rat intestinal epithelial cells. FASEB J 17(10):1310–1312. https://doi.org/10.1096/fj.02-0883fje

di Magliano MP, Logsdon CD (2013) Roles for KRAS in pancreatic tumor development and progression. Gastroenterology 144(6):1220–1229. https://doi.org/10.1053/j.gastro.2013.01.071

Dias-Jácome E, Libânio D, Borges-Canha M, Galaghar A, Pimentel-Nunes P (2016) Gastric microbiota and carcinogenesis: the role of non-helicobacter pylori bacteria - a systematic review. Rev Esp Enferm Dig 108(9):530–540. https://doi.org/10.17235/reed.2016.4261/2016

Dickson RP, Erb-Downward JR, Martinez FJ, Huffnagle GB (2016) The microbiome and the respiratory tract. Annu Rev Physiol 78:481–504. https://doi.org/10.1146/annurev-physiol-021115-105238

Ding HT, Taur Y, Walkup JT (2017) Gut microbiota and autism: key concepts and findings. J Autism Dev Disord 47(2):480–489. https://doi.org/10.1007/s10803-016-2960-9

Favier CF, Vaughan EE, De Vos WM, Akkermans ADL (2002) Molecular monitoring of succession of bacterial communities in human neonates. Appl Environ Microbiol 68(1):219–226. https://doi.org/10.1128/aem.68.1.219-226.2002

Fearon ER (2011) Molecular genetics of colorectal cancer. Annu Rev Pathol 6:479–507. https://doi.org/10.1146/annurev-pathol-011110-130235

Fernandes J, Su W, Rahat-Rozenbloom S, Wolever TM, Comelli EM (2014) Adiposity, gut microbiota and faecal short chain fatty acids are linked in adult humans. Nutr Diabetes 4(6):e121. https://doi.org/10.1038/nutd.2014.23

Finegold SM, Dowd SE, Gontcharova V, Liu C, Henley KE, Wolcott RD, Youn E, Summanen PH, Granpeesheh D, Dixon D, Liu M, Molitoris DR, Green JA 3rd (2010) Pyrosequencing study of fecal microflora of autistic and control children. Anaerobe 16(4):444–453. https://doi.org/10.1016/j.anaerobe.2010.06.008

Flint HJ, Bayer EA, Rincon MT, Lamed R, White BA (2008) Polysaccharide utilization by gut bacteria: potential for new insights from genomic analysis. Nat Rev Microbiol 6(2):121–131. https://doi.org/10.1038/nrmicro1817

Fox JG, Feng Y, Theve EJ, Raczynski AR, Fiala JL, Doernte AL, Williams M, McFaline JL, Essigmann JM, Schauer DB, Tannenbaum SR, Dedon PC, Weinman SA, Lemon SM, Fry RC, Rogers AB (2010) Gut microbes define liver cancer risk in mice exposed to chemical and viral transgenic hepatocarcinogens. Gut 59(1):88–97. https://doi.org/10.1136/gut.2009.183749

Fukuda A, Wang SC, JPt M, Folias AE, Liou A, Kim GE, Akira S, Boucher KM, Firpo MA, Mulvihill SJ, Hebrok M (2011) Stat3 and MMP7 contribute to pancreatic ductal adenocarcinoma initiation and progression. Cancer Cell 19(4):441–455. https://doi.org/10.1016/j.ccr.2011.03.002

Gao R, Gao Z, Huang L, Qin H (2017) Gut microbiota and colorectal cancer. Eur J Clin Microbiol Infect Dis 36(5):757–769. https://doi.org/10.1007/s10096-016-2881-8

García A, Zeng Y, Muthupalani S, Ge Z, Potter A, Mobley MW, Boussahmain C, Feng Y, Wishnok JS, Fox JG (2011) Helicobacter hepaticus--induced liver tumor promotion is associated with increased serum bile acid and a persistent microbial-induced immune response. Cancer Res 71(7):2529–2540. https://doi.org/10.1158/0008-5472.can-10-1975

García-Cañaveras JC, Donato MT, Castell JV, Lahoz A (2012) Targeted profiling of circulating and hepatic bile acids in human, mouse, and rat using a UPLC-MRM-MS-validated method. J Lipid Res 53(10):2231–2241. https://doi.org/10.1194/jlr.D028803

Goodwin AC, Destefano Shields CE, Wu S, Huso DL, Wu X, Murray-Stewart TR, Hacker-Prietz A, Rabizadeh S, Woster PM, Sears CL, Casero RA Jr (2011) Polyamine catabolism contributes to enterotoxigenic Bacteroides fragilis-induced colon tumorigenesis. Proc Natl Acad Sci U S A 108(37):15354–15359. https://doi.org/10.1073/pnas.1010203108

Grosberg RK, Strathmann RR (2007) The evolution of multicellularity: a minor Major transition? Annu Rev Ecol Evol Syst 38(1):621–654. https://doi.org/10.1146/annurev.ecolsys.36.102403.114735

Hold GL (2016) Gastrointestinal microbiota and colon cancer. Dig Dis 34(3):244–250. https://doi.org/10.1159/000443358

Jia L, Li D, Feng N, Shamoon M, Sun Z, Ding L, Zhang H, Chen W, Sun J, Chen YQ (2017) Anti-diabetic effects of clostridium butyricum CGMCC0313.1 through promoting the growth of gut butyrate-producing bacteria in type 2 diabetic mice. Sci Rep 7(1):7046. https://doi.org/10.1038/s41598-017-07335-0

Jones BV, Begley M, Hill C, Gahan CGM, Marchesi JR (2008) Functional and comparative metagenomic analysis of bile salt hydrolase activity in the human gut microbiome. Proc Natl Acad Sci 105(36):13580–13585. https://doi.org/10.1073/pnas.0804437105

Jonsson AL, Bäckhed F (2017) Role of gut microbiota in atherosclerosis. Nat Rev Cardiol 14(2):79–87. https://doi.org/10.1038/nrcardio.2016.183

Kastin AJ, Pan W (2010) Concepts for biologically active peptides. Curr Pharm Des 16(30):3390–3400. https://doi.org/10.2174/138161210793563491

Khan S (2015) Potential role of Escherichia coli DNA mismatch repair proteins in colon cancer. Crit Rev Oncol Hematol 96(3):475–482. https://doi.org/10.1016/j.critrevonc.2015.05.002

Kim J, Park W (2013) Indole inhibits bacterial quorum sensing signal transmission by interfering with quorum sensing regulator folding. Microbiology 159(Pt_12):2616–2625. https://doi.org/10.1099/mic.0.070615-0

Kinross JM, Darzi AW, Nicholson JK (2011) Gut microbiome-host interactions in health and disease. Genome Med 3(3):14. https://doi.org/10.1186/gm228

Kobayashi T, Glatz M, Horiuchi K, Kawasaki H, Akiyama H, Kaplan DH, Kong HH, Amagai M, Nagao K (2015) Dysbiosis and Staphylococcus aureus colonization drives inflammation in atopic dermatitis. Immunity 42(4):756–766. https://doi.org/10.1016/j.immuni.2015.03.014

Kostic AD, Gevers D, Siljander H, Vatanen T, Hyötyläinen T, Hämäläinen AM, Peet A, Tillmann V, Pöhö P, Mattila I, Lähdesmäki H, Franzosa EA, Vaarala O, de Goffau M, Harmsen H, Ilonen J, Virtanen SM, Clish CB, Orešič M, Huttenhower C, Knip M, Xavier RJ (2015) The dynamics of the human infant gut microbiome in development and in progression toward type 1 diabetes. Cell Host Microbe 17(2):260–273. https://doi.org/10.1016/j.chom.2015.01.001

Kumar R, Herold JL, Schady D, Davis J, Kopetz S, Martinez-Moczygemba M, Murray BE, Han F, Li Y, Callaway E, Chapkin RS, Dashwood WM, Dashwood RH, Berry T, Mackenzie C, Xu Y (2017) Streptococcus gallolyticus subsp. gallolyticus promotes colorectal tumor development. PLoS Pathog 13(7):e1006440. https://doi.org/10.1371/journal.ppat.1006440

Lam SY, Yu J, Wong SH, Peppelenbosch MP, Fuhler GM (2017) The gastrointestinal microbiota and its role in oncogenesis. Best Pract Res Clin Gastroenterol 31(6):607–618. https://doi.org/10.1016/j.bpg.2017.09.010

Lane N (2014) Bioenergetic constraints on the evolution of complex life. Cold Spring Harb Perspect Biol 6(5):a015982. https://doi.org/10.1101/cshperspect.a015982

Larsen N, Vogensen FK, van den Berg FWJ, Nielsen DS, Andreasen AS, Pedersen BK, Al-Soud WA, Sørensen SJ, Hansen LH, Jakobsen M (2010) Gut microbiota in human adults with type 2 diabetes differs from non-diabetic adults. PLoS One 5(2):e9085–e9085. https://doi.org/10.1371/journal.pone.0009085

Leloup J, Fossing H, Kohls K, Holmkvist L, Borowski C, Jørgensen BB (2009) Sulfate-reducing bacteria in marine sediment (Aarhus Bay, Denmark): abundance and diversity related to geochemical zonation. Environ Microbiol 11(5):1278–1291. https://doi.org/10.1111/j.1462-2920.2008.01855.x

Ley RE, Bäckhed F, Turnbaugh P, Lozupone CA, Knight RD, Gordon JI (2005) Obesity alters gut microbial ecology. Proc Natl Acad Sci U S A 102(31):11070–11075. https://doi.org/10.1073/pnas.0504978102

Ley RE, Peterson DA, Gordon JI (2006a) Ecological and evolutionary forces shaping microbial diversity in the human intestine. Cell 124(4):837–848. https://doi.org/10.1016/j.cell.2006.02.017

Ley RE, Turnbaugh PJ, Klein S, Gordon JI (2006b) Microbial ecology: human gut microbes associated with obesity. Nature 444(7122):1022–1023. https://doi.org/10.1038/4441022a

Louis P, Hold GL, Flint HJ (2014) The gut microbiota, bacterial metabolites and colorectal cancer. Nat Rev Microbiol 12(10):661–672. https://doi.org/10.1038/nrmicro3344

Lynch SV, Bruce KD (2013) The cystic fibrosis airway microbiome. Cold Spring Harb Perspect Med 3(3):a009738–a009738. https://doi.org/10.1101/cshperspect.a009738

Macfarlane S, Macfarlane GT (2006) Composition and metabolic activities of bacterial biofilms colonizing food residues in the human gut. Appl Environ Microbiol 72(9):6204–6211. https://doi.org/10.1128/aem.00754-06

Maddocks OD, Short AJ, Donnenberg MS, Bader S, Harrison DJ (2009) Attaching and effacing Escherichia coli downregulate DNA mismatch repair protein in vitro and are associated with colorectal adenocarcinomas in humans. PLoS One 4(5):e5517. https://doi.org/10.1371/journal.pone.0005517

Mashima H, Suzuki J, Hirayama T, Yoshikumi Y, Ohno H, Ohnishi H, Yasuda H, Fujita T, Omata M (2008) Involvement of vesicle-associated membrane protein 7 in human gastric epithelial cell vacuolation induced by helicobacter pylori-produced VacA. Infect Immun 76(6):2296–2303. https://doi.org/10.1128/iai.01573-07

Maslowski KM, Vieira AT, Ng A, Kranich J, Sierro F, Yu D, Schilter HC, Rolph MS, Mackay F, Artis D, Xavier RJ, Teixeira MM, Mackay CR (2009) Regulation of inflammatory responses by gut microbiota and chemoattractant receptor GPR43. Nature 461(7268):1282–1286. https://doi.org/10.1038/nature08530

Michaud DS (2013) Role of bacterial infections in pancreatic cancer. Carcinogenesis 34(10):2193–2197. https://doi.org/10.1093/carcin/bgt249

Mohammad S, Ahmad J (2016) Management of obesity in patients with type 2 diabetes mellitus in primary care. Diabetes Metab Syndr 10(3):171–181. https://doi.org/10.1016/j.dsx.2016.01.017

Neish AS (2009) Microbes in gastrointestinal health and disease. Gastroenterology 136(1):65–80. https://doi.org/10.1053/j.gastro.2008.10.080

Nesić D, Hsu Y, Stebbins CE (2004) Assembly and function of a bacterial genotoxin. Nature 429(6990):429–433. https://doi.org/10.1038/nature02532

Nougayrède JP, Taieb F, De Rycke J, Oswald E (2005) Cyclomodulins: bacterial effectors that modulate the eukaryotic cell cycle. Trends Microbiol 13(3):103–110. https://doi.org/10.1016/j.tim.2005.01.002

O'Neill LA, Golenbock D, Bowie AG (2013) The history of toll-like receptors - redefining innate immunity. Nat Rev Immunol 13(6):453–460. https://doi.org/10.1038/nri3446

Odenbreit S, Püls J, Sedlmaier B, Gerland E, Fischer W, Haas R (2000) Translocation of helicobacter pylori CagA into gastric epithelial cells by type IV secretion. Science 287(5457):1497–1500. https://doi.org/10.1126/science.287.5457.1497

Oliver A, Cantón R, Campo P, Baquero F, Blázquez J (2000) High frequency of hypermutable Pseudomonas aeruginosa in cystic fibrosis lung infection. Science 288(5469):1251–1254. https://doi.org/10.1126/science.288.5469.1251

Ouwehand AC, Derrien M, de Vos W, Tiihonen K, Rautonen N (2005) Prebiotics and other microbial substrates for gut functionality. Curr Opin Biotechnol 16(2):212–217. https://doi.org/10.1016/j.copbio.2005.01.007

Pace NR, Sapp J, Goldenfeld N (2012) Phylogeny and beyond: scientific, historical, and conceptual significance of the first tree of life. Proc Natl Acad Sci 109(4):1011–1018. https://doi.org/10.1073/pnas.1109716109

Papastergiou V, Karatapanis S, Georgopoulos SD (2016) Helicobacter pylori and colorectal neoplasia: is there a causal link? World J Gastroenterol 22(2):649–658. https://doi.org/10.3748/wjg.v22.i2.649

Péré-Védrenne C, Cardinaud B, Varon C, Mocan I, Buissonnière A, Izotte J, Mégraud F, Ménard A (2016) The cytolethal distending toxin subunit CdtB of helicobacter induces a Th17-related and antimicrobial signature in intestinal and hepatic cells in vitro. J Infect Dis 213(12):1979–1989. https://doi.org/10.1093/infdis/jiw042

Péré-Védrenne C, Prochazkova-Carlotti M, Rousseau B, He W, Chambonnier L, Sifré E, Buissonnière A, Dubus P, Mégraud F, Varon C, Ménard A (2017) The cytolethal distending toxin subunit CdtB of helicobacter hepaticus promotes senescence and Endoreplication in xenograft mouse models of hepatic and intestinal cell lines. Front Cell Infect Microbiol 7:268. https://doi.org/10.3389/fcimb.2017.00268

Qin J, Li R, Raes J, Arumugam M, Burgdorf KS, Manichanh C, Nielsen T, Pons N, Levenez F, Yamada T, Mende DR, Li J, Xu J, Li S, Li D, Cao J, Wang B, Liang H, Zheng H, Xie Y, Tap J, Lepage P, Bertalan M, Batto J-M, Hansen T, Le Paslier D, Linneberg A, Nielsen HB, Pelletier E, Renault P, Sicheritz-Ponten T, Turner K, Zhu H, Yu C, Li S, Jian M, Zhou Y, Li Y, Zhang X, Li S, Qin N, Yang H, Wang J, Brunak S, Doré J, Guarner F, Kristiansen K, Pedersen O, Parkhill J, Weissenbach J, Antolin M, Artiguenave F, Blottiere H, Borruel N, Bruls T, Casellas F, Chervaux C, Cultrone A, Delorme C, Denariaz G, Dervyn R, Forte M, Friss C, van de Guchte M, Guedon E, Haimet F, Jamet A, Juste C, Kaci G, Kleerebezem M, Knol J, Kristensen M, Layec S, Le Roux K, Leclerc M, Maguin E, Melo Minardi R, Oozeer R, Rescigno M, Sanchez N, Tims S, Torrejon T, Varela E, de Vos W, Winogradsky Y, Zoetendal E, Bork P, Ehrlich SD, Wang J, Meta HITC (2010) A human gut microbial gene catalogue established by metagenomic sequencing. Nature 464(7285):59–65. https://doi.org/10.1038/nature08821

Qin J, Li Y, Cai Z, Li S, Zhu J, Zhang F, Liang S, Zhang W, Guan Y, Shen D, Peng Y, Zhang D, Jie Z, Wu W, Qin Y, Xue W, Li J, Han L, Lu D, Wu P, Dai Y, Sun X, Li Z, Tang A, Zhong S, Li X, Chen W, Xu R, Wang M, Feng Q, Gong M, Yu J, Zhang Y, Zhang M, Hansen T, Sanchez G, Raes J, Falony G, Okuda S, Almeida M, LeChatelier E, Renault P, Pons N, Batto JM, Zhang Z, Chen H, Yang R, Zheng W, Li S, Yang H, Wang J, Ehrlich SD, Nielsen R, Pedersen O, Kristiansen K, Wang J (2012) A metagenome-wide association study of gut microbiota in type 2 diabetes. Nature 490(7418):55–60. https://doi.org/10.1038/nature11450

Queiroz DM, Rocha AM, Rocha GA, Cinque SM, Oliveira AG, Godoy A, Tanno H (2006) Association between helicobacter pylori infection and cirrhosis in patients with chronic hepatitis C virus. Dig Dis Sci 51(2):370–373. https://doi.org/10.1007/s10620-006-3150-y

Reichardt N, Duncan SH, Young P, Belenguer A, McWilliam Leitch C, Scott KP, Flint HJ, Louis P (2014) Phylogenetic distribution of three pathways for propionate production within the human gut microbiota. ISME J 8(6):1323–1335. https://doi.org/10.1038/ismej.2014.14

Ricci V (2016) Relationship between VacA toxin and host cell autophagy in helicobacter pylori infection of the human stomach: a few answers, many questions. Toxins (Basel) 8(7):203. https://doi.org/10.3390/toxins8070203

Ridlon JM, McGarr SE, Hylemon PB (2005) Development of methods for the detection and quantification of 7alpha-dehydroxylating clostridia, Desulfovibrio vulgaris, Methanobrevibacter smithii, and lactobacillus plantarum in human feces. Clin Chim Acta 357(1):55–64. https://doi.org/10.1016/j.cccn.2005.02.004

Ringel Y (2017) The gut microbiome in irritable bowel syndrome and other functional bowel disorders. Gastroenterol Clin N Am 46(1):91–101. https://doi.org/10.1016/j.gtc.2016.09.014

Rosenberg E, Koren O, Reshef L, Efrony R, Zilber-Rosenberg I (2007) The role of microorganisms in coral health, disease and evolution. Nat Rev Microbiol 5(5):355–362. https://doi.org/10.1038/nrmicro1635

Rossi M, Corradini C, Amaretti A, Nicolini M, Pompei A, Zanoni S, Matteuzzi D (2005) Fermentation of Fructooligosaccharides and inulin by Bifidobacteria: a comparative study of pure and fecal cultures. Appl Environ Microbiol 71(10):6150–6158. https://doi.org/10.1128/aem.71.10.6150-6158.2005

Rubinstein MR, Wang X, Liu W, Hao Y, Cai G, Han YW (2013) Fusobacterium nucleatum promotes colorectal carcinogenesis by modulating E-cadherin/β-catenin signaling via its FadA adhesin. Cell Host Microbe 14(2):195–206. https://doi.org/10.1016/j.chom.2013.07.012

Sakr SA, Badrah GA, Sheir RA (2013) Histological and histochemical alterations in liver of chronic hepatitis C patients with helicobacter pylori infection. Biomed Pharmacother 67(5):367–374. https://doi.org/10.1016/j.biopha.2013.03.004

Sampson TR, Debelius JW, Thron T, Janssen S, Shastri GG, Ilhan ZE, Challis C, Schretter CE, Rocha S, Gradinaru V, Chesselet MF, Keshavarzian A, Shannon KM, Krajmalnik-Brown R,

Wittung-Stafshede P, Knight R, Mazmanian SK (2016) Gut microbiota regulate motor deficits and neuroinflammation in a model of Parkinson's disease. Cell 167(6):1469–1480.e12. https://doi.org/10.1016/j.cell.2016.11.018

Samuel BS, Gordon JI (2006) A humanized gnotobiotic mouse model of host–archaeal–bacterial mutualism. Proc Natl Acad Sci 103(26):10011–10016. https://doi.org/10.1073/pnas.0602187103

Sandyk R (1992) L-tryptophan in neuropsychiatry disorders: a review. Int J Neurosci 67(1-4):127–144. https://doi.org/10.3109/00207459208994781

Scanlan PD, Shanahan F, Marchesi JR (2009) Culture-independent analysis of desulfovibrios in the human distal colon of healthy, colorectal cancer and polypectomized individuals. FEMS Microbiol Ecol 69(2):213–221. https://doi.org/10.1111/j.1574-6941.2009.00709.x

Schwiertz A, Taras D, Schäfer K, Beijer S, Bos NA, Donus C, Hardt PD (2010) Microbiota and SCFA in lean and overweight healthy subjects. Obesity (Silver Spring) 18(1):190–195. https://doi.org/10.1038/oby.2009.167

Sekirov I, Russell SL, Antunes LC, Finlay BB (2010) Gut microbiota in health and disease. Physiol Rev 90(3):859–904. https://doi.org/10.1152/physrev.00045.2009

Simrén M, Barbara G, Flint HJ, Spiegel BM, Spiller RC, Vanner S, Verdu EF, Whorwell PJ, Zoetendal EG (2013) Intestinal microbiota in functional bowel disorders: a Rome foundation report. Gut 62(1):159–176. https://doi.org/10.1136/gutjnl-2012-302167

Sonnenburg JL, Xu J, Leip DD, Chen C-H, Westover BP, Weatherford J, Buhler JD, Gordon JI (2005) Glycan foraging in vivo by an intestine-adapted bacterial symbiont. Science 307(5717):1955–1959. https://doi.org/10.1126/science.1109051

Stappenbeck TS, Hooper LV, Gordon JI (2002) Developmental regulation of intestinal angiogenesis by indigenous microbes via paneth cells. Proc Natl Acad Sci U S A 99(24):15451–15455. https://doi.org/10.1073/pnas.202604299

Swidsinski A, Loening-Baucke V, Lochs H, Hale LP (2005) Spatial organization of bacterial flora in normal and inflamed intestine: a fluorescence in situ hybridization study in mice. World J Gastroenterol 11(8):1131–1140. https://doi.org/10.3748/wjg.v11.i8.1131

Tang WHW, Hazen SL (2014) The contributory role of gut microbiota in cardiovascular disease. J Clin Invest 124(10):4204–4211. https://doi.org/10.1172/JCI72331

Tsoi H, Chu ESH, Zhang X, Sheng J, Nakatsu G, Ng SC, Chan AWH, Chan FKL, Sung JJY, Yu J (2017) Peptostreptococcus anaerobius induces intracellular cholesterol biosynthesis in colon cells to induce proliferation and causes dysplasia in mice. Gastroenterology 152(6):1419–1433.e5. https://doi.org/10.1053/j.gastro.2017.01.009

Turnbaugh PJ, Hamady M, Yatsunenko T, Cantarel BL, Duncan A, Ley RE, Sogin ML, Jones WJ, Roe BA, Affourtit JP, Egholm M, Henrissat B, Heath AC, Knight R, Gordon JI (2009) A core gut microbiome in obese and lean twins. Nature 457(7228):480–484. https://doi.org/10.1038/nature07540

van der Crabben SN, Blümer RM, Stegenga ME, Ackermans MT, Endert E, Tanck MW, Serlie MJ, van der Poll T, Sauerwein HP (2009) Early endotoxemia increases peripheral and hepatic insulin sensitivity in healthy humans. J Clin Endocrinol Metab 94(2):463–468. https://doi.org/10.1210/jc.2008-0761

Verstraelen H (2008) Cutting edge: the vaginal microflora and bacterial vaginosis. Verh K Acad Geneeskd Belg 70(3):147–174

Vijay-Kumar M, Aitken JD, Carvalho FA, Cullender TC, Mwangi S, Srinivasan S, Sitaraman SV, Knight R, Ley RE, Gewirtz AT (2010) Metabolic syndrome and altered gut microbiota in mice lacking toll-like receptor 5. Science 328(5975):228–231. https://doi.org/10.1126/science.1179721

Vrieze A, Van Nood E, Holleman F, Salojärvi J, Kootte RS, Bartelsman JF, Dallinga-Thie GM, Ackermans MT, Serlie MJ, Oozeer R, Derrien M, Druesne A, Van Hylckama Vlieg JE, Bloks VW, Groen AK, Heilig HG, Zoetendal EG, Stroes ES, de Vos WM, Hoekstra JB, Nieuwdorp M (2012) Transfer of intestinal microbiota from lean donors increases insulin sensitivity in individuals with metabolic syndrome. Gastroenterology 143(4):913–6.e7. https://doi.org/10.1053/j.gastro.2012.06.031

Vyas U, Ranganathan N (2012) Probiotics, prebiotics, and synbiotics: gut and beyond. Gastroenterol Res Pract 2012:872716. https://doi.org/10.1155/2012/872716

Wang X, Yang Y, Moore DR, Nimmo SL, Lightfoot SA, Huycke MM (2012) 4-hydroxy-2-nonenal mediates genotoxicity and bystander effects caused by enterococcus faecalis-infected macrophages. Gastroenterology 142(3):543–551.e7. https://doi.org/10.1053/j.gastro.2011.11.020

Wang F, Meng W, Wang B, Qiao L (2014) Helicobacter pylori-induced gastric inflammation and gastric cancer. Cancer Lett 345(2):196–202. https://doi.org/10.1016/j.canlet.2013.08.016

Wang X, Xu X, Xia Y (2017) Further analysis reveals new gut microbiome markers of type 2 diabetes mellitus. Antonie Van Leeuwenhoek 110(3):445–453. https://doi.org/10.1007/s10482-016-0805-3

Whitman WB, Coleman DC, Wiebe WJ (1998) Prokaryotes: the unseen majority. Proc Natl Acad Sci U S A 95(12):6578–6583. https://doi.org/10.1073/pnas.95.12.6578

Wong SH, Zhao L, Zhang X, Nakatsu G, Han J, Xu W, Xiao X, Kwong TNY, Tsoi H, Wu WKK, Zeng B, Chan FKL, Sung JJY, Wei H, Yu J (2017) Gavage of fecal samples from patients with colorectal cancer promotes intestinal carcinogenesis in germ-free and conventional mice. Gastroenterology 153(6):1621–1633.e6. https://doi.org/10.1053/j.gastro.2017.08.022

Wu S, Rhee KJ, Albesiano E, Rabizadeh S, Wu X, Yen HR, Huso DL, Brancati FL, Wick E, McAllister F, Housseau F, Pardoll DM, Sears CL (2009) A human colonic commensal promotes colon tumorigenesis via activation of T helper type 17 T cell responses. Nat Med 15(9):1016–1022. https://doi.org/10.1038/nm.2015

Wu H, Xie S, Miao J, Li Y, Wang Z, Wang M, Yu Q (2020) Lactobacillus reuteri maintains intestinal epithelial regeneration and repairs damaged intestinal mucosa. Gut Microbes 11(4):997–1014. https://doi.org/10.1080/19490976.2020.1734423

Xu J, Gordon JI (2003) Honor thy symbionts. Proc Natl Acad Sci 100(18):10452–10459. https://doi.org/10.1073/pnas.1734063100

Yang Y, Weng W, Peng J, Hong L, Yang L, Toiyama Y, Gao R, Liu M, Yin M, Pan C, Li H, Guo B, Zhu Q, Wei Q, Moyer MP, Wang P, Cai S, Goel A, Qin H, Ma Y (2017) Fusobacterium nucleatum increases proliferation of colorectal cancer cells and tumor development in mice by activating toll-like receptor 4 signaling to nuclear factor-κB, and upregulating expression of microRNA-21. Gastroenterology 152(4):851–866.e24. https://doi.org/10.1053/j.gastro.2016.11.018

Yong X, Tang B, Li BS, Xie R, Hu CJ, Luo G, Qin Y, Dong H, Yang SM (2015) Helicobacter pylori virulence factor CagA promotes tumorigenesis of gastric cancer via multiple signaling pathways. Cell Commun Signal 13:30. https://doi.org/10.1186/s12964-015-0111-0

Yu T, Guo F, Yu Y, Sun T, Ma D, Han J, Qian Y, Kryczek I, Sun D, Nagarsheth N, Chen Y, Chen H, Hong J, Zou W, Fang JY (2017) Fusobacterium nucleatum promotes Chemoresistance to colorectal cancer by modulating autophagy. Cell 170(3):548–563.e16. https://doi.org/10.1016/j.cell.2017.07.008

Zhou CB, Fang JY (2018) The regulation of host cellular and gut microbial metabolism in the development and prevention of colorectal cancer. Crit Rev Microbiol 44(4):436–454. https://doi.org/10.1080/1040841x.2018.1425671

Zilber-Rosenberg I, Rosenberg E (2008) Role of microorganisms in the evolution of animals and plants: the hologenome theory of evolution. FEMS Microbiol Rev 32(5):723–735. https://doi.org/10.1111/j.1574-6976.2008.00123.x

The Microbiome Antibiotic Resistome: Significant Strategies Toward Microbiome-Targeted Therapeutic Interventions in Medicine

13

Lakshmi Pethakamsetty, Sudhakar Pola, and Joseph G. Giduthuri

Abstract

Antibiotic resistance/drug resistance occurs in some, or less commonly all, subpopulations of microbes, usually bacteria, which can survive after exposure to one or more antibiotic environments. A microbiome is a community of bacteria, archaea, fungi, protozoa, and viruses that inhabit an ecosystem or an organism. Microorganisms are ubiquitous, existing in, on, and around biotic and abiotic systems, and their habitats are redundant with diversity. They are generally interconnected with their host with a live-in relationship which is often beneficial and essential to both the host and the residential microorganisms. Scientists are just beginning to revise microbiomes to address the risk of emerging antibiotic resistance. The present book chapter deals with the causes for the emergence of drug resistance and the counterstrategies to nullify antibiotic resistance through advanced research ranging from conventional drug therapy to microbiome-targeted therapies (MTT). The ideal way to address the threat of antimicrobial resistance (AMR) with enriched resistomes is to manipulate the microbiome composition via lifestyle alterations and personalized diet recommendations, fine-tuning the microbiota toward a healthy state.

Keywords

Microbiome-targeted therapies (MTT) · Antimicrobial resistance (AMR) · Microbiota · Enriched resistomes

L. Pethakamsetty
Department of Microbiology, AUCST, Andhra University, Visakhapatnam, India

S. Pola (✉)
Department of Biotechnology, AUCST, Andhra University, Visakhapatnam, India

J. G. Giduthuri
Swiss Tropical and Public Health Institute University of Basel, Basel, Switzerland

P. Veera Bramhachari (ed.), *Human Microbiome in Health, Disease, and Therapy*, https://doi.org/10.1007/978-981-99-5114-7_13

13.1 Introduction

Antibiotic resistance (AR) or antimicrobial resistance (AMR) is acknowledged as one of the significant universal health catastrophes of the twenty-first century that jeopardizes the accomplishments of modern medicine by reducing clinical efficacy and increasing treatment costs. Currently, AMR mechanisms are creating havoc on healthcare systems worldwide, contributing to the emergence of epidemic and pandemic superbugs such as *Staphylococcus aureus*, *Neisseria gonorrhoea*, *Clostridium difficile*, *Klebsiella* spp., *Enterobacter* spp., etc. The international food trade and travel have also been responsible for the global emergence and spread of antibiotic resistance, especially for species of the Enterobacteriaceae residing in the human gut leading to cephalosporin (ESBLs) antibiotic pollution worldwide (Geser et al. 2012). *Enterococcus faecium*, *S.aureus*, *Klebsiella pneumoniae*, *Acinetobacter baumannii*, *P. aeruginosa*, and *Enterobacter* displayed antibiotic resistance (carbapenems, fluoroquinolones, chloramphenicol, and aminoglycosides) referred to as ESKAPE pathogens that are responsible for six most fatal hospital-acquired diseases (Burcham et al. 2019).

Understanding microbial diversity and antibiotic resistance may be insightful in disease diagnosis and treatment. It is understood that the treatment with a broad-spectrum antibiotic causes a dramatic loss of in-house microbial diversity with the insurgence of antibiotic-resistant strains and upregulation of antibiotic resistance genes within the host systems (Tanwar et al. 2014). Therefore, human microbiome warrants special attention as possibly the most accessible reservoir of resistance genes due to their high prospects of contact and genetic exchange with potential pathogens (John et al. 2013). Further, microbiome analysis in more diverse populations enables the generation of worldwide databases, which include epidemiological and host genetic information with a better understanding of host-microbe interactions and their crucial role in the origin and progression of diseases. Currently, metagenomics with complementary studies such as metabolomics and metatranscriptomics would stem the mechanistic models of host-microbe communications (Langdon et al. 2016).

13.2 Antimicrobial Resistance (AMR)

Antimicrobial resistance (AMR) is the mechanism by which a drug renders ineffective over a pathogen or a disease caused by the pathogen, which was earlier effective over a range of microorganisms (mostly bacteria). The susceptible microorganisms develop resistance to a drug during the period and will no longer be controlled or killed by the antibiotics. When an organism resists more than one antimicrobial, it is called a multidrug-resistant (MDR) organism/superbugs, a common scenario recorded in nosocomial infections. Microbes acquire antibiotic resistance either genetically or by nongenetic mechanisms. Nongenetic resistance is acquired either by evasion or in the form of L-forms. Some microorganisms, such as mycobacteria, exhibit an evasion or escape mechanism that causes tuberculosis, which persists in the tissues escaping the harsh effects of antibiotics and later reverting to parasitism. Some bacterial strains shed their cell walls temporarily and change to L-forms or wall-deficient forms exhibiting antibiotic resistance (Davies and Davies 2010; Black 2015). Mutations or gene rearrangements frequently occur in bacterial populations (one in $10^7/10^9$), favoring natural selection habitually and resulting in genetic resistance. Auxotrophs (mutants), which resist antimicrobials, multiply over a short period, resulting in the progeny that exhibits AMR genetic resistance being acquired through mutations or horizontal gene transfer (HGT), which is chromosomal or extrachromosomal. Extrachromosomal resistance is conferred by the plasmids (mobile genetic elements), which are autonomous, self-replicating extranuclear bits of DNA that carry the R factor (R-plasmids) (Deschamps et al. 2009).

13.3 Antimicrobial Resistance Mechanisms

Since the dawn of the antibiotic era, resistance to antibiotics has been identified. Paul Ehrlich, the father of modern chemotherapy, demonstrated resistance that, once acquired, would be stably inherited either by substituting the "target" for "chemoreceptor" or inactivating the drug (Fig. 13.1). The following are the mechanisms responsible for AMR:

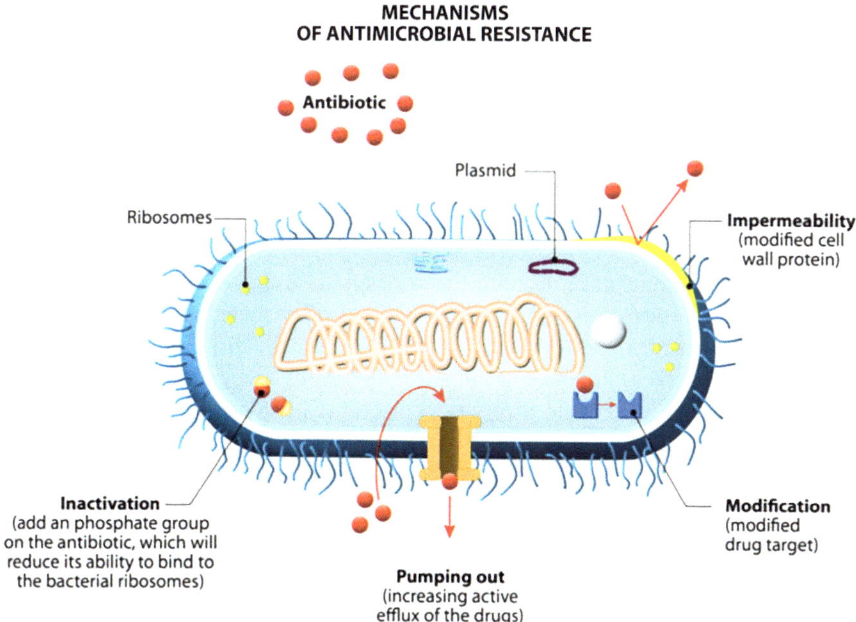

Fig. 13.1 Different mechanisms of AMR

1. *Alteration of targets*: Alteration of the bacterial transcription changing the gene product or target (Eg-PBPs) to which the antibiotic has to bind. Resistance to penicillins, rifamycins, erythromycins, and antimetabolites has been tagged with this mechanism.
2. *Alteration of membrane permeability*: Mutations lead to changes in membrane proteins, altering the membrane permeability with modified pore size and membrane transport mechanisms. This mechanism reduces drug accumulation and enhances the antibiotic's active efflux (pumping out) across the cell surface. Resistance to aminoglycosidic antibiotics (streptomycin, kanamycin), tetracyclines, and quinolones has occurred through this mechanism.
3. *Development of enzymes*: The development of enzymes conferred cross-resistance to many antimicrobials via a common mechanism. Enzymes such as β-lactamases break down the β-lactam ring in penicillin and cephalosporin antibiotics. Chloramphenicol and aminoglycosidic antibiotics are destroyed by developing peptidases and acetyltransferases, respectively.
4. *Alteration of enzymes*: Resistance for sulfonamide drugs is due to the alteration of bacterial enzymes, which shows high affinity toward PABA (para-aminobenzoic acid, a precursor for synthesizing folic acid) and significantly less affinity to sulfonamide, allowing the bacteria to function normally.
5. *Alteration of the metabolic pathway*: This resistance mechanism is identified for sulfonamide drugs. The organisms bypass their common metabolic pathway by acquiring the ability to use folic acid instantaneously instead of developing from PABA.
6. *Ribosome splitting and recycling*: Heat shock proteins (HSP) identified in *Listeria* monocytogenes are responsible for installing ribosomes and successful bacterial translation of functional proteins. This type of mechanism confers resistance to macrolide drugs like lincomycin and erythromycin (Dinos 2017).

13.4 Resistome/Antibiotic Resistance Genes (ARG)

Antibiotic resistance genes (ARG) or antimicrobial genes (AMG) are ubiquitous in community environments, reflecting the widespread usage of drugs over the past 90 years. The resistome constitutes many antibiotic resistance genes (ARGs) from environmental and commensal bacterial communities. Intense selection pressures, such as exposure to high concentrations of antibiotics and their prolonged usage, can enrich the abundance and diversity of ARGs (Cheng et al. 2012; Sultan et al. 2018). Besides the microbiota in the gut, other microbial communities can also act as hubs for ARG exchange in clinical settings, both in the patient (e.g., nasal cavity, vagina, or cystic fibrosis lung) and in the hospital environment (e.g., washbasins and drains). Studying the mechanisms of evolution and the mobility of antibiotic resistance genes into human pathogens can facilitate early surveillance of disease diagnostics and their treatments. The diversity and extent of resistome dissemination within the human population are determined by cultural and sequencing methods, either by stool sampling or postmortem microbiome sampling (Perry et al. 2014).

Resistomes were generally constrained by the phylogenetic diversity of their underlying microbiota. The microbiome resistome can be classified into intrinsic and mobile resistance genes (MGEs). The intrinsic resistance genes are nonmobile resistance genes inherited and provide tolerance to a particular drug without prior exposure. Plasmids, episomes, integrons, transposons, conjugative elements, phages, and genomic islands are mobile DNA elements that can render resistance to bacterial metagenome through horizontal gene transfer (HGT) involving conjugation, transduction, and transformation (Fig. 13.2). The flow of genetic information through HGT might have unfavorable consequences for the gut community since resistant pathogens serve as more venerable adversaries to the residential microbes of the host. Some commensal bacteria can rapidly increase and acquire pathogenic features known as pathobionts, such as *Clostridium difficile* or vancomycin-resistant *Enterococcus*.

Further, the use of a safe antibiotic, avoparcin, in animal feed is found to be responsible for enhancing AMR in human microflora (van Bogaard and Stobberingh 2000; Angulo et al. 2004). The ARGs with versatile diversity in gut microbiota and human-associated environments will likely promote multiple drug resistance in pathogens. However, the dissemination of resistome into clinical environments is still unclear and is essential in understanding their impact on microbiota (Sommer et al. 2010).

MGEs offer bacteria the capacity for niche expansion and functional diversification and can constitute up to 15–20% of prokaryotic genomes. Resistance islands (RIs) are formed by the accumulation of ARGs into particular regions of genomic DNA, leading to multiple drug resistance. The largest RIs (86 kb) of ARGs were recorded in *Acinetobacter baumannii* (Ji Youn Sung et al. 2012). Resistant ESBL genes were identified from *Kluyvera* species (Canton and Coque 2006) and the wide distribution of type A streptogramin acetyltransferases across bacterial species (Peterson and Kaur 2018). Microbial species with the maximum number of ARGs have been identified for *Escherichia coli* with 5386 potential resistance genes, *Enterobacter cloacae with* 2098 resistant genes, *Staphylococcus aureus* with 297,

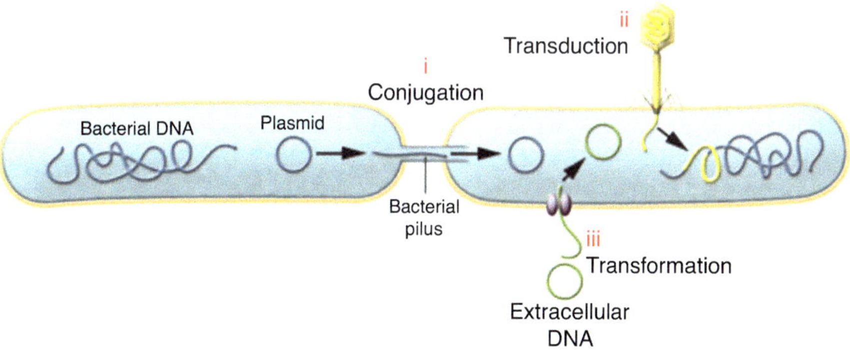

Fig. 13.2 Horizontal gene transfer (https://doi.org/10.1172/JCI72333)

251 in *Bacteroides uniformis*, 229 in *Campylobacter jejuni*, and 222 in *Klebsiella pneumoniae* followed by 214 in *Enterococcus faecalis*, respectively (Raymond et al. 2019). Metagenomic analyses of the human gut microbiome showed a high percentage of ARGs displaying resistance toward mostly tetracycline, macrolides, vancomycin, and bacitracin (Lobanovska and Pilla 2017). Metagenomic analysis of viruses revealed consistent viral populations with lysogenic and integrase genes. Viral metagenomes with CRISPR (clustered regularly interspaced short palindromic repeat) identified variability of virotypes, implying viruses are employed as transducing vehicles rather than predatory agents (Sheetal et al. 2014).

13.5 Human Microbiome

Several microbes encounter the human system within and on the external surface of the cells. The prevalence of microbes is more compared to the body cells. Microbial organisms have beneficial and harmful effects on the system (Sanapala and Pola 2021). Joshua Lederberg first defined the microbiome as the collective genomes of microbes within a community of commensal, symbiotic, and pathogenic microorganisms. The human microbiome, also called the human metagenome, constitutes the communal genomes of all residential microorganisms, including bacteria, fungi, archaea, protozoa, and viruses (Amon and Sanderson 2017). Microbiota is the aggregate of microbes colonized within their tissues and anatomical sites, namely, the skin, gut, viscera, and urogenital tracts. The working of the Human Microbiome Project (2012) (Consortium HMP) has revealed approximately 37.2 trillion microbes with an average occurrence of a 3:1 ratio of microbial cells in the human body to the human cells (Jack A Gilbert et al. 2018). Recent advances in metagenomics have revealed the gut microbiota with a rough estimate of 1000 bacterial species (approximately 2000 genes per species amounting to 2,000,000 genes), 100 times the number of human genes according to HGP (Morgan and Huttenhower 2012).

The human microbiome is dynamic and is influenced by developmental and environmental factors such as age, diet, lifestyle, habitat, and diseases (Gilbert et al. 2018). Comparative studies on microbiome composition, such as microbiome/metagenome-wide association studies (MWAS), in parallel to genome-wide association studies (GWAS), are carried out to discover potential mechanisms under the origin, progression, and effects of the disease based on the monitoring or modulation of key elements of the microbiota (Crofts et al. 2017).

13.6 Types of Microbes in the Human Microbiome

Microbiomes are biological systems with functional gene diversity to adapt and respond to environmental changes, thereby playing a critical role in fighting debilitating and chronic diseases. The most remarkable changes in microbiome composition occur in infancy and early childhood, which is alleged to be imperative in maintaining homeostasis with the host's immune system that impacts health in later years (Nogueira et al. 2019). Normal-born babies are colonized predominantly by *Lactobacillus*, *Prevotella*, or *Sneathia* genera from the mother's vaginal microbiota. Cesarean-born babies with bacterial communities originated from the skin *Staphylococcus*, *Corynebacterium*, and *Propionibacterium* genera (Dominguez-Bello et al. 2010). The microbiome adapts during the infant's early years, stabilizing at around 3–4 years, with a stable adult microbial configuration. The microbiome comprises different cluster profiles called enterotypes (*Bacteroides*, *Prevotella*, and *Ruminococcus*), representing the abundance of one or a few organisms that remain stable over time (Arumugam et al. 2011). However, the concept of enterotypes is unclear, and researchers prefer a cluster-centric approach to microbial communities with defined boundaries (Costea et al. 2018).

Microbial populations invade dermal and mucosal surfaces of the human body with distinctive microbial communities restricted to different regions. The skin and genital sites show less diversity compared to the mouth and gut microbiota. The human microbiome lodges 2–7 microbial community types (Fig. 13.3), with relative abundances of at least 63 bacterial genera, besides various species of archaea, fungi, protozoa, viruses, and other microorganisms. However, summative knowledge provides evidence of about 55 bacterial divisions in our body, including mainly Bacteroidetes (48%) and Firmicutes (51%), with the remaining 1% of phylotypes

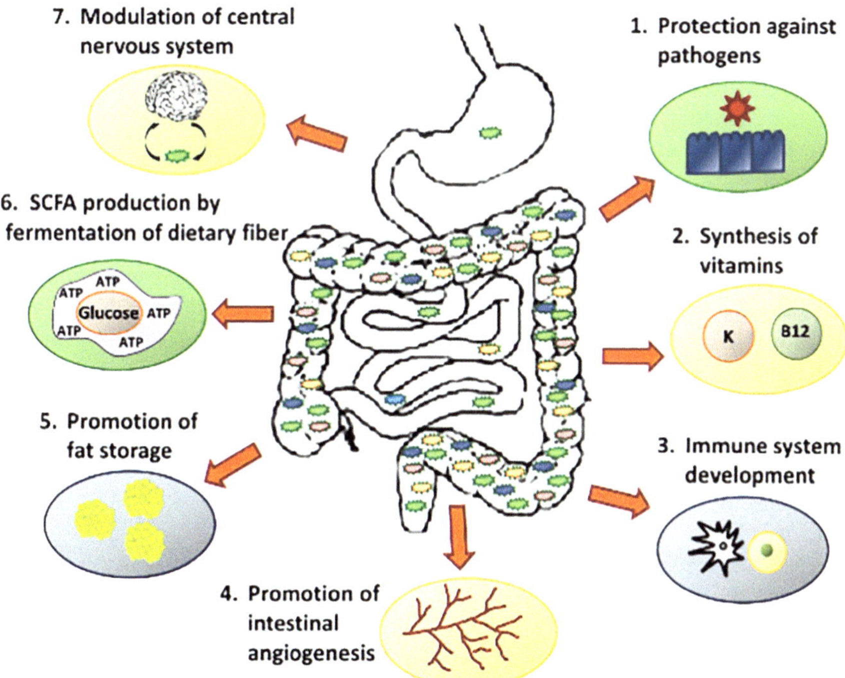

Fig. 13.3 The human microbiome is critical in monitoring vital homeostatic mechanisms in the body (Amon and Sanderson 2017)

comprising Proteobacteria, Verrucomicrobia, Fusobacteria, Cyanobacteria, Actinobacteria, and Spirochaetes (Andersson et al. 2008). Archea domain population is also recorded in the microbiome dominated by methanogenic bacteria such as *Methanobrevibacter smithii* and *Methanosphaera stadtmanae* (Nkamga et al. 2017). The mycobiota constitutes yeasts such as *Malassezia* species in sebaceous glands, and *Acremonium* is dominant on dandruff-afflicted scalps (Kamamoto et al. 2017) particularly. Viruses, especially bacteriophages, reside in the skin, gut, lungs, and oral cavity (Manrique et al. 2016). Bacterial species in the mouth, such as *Actinomyces viscosus* and *A. naeslundii*, form apart of a sticky substance called plaque. In contrast, vaginal microflora consists of various *Lactobacillus* species such as *Lactobacillus acidophilus*, *L. iners*, *L. crispatus*, *L. jensenii*, *L. delbrueckii*, and *L. gasseri* (Ravel et al. 2011). Conjunctival microflora includes bacterial gram-positive cocci (e.g., *Staphylococcus* and *Streptococcus*) and gram-negative rods and cocci (e.g., *Haemophilus* and *Neisseria*) and fungal organisms (*Candida*, *Aspergillus*, and *Penicillium)* (Suto et al. 2012).

13.7 Host-Microbiome Interactions

The human microbiome has widespread functions like defense against pathogens; development of immunity; host nutrition aiding in energy metabolism by short-chain fatty acids, synthesis of vitamins (vitamins B12, thiamine and riboflavin, and vitamin K), and fat storage; as well as influence on human behavior. Intestinal microflora utilizes dietary components to produce energy and metabolites, many of which are taken up and further metabolized or affect the host metabolism (Kinross et al. 2011). The effects of bacterial metabolites on host metabolism can be beneficial and harmful. For example, short-chain fatty acids (SCFAs), derived from otherwise undigestible fiber, have generally beneficial effects on the host, including anti-obesity and antidiabetic actions (Siegfried Ussar et al. 2016). On the other hand, N-nitroso compounds, ammonia, and hydrogen sulfide derived by bacteria from dietary protein can induce reactive oxygen species (ROS) and DNA damage activating inflammatory pathways. Deoxycholic acid, a secondary bile acid produced by the gut microbiota, promotes the development of hepatocellular carcinoma (Dzutsev et al. 2015). Trimethylamine-N-oxide (TMAO), an end metabolite of dietary choline, has been shown to promote arteriosclerosis and correlate with cardiovascular disease (CVD), stroke, and death. Microbiomes with Bifidobacteria are believed to be protective, while Proteobacteria has a reported risk factor (Sanz et al. 2015).

The Bacteroidota phylum includes several bacterial species that protect the host against pathogens. For example, *Bacteroides thetaiotaomicron* confers protection against gram-positive (C-type lectins REGIIIγ and REGIIIβ) and viral infections (type I IFN-induced GTPases). In the Firmicutes phylum, *Lactobacillus* spp. maintains intestinal colonization by producing short-chain fatty acids (SCFAs) and aiding bile acid metabolism. The function of SCFAs relies on their capacity to suppress histone deacetylase activity, indicating the presence of epigenetic regulation.

Different receptors for SCFAs with an essential role in immune regulation and metabolism have been identified in intestinal epithelial cells, immune cells, and adipocytes. Further, the production of SCFAs is one of the protective mechanisms employed by the endogenous microbiota to prevent the attachment and invasion of enteric pathogens to the intestinal epithelium (Riiser 2015). Besides influencing localized immune responses, the microbiota has broader effects on innate and adaptive immunity at multiple levels (Nikoopour and Bhagirath 2014).

13.8 Human Microbiome: Analysis

The composition of the microbiome is studied by essentially identifying the members of the microbial community based on DNA/RNA or protein analysis either by amplicon studies or shotgun metagenomic approaches (culture, cloning and sequencing, pyrosequencing, and NGS) (Andersson et al. 2008; Li et al. 2012; Hu et al. 2013; Gupta et al. 2014). The amplicon studies involve specific known marker genes to designate the microbial population taxonomically, whereas metagenomic studies reveal the functional potentialities of the microbial community (Martiny et al. 2011). The screening strategies rely on identifying particular microorganisms on selective media, followed by statistical analysis with bioinformatic pipeline databases. The more significant challenge in the genomic analysis of microbiota is the complete core analysis of microbial community, which is highly variable not only from person to person but also from different sites within the same person and further by carefully avoiding the host DNA in the studies. Phylogenetic analysis is based on ribosomal editing (16s rRNA) or oligotyping to discriminate closely related distinct taxa in the microbiota (Dethlefsen et al. 2008).

Both 16S and WGS approaches have been extensively used to study the human gut microbiome (MetaHIT) (Qin et al. 2010) and HMP (Human Microbiome Project) developed by the National Human Genome Research Institute (NHGRI) to understand the role of the microbiota in health, immunity, nutrition, and diseases. Assembling metagenomes requires sophisticated biocomputational tools and software for annotating the target genes and resolving their functional potentialities by creating genomic data banks (Qin et al. 2010). Integrated Microbial Genomes-Expert Review (IMG/ER) system and the National Microbial Pathogen Data Resources' RAST (Rapid Annotation using Subsystems Technology) server, JCVI (J. Craig Venter Institute), are annotation servers for microbiome analysis. In this line, the use of metatranscriptomics and metabolomics approaches can complement metagenomic approaches. Metatranscriptomics is the sequencing-based analysis of expressed transcripts in a sample, which provides information on the active genes during the experiment. Metatranscriptomics can help to elucidate biological functions underlying microbial dysbiosis associated with multiple diseases (Franzosa et al. 2014) and its relationship with disorders such as inflammatory bowel disease (IBD) (Ahmed et al. 2016).

Metagenomics enabled a complementary view of a bacterial community, including its ARGs, through alignment-based homology searches against an ARG reference database such as ARDB, SARG, CARD, Resfams, P.C.M., and ResFinder (Yin et al. 2018). Although these reference databases reveal thousands of characterized ARGs, they only reflect a small proportion of the total resistome (Gupta et al. 2014). Recently, deepARG was recently published to find novel ARGs directly from shotgun metagenomic data based on artificial neural networks, followed by fARGene (https://github.com/fannyhb/fargene) based on probabilistic HMM gene models optimized to accurately identify previously uncharacterized resistance genes (Arango-Argoty et al. 2018; Berglund et al. 2019).

13.9 The Human Microbiome and Diseases

The main criteria in studying the microbiome is understanding the microbial imbalance or maladaptation that results in disease, sometimes termed "dysbiosis" (Pflughoeft and Versalovic 2012). Gut microbes and their metabolites showed a major influence on the host immune response, thereby activating pro-inflammatory mediators such as cytokines, interleukin-6 (IL-6), and tumor necrosis factor-α (TNF-α), which damage epithelial cells inducing dysfunction (Cho and Blaser 2012; Rooks and Garrett 2016). Dysbiosis-related inflammation and the biosynthesis of chemical carcinogens (e.g., acetaldehyde, N-nitroso compounds) by microbes are among several possible mechanisms through which the microbiota may have a role in carcinogenesis. Enterotoxin-producing bacteria such as *Enterococcus faecalis* produce toxins that can trigger inflammatory responses damaging the gut cells. Further, some gut biota, *Bacteroides fragilis*, produce ROS that damage the oxidative DNA, activating β-catenin nuclear signaling and leading to cellular proliferation. Similarly, *Fusobacterium nucleatum* induces inflammatory changes by adhering to colonic epithelial cells by the FadA surface protein, which interacts with E-cadherin to mediate changes in β-catenin and Wnt signaling (Lynch and Pedersen 2016; Sofia and Elena 2020).

The colon is the most heavily colonized section of the digestive tract, with approximately 70% of the estimated microflora. It has been identified that more than 20% of the cancer burden worldwide is attributed to known intestinal infectious gut microbiota (Rubinstein et al. 2013). The gut microbiome helps uphold mucosal homeostasis and epithelial barrier function, compartmentalizing bacteria to the lumen. However, perturbations in gut barrier function lead to increased "intestinal permeability," which is shown to be coupled with a variety of gastrointestinal disorders and diseases (Table 13.1), such as IBD (inflammatory bowel disease), irritable bowel syndrome (IBS), necrotizing enterocolitis (NEC), celiac disease, chronic inflammation, and colorectal cancer development (Jalanka-Tuovinen et al. 2011; Grishin et al. 2013; Vancheswaran Gopalakrishnan et al. 2018). Atopic diseases such as eczema, asthma, and food allergies are increasing, often linked to the lack of early-life exposure to microbial antigens in hygienic developed countries. The microbiota plays a crucial role in human pathogenesis, although other environmental factors such as diet, stress, genetic susceptibility, habits, and habitats also

Table 13.1 List of some of the diseases recorded due to altered microbiome

S. no.	Diseases	Organ/ body	Altered microbiome	References
1	Acne, eczema, allergy	Skin	Dysregulation of immune system due to increased pathogenic strains	
2	Inflammatory bowel syndrome, irritable bowel syndrome, and gut infections	Gut	Altered mucosal barrier and dysregulated immune response	Grishin et al. (2013)
3	Cardiovascular disease	Heart	Production of pro-inflammatory metabolites	
4	Autism spectrum disorder	Brain	Abundance of bacterial toxins/disrupted fermentation	
5	Asthma and cystic fibrosis	Lung	Reduced immunological tolerance and altered gene expression	Riiser (2015)
6	Nonalcoholic fatty liver disease	Liver	Bile acid metabolism altered	
7	Diabetes type 1 and type 2	Pancreas	Reduction in insulin sensitivity	Sanz et al. (2015)
8	Metabolic syndrome or obesity	Adipose tissue	Reduced intestinal gluconeogenesis and insulin resistance	Ahmed et al. (2016)

contribute to diseases. However, research on the human microbiome concerning diseases is still vague and inconsistent.

13.10 Microbiome and Disease Diagnosis

Recent advances in metagenomic studies have made it possible to utilize microbiome signatures in cancer diagnostics where a specific species or sets of species could act as key indicators (Nishiumi et al. 2012). *F. nucleatum* has been consistently found to be associated with colorectal cancers serving as indicative species (Chung et al. 2017). Culture-based analysis of AMR showed *E. coli* and *Niesseria gonorrhea* as indicator microorganisms for gut and urinary tract infections, respectively (Okeke et al. 2011). Proteobacteria are natural residents of the gut homeostatic microbiota that serve as potential diagnostic signature organisms of dysbiosis (Shin et al. 2015, Gorvitovskaia et al. 2016).

13.11 Antibiotics and Microbiome

Antibiotics are vital paraphernalia in modern medicine, essential for treating infectious diseases and as a critical prop-up therapy in key medical interventions such as surgery and cancer chemotherapy. The misuse of antibiotics plays a significant role

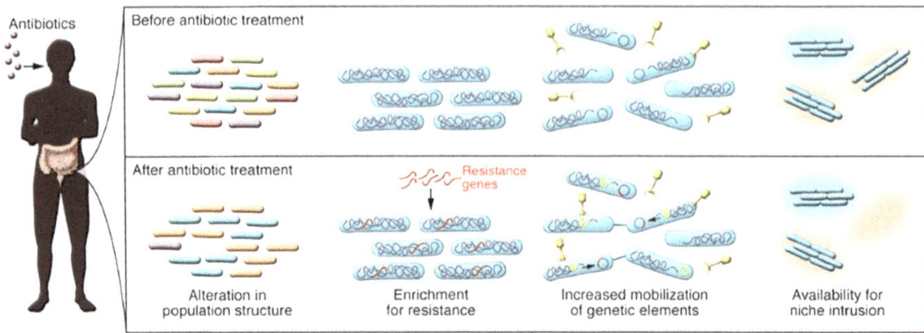

Fig. 13.4 Effects of antibiotics on the gut microbiome (J Clin Invest. 2014; 124(10): 4212–4218. https://doi.org/10.1172/JCI72333)

in the pathogenesis of several diseases associated with the destruction of microbiota (Fig. 13.4). However, genetic mutations in the microbiome can confer biomedically relevant traits, such as the ability to extract nutrients from food, metabolize drugs, evade antibiotics, and communicate with the host immune system (Kim et al. 2011; Nandita and Katherine 2020). "Genome-mining" approaches have revealed the abundance and diversity of antibiotic resistance genes, which might be insightful for novel drug developments (Langdon et al. 2016). Antibiotics, such as clindamycin, clarithromycin, metronidazole, and omeprazole, have typically caused the longest-lasting effects on pharyngeal and gut community composition (Ferrer et al. 2017).

Dethlefsen et al. identified ciprofloxacin to decrease taxonomic richness in the gut biota (2008). Aminoglycosidic resistance was recorded in patients admitted into the intensive care unit (ICU), and the administration of cephalosporin, cefprozil, increased *Lachnoclostridium bolteae* and *Enterobacter cloacae* (Climent et al. 2018; Raymond et al. 2019). High-level azithromycin and ceftriaxone-resistant *Neisseria gonorrhea* were identified in March 2018 by the Public Health England Reference Laboratory. Aminosalicylate drugs have been reported to mainly influence only members of Firmicutes where polymyxins, polyenes, and sulfonamides seem to affect only Bacteroidetes and Firmicutes. Quinolones have been reported to affect only Firmicutes and Proteobacteria, while beta-lactams (amoxicillin), lincosamides (clindamycin), and phosphoglycolipids (flavomycin) affected Fusobacteria, and cephalosporin (ceftriaxone) affect Verrucomicrobiota/*Akkermansia*, respectively (Buffie and Pamer 2013; Ferrer et al. 2017). Table 13.2 lists some antibiotics susceptible to microbial pathogens and their AMR mechanisms.

The administration of antibiotics may contribute to dysbiosis by directly eliminating the bacterial populations that confer colonization resistance to the intestinal microbiome antibiotics. Colonization resistance develops either by:

(a) Direct mechanisms, through direct competition, nutrient depletion, and secretion of bacteriocins.

Table 13.2 List of some of the common antibiotics with resistant microorganisms and their antimicrobial mechanisms

Name of antibiotic	Mode of action of antibiotic	List of resistant microorganisms	AMR mechanisms	References
Antibacterial penicillin	Binds to PBPs (penicillin-binding proteins), inhibiting peptidoglycan synthesis	*Staphylococcus aureus, E. coli, H. influenzae, N. gonorrhoeae*, Streptococcus, *K. pneumoniae*, methicillin-resistant *Staphylococcus aureus* (MRSA)	Plasmid-encoded beta-lactamase, breaking the beta-lactam ring and inactivating it (e.g., TEM-1, TEM-2, and SHV-1)	Mariya and Giulia (2017)
Cephalosporins (ceftizoxime, cefotaxime, ceftriaxone, cephamycins, cefoxitin, or cefotetan and ceftazidime)	Inhibits peptidoglycan and cell wall synthesis	*E. cloacae, C. freundii, S. marcescens*, and *P. aeruginosa, K. pneumoniae, Salmonella* spp., *P. mirabilis, Campylobacter* spp.	Chromosomal class C β-lactamase production (TEM- or SHV-type ESBLs)	Adesoji et al. (2016)
Ampicillin	Inhibits peptidoglycan synthesis	*E. coli, H. influenzae*, and *N. gonorrhoeae* and *K. pneumoniae*	The amino acid substitutions (TEM1)	Kaczmarek et al. (2004)
Aminoglycosides (streptomycin, gentamicin)	Inhibits protein synthesis by binding to 30S ribosome subunit	*E. coli, Pseudomonas aeruginosa, K. pneumoniae, Mycobacterium tuberculosis* (MDR TB)	Plasmid-encoded enzymes (e.g., *mexAB-oprM, mexXY)* that chemically inactivate the drug (acetylation or phosphorylation), S12 ribosomal mutations	Sylvie and Kristin(2016)
Macrolides (erythromycin, azithromycin, clindamycin)	Inhibits protein synthesis by binding to 50S ribosome subunit	*Campylobacter, Streptococcus pyogenes, Staphylococcus aureus*, methicillin-resistant *Staphylococcus aureus* (MRSA), *Neisseria gonorrhoeae, S. pneumoniae, H. influenzae*	Resistance occurs through SNPs in the 23S rRNA gene mutations and modifying the 50S ribosome subunit	Dinos (2017)

(continued)

Table 13.2 (continued)

Name of antibiotic	Mode of action of antibiotic	List of resistant microorganisms	AMR mechanisms	References
Fluoroquinolones (ciprofloxacin, moxifloxacin, and levofloxacin)	Inhibits DNA synthesis by binding to DNA topoisomerase	*Clostridium difficile*, *Escherichia coli* and *Salmonella*, *Mycoplasma genitalium*	Alteration of DNA topoisomerase enzyme and also widespread usage of antibiotics in livestock	Kim and Hooper(2014)
Tetracycline (doxycycline, tigecycline)	Inhibits protein synthesis by blocking tRNA	*Staphylococcus aureus*, methicillin-resistant *Staphylococcus aureus* (MRSA), *S. dysenteriae*, *Mycoplasma genitalium*, *E. coli*, *Shigella dysenteriae*	Epigenetic inheritance, alteration of membrane permeability	JanaMarkeley and Timothy Wencewicz (2018)
Carbapenems (imipenem, meropenem)	Inhibition of cell wall synthesis and transpeptidases	*Enterobacter* spp. (carbapenem-resistant Enterobacteriaceae (CRE)), *K. pneumoniae*, *Serratia marcescens*, *Acinetobacter* spp.	Carbapenemase-producing Enterobacteriaceae (CPE), NDM-1 (*New Delhi metallo-beta-lactamase 1*), *Klebsiella pneumoniae* carbapenemase (KPC)	Gupta et al. (2011)
Linezolid	Protein synthesis inhibitor; prevents the initiation step	*Enterococcus faecalis*, linezolid-resistant *Enterococcus*	23S RNA mutations (Cfr methyltransferase)	Brickner et al. (2008)
Polymyxins	Affects cytoplasmic membrane permeability	*Klebsiella pneumoniae*, *Pseudomonas aeruginosa*, and *Acinetobacter baumannii*	Plasmid-mediated resistance (MCR-1) and alteration of cell membrane integrity	Falagas et al. (2010)
Rifampin	Inhibits RNA synthesis by binding to the RNA polymerase	*Mycobacterium tuberculosis* (MDR TB)	Alteration of polymerase enzyme	Goldstein (2014)

(continued)

Table 13.2 (continued)

Name of antibiotic	Mode of action of antibiotic	List of resistant microorganisms	AMR mechanisms	References
Trimethoprim	Inhibit the folic acid pathway	*Klebsiella pneumoniae*, *P. aeruginosa*, *Serratia marcescens*, *S. aureus*, *Staphylococcus haemolyticus*, *Campylobacter jejuni*, and *Helicobacter pylori*	Alteration of enzymes and antimetabolites	George and Huovinen (2001)
Chloram-phenicol	Inhibits protein synthesis by binding to 50S ribosome subunit and translocation by inhibiting the formation of peptide bonds	*S. aureus* (MRSA)*S. epidermidis (MRSE) CoNSStreptococcus-Pseudomonas* Corynebacteria	Plasmid-encoded enzyme that acetylates the drug	Čivljak et al. (2014)
Isoniazid	Inhibition of mycolic acid synthesis	*Mycobacterium tuberculosis* (MDR TB)	Resistance is primarily mediated by mutations in *katG* and *inhA*, *ndh*, *etc.*	Catherine et al. (2014)
Daptomycin	Binds to the membrane and causes rapid depolarization, resulting in a loss of membrane potential leading to inhibition of protein, DNA, and RNA synthesis	MRSA, *Enterococcus faecalis*	Cell wall thickening	William et al. (2016)
Vancomycin	Inhibits peptidoglycan synthesis	Vancomycin-resistant *Staphylococcus aureus*, vancomycin-resistant *Enterococcus faecium*	Plasmid-mediated antibiotic resistance, D-Ala-D-Ala replacement	Howden et al. (2010)

(continued)

Table 13.2 (continued)

Name of antibiotic	Mode of action of antibiotic	List of resistant microorganisms	AMR mechanisms	References
Antifungal fluconazole, caspofungin, amphotericin B, azoles, and echinocandins	Disruption of cell membrane integrity	*Candida*, *Cryptococcus neoformans*, and *Aspergillus fumigatus*	Point mutations within the *ERG11* gene *CYP51A* gene cause microbiologic resistance	Nathan (2017)
Antiprotozoal artemisinin, pentamidine, suramin, benznidazole, and nifurtimox	Interferes with metabolic processes, reproduction, and larval physiology and neuromuscular physiology of parasites	*Malarial parasitr---Plasmodium* spp., *Trypanosoma* spp., and *Leishmania* spp. and *Entamoeba* spp.	Export and decrease of drug uptake and alteration of drug targets. The amino acid substitutions	Capela et al. (2019)
Antivirals oseltamivir, acyclovir, amantadine, 3'-azido-3'-deoxythymidine (AZT), etc.	Neuraminidase, protease and nucleic acid analogues, anti-HIV drugs, including NRTIs and NNRTIs, which target reverse transcriptase	HIV, hepatitis B, hepatitis C, influenza, herpes viruses including varicella zoster virus, *Cytomegalovirus*, and Epstein-Barr virus	Resistance acquired through mutations in the genes that encode the protein targets of the drugs. Mutations are in viral thymidine kinase gene, HIV-1 reverse transcriptase, etc.	Warnke et al. (2007) and Kristen et al. (2016)

(b) Indirect mechanisms involve the activation of innate immune defenses in the mucosa and the production of protective secondary metabolites such as secondary biliary acids, antimicrobial peptides, and short-chain fatty acids.

The ability of the microbiota to restore a healthy microbiome ecosystem ("resilience") depends on age, diet, class of antibiotic, pharmacokinetics, pharmacodynamics, range of action, dosage, duration, and administration route (Yang et al. 2017; Climent et al. 2018). Antibiotic resistance genes and their evolutionary mechanisms are mostly unexplored in humans and, therefore, necessary to account for the entire microbial ecosystem within the human landscape (Burcham et al. 2019). National and international antibiotic stewardship programs for monitoring and mapping AMR were carried out globally. ResistanceOpen is an online global mapping program displaying AMR aggregate data. ResistanceMap is a website developed by the Center for Disease Dynamics, Economics & Policy, which provides worldwide data on AMR WHO has organized a global action plan to tackle the mounting problem of AMR and has promoted the first World Antimicrobial Awareness Week from November 16 to 22, 2015.

13.12 Recent Advances in Limiting Drug Resistance Through MTT

Recent advances in microbiome research suggest manipulating the microbiota as a therapeutic tool in limiting drug resistance. Microbiome-targeted therapies (MTT) are recently exploited by managing microbes and their metabolism, ranging from simple dietary manipulation to high precision (Hartstra et al. 2015). Enrichment of microbiota through probiotics, prebiotics, phage therapies, and fecal microbiota transplantation restores the beneficial microbiota reversing dysbiosis (Schmidt et al. 2018). Immunologic stimulation by *Bacteroides* spp. and *Bifidobacterium* spp. was found to have a profound effect on therapy efficacy (Gorvitovskaia et al. 2016). Further, imbalance in dietary sphingolipids significantly impacted the therapeutic efficacy of chemotherapy and radiation (Camp et al. 2017). Dietary approaches such as foods rich in polyphenols, including fiber, and moderation of high-fat foods or specific food additives that preserve beneficial microbial communities are potentially effective long-term preventive strategies (Maruvada et al. 2017).

Prebiotics and probiotics can benefit the microbiome's development by introducing nutrients that benefit the existing and new benign/symbiotic bacteria (Kashima et al. 2015). Administration of bacterial consortia or "designer probiotics" could also provide a more feasible method of microbial manipulation in the clinical setting (Conlon and Bird 2015). Probiotics help to preserve healthy microbiota by regulating pathogenic bacteria and immune system response, which may reduce blood cholesterol and colitis and prevent chronic inflammations and cancers (Fernández et al. 2016). Various prebiotics can prevent cancers by different mechanisms: releasing detoxifying agents, anti-inflammatory factors, anticancer compounds (antiangiogenesis, promoting anti-PDL1 drugs), and short-chain fatty acids (SCFA) that improve the intestinal barrier function (Pandey et al. 2015).

It has been reported that the butyrate-producing species *Clostridium butyricum* and *Bacillus subtilis* may have an antitumor effect in a colorectal cancer mouse model (Chen et al. 2015). *L. casei* has been reported to produce ferrichrome, which inhibits colon cancer progression through apoptosis mediated by utilizing the c-Jun N-terminal kinase pathway (Konishi et al. 2016). Another *L. casei* strain (variety rhamnosus, Lcr35) showed to prevent induced intestinal mucositis in CRC-bearing mice (Chang et al. 2018). Recent clinical studies also revealed that *Bifidobacterium* probiotics restore the equilibrium of gut dysbiosis and reduce intestinal perturbations (Liang et al. 2017). Synbiotic therapies/synbiotics are the amalgamations of prebiotics and probiotics in dietary supplements such as *Bifidobacterium animalis* subsp. *Lactis* spp. combinations or oat fiber/*L. plantarum* and FOS/*L. sporogenes* formulations can modulate host immunity by regulating cytokine production and the proliferation and differentiation of macrophages, lymphocytes, and intestinal epithelial cells (Bozkurt et al. 2019).

13.12.1 Fecal Microbiota Transplantation (FMT)

FMT is a medical practice for restoring and reestablishing colonization resistance and other functions associated with the normal intestinal microbiota. FMT was first described in 1958 by Eiseman et al. while treating *Clostridium difficile* infection (CDI). Fecal microbial transplantations are possible through gastroscopy/colonoscopy or oral administration of lyophilized pills (Lee et al. 2016). Presently, FMT has been reevaluated as a hopeful therapeutic method to treat other disorders involving gut dysbiosis, such as colorectal cancer, ulcerative colitis, IBD, IBS, metabolic syndrome, types 1 and 2 diabetes, atopy, obesity, multiple sclerosis, and autism (Filip et al. 2018).

13.13 Phage Therapy

Conventionally, phage therapy involves using naturally occurring phages to infect and lyse bacteria at the site of infection. Biotechnological advances have extended potential phage therapeutics' repertoire to include novel strategies using bioengineered phages and purified phage lytic proteins to restore a desired bacterial equilibrium (Paule et al. 2018). Research on phages and their lytic proteins, specifically against multidrug-resistant bacterial infections, is regarded as the potential therapy to supplement antibiotic resistance (Derek et al. 2017).

13.14 Conclusions and Future Perspectives

It is understood from the rate of emergence of AMR that the microbes naturally evolve to resist the drugs gradually. Yet, unfortunately, the abuse or misuse of antibiotics in hospitals and clinics is speeding up their evolution resulting in the speciation of superbugs. It was predicted by the UK government that by 2050, the mortality rates through AMR infections will be more than the current toll from cancer. AMR infections are associated with 23,000 deaths and 2 million illnesses in the USA annually (Fortman and Mukhopadhyay 2016). Therefore, there is an immediate need to circumvent the undesirable effects on the microbiota, resulting in NCDs (noncommunicable diseases) such as colon cancer, autism, and obesity. Microbiome research is poised to influence human health by developing low-risk probiotics, prebiotics, phage therapy, and FMT treatments to fight infections, malnourishment, metabolic disorders, and vaccine efficacy. Research quests in these areas might lead to the design of biologically inspired therapeutics, such as consortia-based solutions/multi-organism cocktails to reboot dysbiotic ecosystems.

The emergence of resistance to the "last line" of drugs such as methicillin and vancomycin is a significant public health threat. It is, therefore, critical to understand AMR in medical environments. Shortly, the capacity of microbes to develop resistance outruns the production of new and novel drugs, provoking a need for alternate strategies to be implemented by governments and pharmaceutical industries. The discovery of new antimicrobials is important, but in reality, the major solution to address the problem of AMR is the rational usage of existing antibiotics and encouraging MTT. The surveillance of ARG dissemination across human populations would allow source tracking, outbreak preparation, and treatment alternatives which helps to focus on preventative measures instead of reactive medicine.

References

Adesoji AT, Onuh JP, Okunye OL (2016) Bacteria resistance to Cephalosporins and its implication to public health. J Bacteriol Mycol 3(1):1021

Ahmed I, Roy BC, Khan SA et al (2016) Microbiome, metabolome and inflammatory bowel disease. Microorganisms 4:E20

Amon P, Sanderson I (2017) What is the microbiome? Arch Dis Child Educ Pract Ed 102:258–261. https://doi.org/10.1136/archdischild-2016-311643

Andersson AF, Lindberg M, Jakobsson H, Backhed F, Nyren P, Engstrand L (2008) Comparative analysis of human gut microbiota by barcoded pyrosequencing. PLoS One 3:e2836. https://doi.org/10.1371/journal.pone.0002836

Angulo FJ, Nargund VN, Chiller TC (2004) Evidence of an association between use of antimicrobial agents in food animals and antimicrobial resistance among bacteria isolated from humans and the human health consequences of such resistance. J Veterinary Med Ser B 51:374–379

Arango-Argoty G, Garner E, Pruden A, Heath LS, Vikesland P, Zhang L (2018) DeepARG: a deep learning approach for predicting antibiotic resistance genes from metagenomic data. Microbiome 6:23

Arumugam M, Raes J, Pelletier E, Le Paslier D, Yamada T, Mende DR et al (2011) Enterotypes of the human gut microbiome. Nature 473:174–180

Berglund F, Österlund T, Boulund F, Marathe NP, Larsson DGJ, Kristiansson E (2019) Identification and reconstruction of novel antibiotic resistance genes from metagenomes. Microbiome 7:52. https://doi.org/10.1186/s40168-019-0670-1

Black JG (2015) Microbiology, principles and explorations, 8th edn index. ISBN 978-0-470-10748-5

Bozkurt HS, Quigley EM, Kara B (2019) Bifidobacterium animalis subspecies lactis engineered to produce mycosporin-like amino acids in colorectal cancer prevention. SAGE Open Med 7:2050312119825784. https://doi.org/10.1177/2050312119825784

Brickner SJ, Barbachyn MR, Hutchinson DK, Manninen PR (2008) Linezolid (ZYVOX), the first member of a completely new class of antibacterial agents for treatment of serious gram-positive infections. J Med Chem 51:1981–1990

Buffie CG, Pamer EG (2013) Microbiota-mediated colonization resistance against intestinal pathogens. Nat Rev Immunol 13(11):790–801

Burcham ZM, Schmidt CJ, Pechal JL, Brooks CP, Rosch JW, Eric Benbow M, Jordan HR (2019) Detection of critical antibiotic resistance genes through routine microbiome surveillance. PLoS One 14(3):e0213280. https://doi.org/10.1371/journal.pone.0213280

Camp ER, Patterson LD, Kester M, Voelkel-Johnson C (2017) Therapeutic implications of bioactive sphingolipids: a focus on colorectal cancer. Cancer Biol Ther 18:640–650. https://doi.org/10.1080/15384047.2017.1345396

Canton R, Coque TM (2006) The CTX-M β-lactamase pandemic. Curr Opin Microbiol 9:466–475

Capela R, Moreira R, Lopes F (2019) An overview of drug resistancein protozoal diseases. Int J Mol Sci 20:5748

Catherine V, William R, Jacobs JR (2014) Resistance to isoniazid and ethionamide in *Mycobacterium tuberculosis*: genes, mutations, and causalities. Microbiol Spectr 2(4):MGM2-0014-2013. https://doi.org/10.1128/microbiolspec.MGM2-0014-2013

Chang C-W, Liu C-Y, Lee H-C, Huang Y-H, Li L-H, Chiau J-SC, Wang T-E, Chu C-H, Shih S-C, Tsai T-H, Chen Y-J (2018) Lactobacillus casei variety rhamnosus probiotic preventively attenuates 5-fluorouracil/oxaliplatin-induced intestinal injury in a syngeneic colorectal cancer model. Front Microbiol 9:983. https://doi.org/10.3389/fmicb.2018.00983

Chen Z-F, Ai L-Y, Wang J-L, Ren L-L, Ya-Nan Y, Jie X, Chen H-Y, Jun Y, Li M, Qin W-X, Ma X, Shen N, Chen Y-X, Hong J, Fang J-Y (2015) Probiotics clostridium butyricum and bacillus subtilis ameliorate intestinal tumorigenesis. Future Microbiol 10(9):1433–1445. https://doi.org/10.2217/fmb.15.66. Epub 2015 Sep 7

Cheng G, Hu Y, Yin Y, Yang X, Xiang C, Wang B et al (2012) Functional screening of antibiotic resistance genes from human gut microbiota reveals a novel gene fusion. FEMS Microbiol Lett 336:11–16

Cho I, Blaser MJ (2012) The human microbiome: at the interface of health and disease. Nat Rev Genet 13:260–270. https://doi.org/10.1038/nrg3182

Chung Y, Guo F, Yanan Y et al (2017) Fusobacterium nucleatum promotes chemoresistance to colorectal cancer by modulating autophagy. Cell 170:548–563

Čivljak R, Giannella M, Di Bella S, Petrosillo N (2014) Could chloramphenicol be used against ESKAPE pathogens? A review of in vitro data in the literature from the 21st century. Expert Rev Anti Infect Ther 12:249–264. https://doi.org/10.1586/14787210.2014.878647

Climent CP, Vergara A, Vila J (2018) Intestinal microbiota and antibiotic resistance: Perspectives and solutions. Hum Microb J 9:11–15

Conlon M, Bird A (2015) The impact of diet and lifestyle on gut microbiota and human health. Nutrients 7:17–44

Costea PI, Hildebrand F, Manimozhiyan A, Bäckhed F, Blaser MJ, Bushman FD, Knight R (2018) Enterotypes in the landscape of gut microbial community composition. Nat Microbiol 3(1):8–16

Crofts TS, Gasparrini AJ, Dantas G (2017) Next-generation approaches to understand and combat the antibiotic resistome. Nat Rev Microbiol 15:422–434

Davies J, Davies D (2010) Origins and evolution of antibiotic resistance. Microbiol Mol Biol Rev 74(3):417–433. https://doi.org/10.1128/MMBR.00016-10

Derek ML, Koskella B, Henry CL (2017) Phage therapy: an alternative to antibiotics in the age of multi-drug resistance. World J Gastrointest Pharmacol Ther 8(3):162–173. Published online 2017 Aug 6. https://doi.org/10.4292/wjgpt.v8.i3.162

Deschamps C, Clermont O, Hipeaux MC, Arlet G, Denamur E, Branger C (2009) Multiple acquisitions of CTX-M plasmids in the rare D2 genotype of *Escherichia coli* provide evidence for convergent evolution. Microbiology 155:1656–1668

Dethlefsen L, Huse S, Sogin ML, Relman DA (2008) The pervasive effects of an antibiotic on the human gut microbiota, as revealed by deep 16S rRNA. PLoS Biol 6:e280. https://doi.org/10.1371/journal.pbio.0060280

Dinos GP (2017) The macrolide antibiotic renaissance. Br J Pharmacol 174:2967–2983

Dominguez-Bello MG, Costello EK, Contreras M, Magris M, Hidalgo G, Fierer N et al (2010) Delivery mode shapes the acquisition and structure of the initial microbiota across multiple body habitats in newborns. Proc Natl Acad Sci U S A 107:11971–11975

Dzutsev A, Goldszmid RS, Viaud S, Zitvogel L, Trinchieri G (2015) The role of the microbiota in inflammation, carcinogenesis, and cancer therapy. Eur J Immunol 45:17–31. https://doi.org/10.1002/eji.201444972

Eiseman B, Silen W, Bascom GS, Kauvar AJ (1958) Fecal enema as an adjunct in the treatment of pseudomembranous enterocolitis. Surgery 44:854–859

Falagas ME, Rafailidis PI, Matthaiou DK (2010) Resistance to polymyxins: mechanisms, frequency and treatment options. Drug Resist Updat 13:132–138. https://doi.org/10.1016/j.drup.2010.05.002

Fernández R-B, Gutiérrez-del-Río EM, Miguélez CJ, Villar FL (2016) Colon microbiota fermentation of dietary prebiotics towards short-chain fatty acids and their roles as anti-inflammatory and antitumour agents: a review. J. Funct. Foods 25:511–522. https://doi.org/10.1016/j.jff.2016.06.032

Ferrer M, Méndez-García C, Rojo D, Barbas C, Moya A (2017) Antibiotic use and microbiome function. Biochem Pharmacol 134:114–126

Filip M, Tzaneva V, Dumitrascu DL (2018) Fecal transplantation: digestive and extradigestive clinical applications. Clujul Med 91:259–265. https://doi.org/10.15386/cjmed-946

Fortman JL, Mukhopadhyay A (2016) The future of antibiotics: emerging technologies and stewardship. Trends Microbiol 24:515–517

Franzosa EA, Morgan XC, Segata N, Waldron L, Reyes J, Earl AM, Giannoukos G, Boylan MR, Ciulla D, Dirket al. (2014) Relating the metatranscriptome and metagenome of the human gut. PNAS 111(22):E2329–E2338. https://doi.org/10.1073/pnas.1319284111

George ME, Huovinen P (2001) Resistance to trimethoprim-sulfamethoxazole. Clin Infect Dis 32(11):1608–1614. https://doi.org/10.1086/320532

Geser N, Stephan R, Korczak BM, Beutin L, Hachler H (2012) Molecular identification of extended-spectrum-beta-lactamase genes from enterobacteriaceae isolated from healthy human carriers in Switzerland. Antimicrob Agents Chemother 56:1609–1612

Gilbert JA, Blaser MJ, Gregory Caporaso J, Jansson JK, Lynch SV, Knight R (2018) Current understanding of the human microbiome. Nat Med 24:392–400

Goldstein BP (2014) Resistance to rifampicin: a review. J Antibiot 67:625–630

Gopalakrishnan V, Helmink BA, Spencer CN, Reuben A, Wargo JA (2018) The influence of the gut microbiome on cancer, immunity, and cancer immunotherapy. Cancer Cell 33(4):570–580. https://doi.org/10.1016/j.ccell.2018.03.015

Gorvitovskaia A, Holmes SP, Huse SM (2016) Interpreting prevotella and bacteroides as biomarkers of diet and lifestyle. Microbiome 4:15

Grishin A, Papillon S, Bell B et al (2013) The role of the intestinal microbiota in the pathogenesis of necrotizing enterocolitis. Semin Pediatr Surg 22:69–75

Gupta N, Limbago BM, Patel JB, Kallen AJ (2011) Carbapenem-resistant enterobacteriaceae: epidemiology and prevention. Clin Infect Dis 53:60–67. https://doi.org/10.1093/cid/cir202

Gupta SK, Padmanabhan BR, Diene SM, Lopez-Rojas R, Kempf M, Landraud L, Rolain JM (2014) ARG-ANNOT, a new bioinformatic tool to discover antibiotic resistance genes in bacterial genomes. Antimicrob Agents Chemother 58:212–220

Hartstra AV, Bouter KEC, Bäckhed F, Nieuwdorp M (2015) Insights into the role of the microbiome in obesity and type 2 diabetes. Diabetes Care 38(1):159–165. https://doi.org/10.2337/dc14-0769

Howden BP, Davies JK, Johnson PD, Stinear TP, Grayson ML (2010) Reduced vancomycin susceptibility in staphylococcus aureus, including vancomycin-intermediate and heterogeneous vancomycin-intermediate strains: resistance mechanisms, laboratory detection, and clinical implications. Clin Microbiol Rev 23(1):99–139. https://doi.org/10.1128/CMR.00042-09

Hu Y, Yang X, Qin J, Lu N, Cheng G, Wu N et al (2013) Metagenome-wide analysis of antibiotic resistance genes in a large cohort of human gut microbiota. Nat Commun 4:2151

Human Microbiome Project Consortium (2012) Structure, function and diversity of the healthy human microbiome. Nature 486:207–214. https://doi.org/10.1038/nature11234

Jalanka-Tuovinen J, Salonen A, Nikkila J, Immonen O, Kekkonen R, Lahti L et al (2011) Intestinal microbiota in healthy adults: temporal analysis reveals individual and common core and relation to intestinal symptoms. PLoS One 6:e23035. https://doi.org/10.1371/journal.pone.0023035

JanaMarkeley and Timothy Wencewicz (2018) Tetracycline-inactivating enzymes. Front. Microbiol 9:1058. https://doi.org/10.3389/fmicb.2018.01058

John P, Stobberingh EE, Savelkoul PHM, Wolffs PFG (2013) The human microbiome as a reservoir of antimicrobial resistance. Front. Microbiol 4:87. https://doi.org/10.3389/fmicb.2013.00087

Kaczmarek FS, Gootz TD, Dib-Hajj F, Shang W, Hallowell S, Cronan M (2004) Genetic and molecular characterization of beta-lactamase-negative ampicillin-resistant *Haemophilus influenzae* with unusually high resistance to ampicillin. Antimicrob Agents Chemother 48:1630–1639

Kamamoto CSL, Nishikaku AS, Gompertz OF, Melo H, Bagatin E (2017) Cutaneous fungal microbiome: *Malassezia* yeasts in seborrheic dermatitis scalp in a randomized, comparative and therapeutic trial. Dermatoendocrinol 9(1):e1361573. https://doi.org/10.1080/19381980.2017.1361573

Kashima S et al (2015) Polyphosphate, an active molecule derived from probiotic *lactobacillus brevis*, improves the fibrosis in murine colitis. Transl Res 166:163–175
Kim ES, Hooper DC (2014) Clinical importance and epidemiology of quinolone resistance. Infect Chemother 46:226–238
Kim SM, Kim HC, Lee SW (2011) Characterization of antibiotic resistance determinants in oral biofilms. J Microbiol 49:595–602
Kinross JM, Darzi AW, Nicholson JK (2011) Gut microbiome-host interactions in health and disease. Genome Med 3:14
Konishi H, Fujiya M, Tanaka H et al (2016) Probiotic-derived ferrichrome inhibits colon cancer progression via JNK-mediated apoptosis. Nat Commun 7:12365. https://doi.org/10.1038/ncomms12365
Kristen KI, Renzette N, Kowalik TF, Jeffrey D (2016) Antiviral drug resistance as an adaptive process. Virus Evol 2(1):vew014. https://doi.org/10.1093/ve/vew014
Langdon A, Crook N, Dantas G (2016) The effects of antibiotics on the microbiome throughout development and alternative approaches for therapeutic modulation. Genome Med 2(4):39. https://doi.org/10.1186/s13073-016-0294-z
Lee CH, Steiner T, Petrof EO et al (2016) Frozen vs fresh faecal microbiota transplantation and clinical resolution of diarrhoea in patients with recurrent clostridium difficile infection: a randomized clinical trial. JAMA 315:142–149
Li LL, Norman A, Hansen LH, Sorensen SJ (2012) Metamobilomics – expanding our knowledge on the pool of plasmid encoded traits in natural environments using high-throughput sequencing. Clin Microbiol Infect 18(Suppl. 4):5–7
Liang Q, Chiu J, Chen Y, Huang Y, Higashimori A, Fang J, Brim H, Ashktorab H, Ng SC, Ng SSM, Zheng S, Chan FKL, Sung JJY, Yu J (2017) Fecal bacteria act as novel biomarkers for noninvasive diagnosis of colorectal cancer. Clin Cancer Res 23:2061–2070. https://doi.org/10.1158/1078-0432.CCR-16-1599
Lynch SV, Pedersen O (2016) The human intestinal microbiome in health and disease. N Engl J Med 375:2369–2379. https://doi.org/10.1056/nejmra1600266
Manrique P, Bolduc B, Walk ST, van der Oost J, de Vos WM, Young MJ (2016) Healthy human gut phageome. Proc Natl Acad Sci 113(37):10400–10405. https://doi.org/10.1073/pnas.1601060113
Mariya L, Giulia P (2017) Penicillin's discovery and antibiotic resistance: lessons for the future? Yale J Biol Med 90(1):135–145
Martiny AC, Martiny JB, Weihe C, Field A, Ellis JC (2011) Functional metagenomics reveals previously unrecognized diversity of antibiotic resistance genes in gulls. Front Microbiol 2:238. https://doi.org/10.3389/fmicb.2011.00238
Maruvada P, Leone V, Kaplan LM, Chang EB (2017) The human microbiome and obesity: moving beyond associations. Cell Host Microbe 22:589–599. https://doi.org/10.1016/j.chom.2017.10.005
Morgan XC, Huttenhower C (2012) Chapter 12: human microbiome analysis. PLoS Comput Biol 8:e1002808. https://doi.org/10.1371/journal.pcbi.1002808
Nandita RG, Katherine SP (2020) Population genetics in the human microbiome. Trends Genet 36(1):53–67. https://doi.org/10.1016/j.tig.2019.10.010
Nathan PW (2017) Antifungal resistance: current trends and future strategies to combat. Infect Drug Resist 10:249–259. https://doi.org/10.2147/IDR.S124918
Nikoopour E, Bhagirath S (2014) Reciprocity in microbiome and immune system interactions and its implications in disease and health. Inflamm Allergy Drug Targets 13(2):94–104. https://doi.org/10.2174/1871528113666140330201056
Nishiumi ST, Kobayashi AI, Yoshie T, Kibi M, Izumi Y, TatsuyaOkuno NH, Kawano S, TadaomiTakenawa TA, Yoshid M (2012) A novel serum metabolomics-based diagnostic approach for colorectal cancer. PLoS One 7:e40459
Nkamga VD, Henrissat B, Drancourt M (2017) Archaea: essential inhabitants of the human digestive microbiota. Hum Microbiome J 3:1–8. https://doi.org/10.1016/j.humic.2016.11.005

Nogueira T, David PHC, Pothier J (2019) Antibiotics as both friends and foes of the human gut microbiome: the microbial community approach. Drug Dev Res 80:86–97. https://doi.org/10.1002/ddr.21466

Okeke IN, Peeling RW, Goossens H, Auckenthaler R, Olmsted SS, De Lavison JF et al (2011) Diagnostics as essential tools for containing antibacterial resistance. Drug Resist Updat 14:95–106

Pandey KR, Naik SR, Vakil BV (2015) Probiotics, prebiotics and synbiotics- a review. J Food Sci Technol 52:7577–7587. https://doi.org/10.1007/s13197-015-1921-1

Paule A, Frezza D, Edeas M (2018) Microbiota and phage therapy: future challenges in medicine. Med Sci 2018:6. https://doi.org/10.3390/medsci6040086

Perry JA, Westman EL, Wright GD (2014) The antibiotic resistome: what's new? Curr Opin Microbiol 21:45–50

Peterson E, Kaur P (2018) Antibiotic resistance mechanisms in bacteria: relationships between resistance determinants of antibiotic producers, environmental bacteria, and clinical pathogens. Front Microbiol 9:2928. https://doi.org/10.3389/fmicb.2018.02928

Pflughoeft KJ, Versalovic J (2012) Human microbiome in health and disease. Annu Rev Pathol 7:99–122

Qin J, Li R, Raes J, Arumugam M, Burgdorf KS, Manichanh C et al (2010) A human gut microbial gene catalogue established by metagenomic sequencing. Nature 464:59–65

Ravel J, Gajer P, Abdo Z, Schneider GM, Koenig SS, Mcculle SL et al (2011) Vaginal microbiome of reproductive-age women. Proc Natl Acad Sci U S A 108(Suppl. 1):4680–4687

Raymond MB, Ouameur AA, Déraspe M, Plante P-L et al (2019) Culture-enriched human gut microbiomes reveal core and accessory resistance genes. Microbiome 7:56. https://doi.org/10.1186/s40168-019-0669-7

Riiser A (2015) The human microbiome, asthma, and allergy. Allergy Asthma Clin Immunol 11:35

Rooks MG, Garrett WS (2016) Gut microbiota, metabolites and host immunity. Nat Rev Immunol 16:341–352

Rubinstein X, Wang W, Liu Y, Hao G, Cai YWH (2013) Fusobacterium nucleatum promotes colorectal carcinogenesis by modulating E-cadherin/β-catenin signaling via its FadA Adhesin. Cell Host Microbe 14:195–206. https://doi.org/10.1016/j.chom.2013.07.012

Sanapala P, Pola S (2021) Modulation of systemic immune responses through genital, skin, and oral microbiota: unveiling the fundamentals of human microbiomes. In: Bramhachari PV (ed) Microbiome in human health and disease. Springer, Singapore, pp 13–34

Sanz Y, Olivares M, Moya-Pérez Á et al (2015) Understanding the role of gut microbiome in metabolic disease risk. Pediatr Res 77:236–244

Schmidt TSB, Raes J, Bork P (2018) The human gut microbiome: from association to modulation. Cell 172:1198–1215. https://doi.org/10.1016/j.cell.2018.02.044

Sheetal RM, Collins JJ, Relman DA (2014) Antibiotics and the gut microbiota. J Clin Invest 124(10):4212–4218. https://doi.org/10.1172/JCI72333

Shin NR, Whon TW, Bae JW (2015) Proteobacteria: microbial signature of dysbiosis in gut microbiota. Trends Biotechnol 33:496–503

Sofia AT, Elena IK (2020) Microbiota and cancer: host cellular mechanisms activated by gut microbial metabolites. Int J Med Microbiol 310(4):151425. https://doi.org/10.1016/j.ijmm.2020.151425

Sommer MO, Church GM, Dantas G (2010) The human microbiome harbors a diverse reservoir of antibiotic resistance genes. Virulence 1:299–303

Sultan I, Rahman S, Jan AT, Siddiqui MT, Mondal AH, RizwanulHaq Q (2018) Antibiotics, Resistome and resistance mechanisms: a bacterial perspective. Front Microbiol 9:2066. https://doi.org/10.3389/fmicb.2018.02066

Sung JY, Koo SH, Cho HH, Kwon KC (2012) AbaR7 a genomic resistance Island found in multidrug-resistant *Acinetobacter baumannii* isolates in Daejeon, Korea. Ann Lab Med 32(5):324–330. https://doi.org/10.3343/alm.2012.32.5.324

Suto C, Morinaga M, Yagi T, Tsuji C, Toshida H (2012) Conjunctival sac bacterial flora isolated prior to cataract surgery. Infect Drug Resist 5:37–41

Sylvie G-T, Kristin JL (2016) Mechanisms of resistance to aminoglycoside antibiotics: Overview and Perspectives. Medchemcomm 7(1):11–27. https://doi.org/10.1039/C5MD00344J

Tanwar J, Das S, Fatima Z et al (2014) Multidrug resistance: an emerging crisis. Interdiscip Perspect Infect Dis 2014:541340

Ussar S, Shiho F, Ronald Kahn C (2016) Interactions between host genetics and gut microbiome in diabetes and metabolic syndrome. Mol Metab 5(9):795–803. https://doi.org/10.1016/j.molmet.2016.07.004

van Bogaard AE, Stobberingh DE (2000) Epidemiology of resistance to antibiotics links between animals and humans. Int J Antimicrob Agents 14:327–335

Warnke D, Barreto J, Temesgen Z (2007) Antiretroviral drugs. J Clin Pharmacol 47:1570–1579. https://doi.org/10.1177/0091270007308034

William RM, Bayerand AS, Cesar A (2016) Arias mechanism of action and resistance to daptomycin in staphylococcus aureus and enterococci. Cold Spring Harb Perspect Med 6(11):a026997. https://doi.org/10.1101/cshperspect.a026997

Yang JH, Bhargava P, McCloskey D, Mao N, Palsson BO, Collins JJ (2017) Antibiotic-induced changes to the host metabolic environment inhibit drug efficacy and alter immune function. Cell Host Microbe 22:757.e3–765.e3

Yin X, Jiang XT, Chai B, Li L, Yang Y, Cole JR et al (2018) ARGs-OAP v2.0 with an expanded SARG database and hidden markov models for enhancement characterization and quantification of antibiotic resistance genes in environmental metagenomes. Bioinformatics 34:2263–2270. https://doi.org/10.1093/bioinformatics/bty053

New Paradigms on Microbiome Diagnostic Design and Engineering

14

Manohar Babu Vadela, Satyanagalakshmi Karri, and Vijay A. K. B. Gundi

Abstract

A vast microbial diversity is associated with different ecologies, including humans, animals, and plants, challenging the microbiologists in the perspective of their identification, potential applications, and clinical diagnostics. Microorganisms, including archaea, bacteria, fungi, and viruses, harbored and colonized on humans' skin and gastrointestinal and gut, play a key role in various diseases. Identifying and characterizing this vast, diverse microbiome and diagnosing clinically important pathogens have triggered the establishmentof massive DNA sequencing technologies.In the modern era, these new sequencing technologies have allowed scientists to precisely identify any organism's taxonomic status. This chapter described a few pathogenic microbiomes and their identification approaches, like in situ microbiome engineering, microfluidic systems, engineered organoids, and single-cell imaging approaches.

Keywords

Diagnostic methods · Microorganisms · Biosensor · Engineered microbes · Microfluidic diagnostic · Antimicrobial peptides

14.1 Introduction

Microbial diversity is a core factor and influences the stability of microbes and their related health of the population (Coyte et al. 2015). The alterations in microbiota in an ecosystem have been associated with pathological conditions such as metabolic dysfunction, neurodegenerations, and cancer. The finding of functional diversity of

M. B. Vadela · S. Karri · Vijay A. K. B. Gundi (✉)
Department of Biotechnology, Vikrama Simhapuri University, Nellore, Andhra Pradesh, India

P. Veera Bramhachari (ed.), *Human Microbiome in Health, Disease, and Therapy*, https://doi.org/10.1007/978-981-99-5114-7_14

microbes rather than taxonomic diversity is a most important and emerging state in the future (Li and Convertino 2019). At present, microbiome research related to the effect on human health is more concerned (Martin et al. 2018). A healthy environmental microbiome governs a healthy human microbiome (Lloyd-Price et al. 2016). Hence, it is important to study microbiomes and the ecosystem where they exist. Diagnostic and therapeutic study of the microbiome is an emerging need regarding their potential metabolic use and contribution to human health.

Understanding the microbiome and its biological process in host cells can help us use microbes as a low-cost diagnostic tool and microbiome as an evaluative tool. In the last decade, microbiomes have been used for disease diagnosis and prognosis tools, for example, inflammatory bowel disease diagnosis (Zhou et al. 2018) and progression of diabetes (Leustean et al. 2018). Among microbial diseases, bacterial diseases have more considerable morbidity and mortality, representing a significant public health threat in developed and developing countries. Over 2.5 million deaths occur yearly due to water-associated diseases worldwide (Prieto et al. 2015). In this context, developing diagnostic tools is extremely important to prevent bacterial disease spread and public health. It can also prevent millions of deaths caused by the lack of these facilities (Yager et al. 2006). Clinical microbiology deals with diagnostics, identifying pathogens from clinical samples, treatment strategies for patients, and surveys and monitoring in public for outbreaks. The traditional methods are reliable, isolating and culturing pathogens from specimens and detecting pathogen-specific antibodies (serology) or antigens. Still, it isn't easy to maintain in the laboratory every time. However, gene-based molecular identification via PCR targets only a limited number of pathogens using specific primers or probes. The metagenomic approaches illustrate all nucleic acid material in a sample, enabling analysis of the entire microbiome and the host genome in patient samples. Still, it is an expensive method (Chiu and Miller 2019). The diagnostic method should be economical, and importantly, the test should be rapid, simple to use, easily interpretable, and stable when transported and stored under extreme conditions. Diseases caused by multiple causative agents, for example, acute lower respiratory infections and diarrheal diseases, have to be diagnosed using multiplex tests and individual tests for emerging and reemerging diseases. Firstly, we emphasized to provide imperative knowledge on molecular-based detection methods; modern diagnostic inventions like impedance spectroscopy, electrochemical, and combined dielectrophoresis and impedance methods; biosensor-based designs; and advanced microfluidic methods in microbial identification. Secondly, the development of engineered microbes in diagnostic and therapeutic application has been discussed.

14.2 Molecular Methods for Microbial Diagnostics

The advantages of PCR-based methods are that they do not require intensive microbial biomass cultivation and stable genotypic characteristics. The organism's pathogenicity coded by the virulence factors or toxin loci in variable regions can be used to detect pathogens. After the whole-genome sequencing approach was established, advanced methods, including nucleic acid hybridization (Procop 2007), ligase chain reaction (Drancourt et al. 2000), strand displacement amplification (Walker et al. 1992), and transcription-based amplification (Compton 1991; Fahy et al. 1991) methods, have been implemented to detect and for genotyping of both common and uncommon microorganisms. Microarray-based assay and universal broad-range 16S rDNA PCR methods are the most commonly used methods of detection of microbes. Moreover, next-generation sequencing technologies are also effectively employed for pathogen detection and discovery.

14.3 Broad-Range PCR

Clinical diagnostic labs are using molecular approaches because of rapid detection, the potentiality to bypass the time-consuming culture techniques and the potential to detect critical culture-negative pathogens as it detects and amplifies the target nucleic acid region in the sample. Broad-range PCR is nonselective and extensively used to detect multiple organisms simultaneously rather than performing multiple monoplex PCR approaches in clinical samples. However, this approach must clone the amplicons for accurate detection before sequencing (Grahn et al. 2003). This approach has been widely used to diagnose various pathogenic microbes, including invasive bacterial and fungal organisms, culture-negative endocarditis and meningitis, and pathogens responsible for inflammatory diseases (Sibley et al. 2012). Despite its application in the diagnostics of various pathogens and sensitivity of detection rates, this approach has its limitations with DNA contamination. The universality of the 16S rRNA gene can be a limitation with the high sensitivity of PCR. A small amount of DNA from a test sample can give false-positive results after amplification (Vandecasteele et al. 2002).

14.4 Combined PCR and Electrospray Ionization Mass Spectrometry (PCR/ESI-MS)

Broad PCR and electrospray ionization mass spectrometry (PCR/ESI-MS) has been developed. It has become a promising diagnostic approach for a broad range of pathogens with prior nucleic acid sequence information knowledge. It can detect pathogens by targeting several genes and their products analyzed by electrospray ionization mass spectrometry (Baldwin et al. 2009). The PCR/ESI-MS provides information not only on the taxonomical status of the organism but also on its genotype, resistance, and virulence factors. This method is widely used in epidemiological and genomic studies because of its diagnostic application and ability to detect and characterize organisms from nearly 300 samples within 24 h (Sibley et al. 2012). After developing PCR/ESI-MS, several studies have demonstrated its potentiality in human pathogen diagnostic applications, bacteremia, and identification and typing of a wide range of human pathogens, including *S. aureus, Mycobacterium* spp., *Candida* spp. and *Ehrlichia* spp., and respiratory-related influenza and pan-orthopoxviruses (Kaleta et al. 2011; Hall et al. 2009).

14.5 Microarrays

Microarray approaches are the most advanced and powerful tools for identifying and characterizing many gene sequences derived from an organism that can reveal a hundred thousand genes compared with other molecular approaches. This genotyping approach can also potentially detect new genes or identify the polymorphism by oligonucleotide hybridization (Miller and Tang 2009). Microarray consists of a solid surface called a matrix with predesigned oligonucleotide probes immobilized in millions of numbers complementary to target nucleic acids. In this method, target nucleic acids extracted from clinical specimens or pathogenic organisms are labeled with fluorescence and hybridized with immobilized probes on the array matrix. The fluorescence can be scanned and detected by fluorescent scanners when the target DNA is hybridized with a complimentary probe nucleotide emits fluorescence. The data generated from microarrays can also be analyzed using sophisticated bioinformatic tools and algorithms (Loewe and Nelson 2011).

Several commercial suppliers have developed microarrays based on several purposes, such as pathogen diagnostic applications, customer-specified assays, and detection of a large panel of respiratory viruses, including influenza A and B and FDA-approved xTAG Respiratory Viral Panel FAST (Pabbaraju et al. 2008). Later on, microarray methods have also been used in the detection and genotyping of human papillomavirus (HPV), a causative agent of cervix cancer and cervical dysplasia (Albrecht et al. 2006; Cho et al. 2011).

14.6 Foodborne Pathogen Detection Using Impedimetric Biosensors

The outbreaks of foodborne epidemics in both developing and developed countries are seriously considered public health issues. Over 250 known diseases transmit through foodborne pathogens such as *Staphylococcus aureus*, *Salmonella* spp., and *Listeria monocytogenes* (Singh et al. 2013). On a priority basis, rapid detection of foodborne pathogens is important to control and prevent epidemics in humans and reduce mortality rates drastically. Different diagnostic methods such as (1) microbiological culturing, (2) immunological technologies, (3) DNA/RNA-based methods, (4) loop-mediated isothermal amplification, (5) rolling circle amplification methods, and (6) biosensor detection methods are available at present for controlling several foodborne pathogens.

Microbiological culture methods are known as traditional culture methods utilizing the chromogenic medium to detect foodborne pathogens. These are considered highly accurate methods to detect pathogens. However, these methods' main disadvantages include requiring 5–7 days of incubation, laborious, and poor sensitivity and specificity (Ngwa et al. 2013). Immunological methods such as ELISA, immunomagnetic separation (IMS), and immune colloidal gold technique (GICT) are commonly used in foodborne pathogen detection. It is well known as a rapid method requiring 4 h to get the results, with relatively high sensitivity and specificity. The important disadvantage of this method is its false-positive rate and poor stability (Jin et al. 2012; Chen et al. 2015). DNA/RNA-based methods mainly rely on gene amplification and quantification by PCR and real-time PCR, respectively, and are recognized as rapid methods requiring only ≤2 h and are relatively sensitive and can be used to detect multiple pathogens in several samples. However, the drawback of these methods is that they are expensive and well-trained personnel should perform the tests (Kordas et al. 2016). Isothermal amplification assays such as LAMP, rolling circle amplification (RCA), and saltatory rolling circle amplification (SRCA) are considered advanced technologies with high sensitivity and selectivity and require only ≤2 h for the detection of pathogens. However, these methods are not suitable for the on-site detection of organisms (Wang et al. 2018). Among these diagnostic methods, biosensor strategies are highly prevalent and useful with their very fast responsiveness, robustness, and cost-effectiveness. They are known to produce results with high sensitivity and selectivity. Also, these biosensors can detect pathogens on-site in real time with minimal sample preparation (Arora et al. 2011).

14.7 Biosensors

Biosensors are devices used for disease monitoring and drug discovery and also for detecting pollutants, disease-causing microorganisms, and markers that are indicators of disease in bodily fluids (blood, urine, saliva, sweat). Biosensor acts as an indicator in analyzing a single pathogen or multiple pathogens. Generally, these methods depend on DNA/RNA components, substrate utilization by pathogens, and interaction with eukaryotic cells and antibodies. Different biosensors are developed using physical and genetic properties (Arora et al. 2011). Biosensors are commonly composed of two parts, such as a bioreceptor and transducer, including others as mentioned below. The types and applications of biosensors are depicted in Table 14.1:

- *Analyte* is a target component that has to be detected in the biosensor.
- *Bioreceptor* is a molecule that recognizes the target component/molecule—for example, enzymes, aptamers, antibodies, and nucleic acid material. When the analyte interacts with a bioreceptor and generates, a signal is known as bio-recognition.
- *Transducer* is an element that converts the bio-recognition energy into measurable energy; the whole process is known as signalization. Most of the transducers are of optical or electrical signals.
- *Electronics* are a part of biosensors, which help to generate signals in analog amplification and convert them into digital form.

Table 14.1 A partial list of modern diagnostic methods and their target pathogenic bacteria

Method	Pathogen recognizing agent	Name of the agent AMP/ Aptamer	Pathogen targeted (bacteria/virus/ parasite)	Reference
Impedance-metry	Antimicrobial peptides (AMPs)	Magainin I	*E. coli O157:H7*, *S. typhimurium*	Mannoor et al. (2010)
			E. coli K12, *B. subtilis*, *S. epidermis*	Li et al. (2014)
		Clavanin A	*K. pneumoniae*, *E. faecalis*, *E. coli*, *B. subtilis*, *S. typhimurium*, *S. aureus*	Andrade et al. (2015) and de Miranda et al. (2017)
		Human lactoferrin (residues 1–11)	*S. sanguinis*	Hoyos-Nogués et al. (2016)
		Cecropin (A, B, and P) parasin, magainin I Polymyxin (B and E) Melittin, bactenecin	*C. burnetii*, *B. melitensis*, *VEE*, *vaccinia virus*	Kulagina et al. (2007)
		Leucocin A	*L. monocytogenes*, *S. aureus*	Etayash et al. (2014)
	Aptamers	H63SL2-M6	*Tuberculosis*	Lavania et al. (2018)
		Thiolated MPT64	*Tuberculosis*	Sypabekova et al. (2019)
		Apt-E. coli, Apt-S. typ	*Salmonella*	Li et al. (2018)
		Chimeric *S. aureus* aptamer	*Staphylococcus aureus*	Cai et al. (2019)
		LM6-116	*Listeria monocytogenes*	Suh et al. (2018)
		EcA5-27	Urethropathogenic *E. coli*	Savory et al. (2014)
		Antitat	*Human immunodeficiency virus (HIV)*	Caglayan and Üstündağ (2020)

(continued)

Table 14.1 (continued)

Method	Pathogen recognizing agent	Name of the agent AMP/ Aptamer	Pathogen targeted (bacteria/virus/ parasite)	Reference
		A thiol-modified aptamer against HBsAg	*Hepatitis B virus*	Xi et al. (2018)
		HPV-07	*Human papillomavirus*	Trausch et al. (2017)
		ssDNA	*SARS-CoV*	Zhou et al. (2020)
		Apt68	*Trypanosoma cruzi*	Babamiri et al. (2018)
		P38 or NG3	*Plasmodium for malaria*	Xi et al. (2018)
Electrochemical impedance spectroscopy	Antibodies	Anti-E. coli (PA1-7213)	*E. coli O157:H7*	Barreiros dos Santos et al. (2013)
	Nucleic acids	DNA (acpcPNA)	*M. tuberculosis*	Teengam et al. (2018)
		ssDNA	*Salmonella* spp.	Ma et al. (2014)
		DNA (NS5)	*Zika virus*	Faria and Zucolotto (2019)
Electrochemical	Graphene quantum Dots, GQD	Antibodies	*Yersinia enterocolitica*	Savas and Altintas (2019)
Fluorescence quenching	Graphene quantum Dots, GQD	Gold nanoparticles (AuNPs)	*E. coli O157:H7*	Saad and Abdullah (2019)
		AuNP	*S. typhimurium* *S. enteritidis*	Wu et al. (2018)
Combined dielectrophoresis and impedance	Polyclonal antibodies		*E. coli* *Strain K12*	Suehiro et al. (2003)
	Fluorescent beads (2 μm)		*B. subtilis* spores	Sabounchi et al. (2008)
Microfluidics			*Escherichia coli* *Staphylococcus aureus*	D'Amico et al. (2017)
			Escherichia coli, *Staphylococcus aureus*, *Pseudomonas aeruginosa*	Yoon et al. (2019)

- *The display* consists of hardware and software that generates the biosensor results. The output signal on the display can be numeric, graphic, tabular, or an image, depending on the requirement of the end user.

Bioreceptors interact with the target molecule, whereas transducers convert this interaction into electric signals. Based on the receptor type, biosensors are categorized into different types, such as antibody biosensors, DNA biosensors, enzyme biosensors, whole-cell biosensors, and phage biosensors. In contrast, according to transducers, these can be classified into electrochemical biosensors, piezoelectric biosensors, calorimetric biosensors, and optical biosensors, etc. (Wu et al. 2019).

14.7.1 Biosensor-Based Diagnostic Methods

Based on the bioreceptor type and transducer, the analyte biosensors can be classified into different types. Some biosensor types are discussed below (Arora et al. 2011).

14.7.2 Optical Biosensor

Based on the selectivity and sensitivity of this method, it is being used widely in pathogen detection. Fiber-optic biosensors are commercially available. The basic principle is that the pathogen or analyte is labeled with the fluorescence material. When the analyte hits, the biosensor gets excited by the laser waves at 635 nm. The fluorescent compound in conjugation with antibodies can be used to detect pathogenic bacteria. In this method, fluorescein isothiocyanate and some lanthanides are used as markers. Very useful methods were developed using fiber-optic biosensors and antibodies for detecting botulinum toxin, staphylococcal enterotoxin, *Listeria*, and *Salmonella*.

14.7.3 Surface Plasmon Resonance Biosensor

SPR biosensors are label-free techniques for monitoring biomolecular interactions in real time. It determines the specificity and affinity during the bond formation between protein-protein (Madeira et al. 2011), protein-nucleic acids (Majka and Speck 2007), enzyme-substrates (Fong et al. 2002), receptor-drug (Salamon et al. 2000), protein-lipid membrane/polysaccharides (Erb et al. 2000; Beccati et al. 2005), and cell/virus-protein (Miyoshi et al. 2006; Zhang et al. 2014a). This optical technique measures the changes in refractive index near thin metal layers (i.e., gold, silver, or aluminum films) in response to biomolecular interactions. When light hits a metal surface at a certain angle, the photon couples with an electron in the metal surface layer, which then move to excitation state, and the movement of electron is known as plasmon. This oscillation generates as electric field between metal surface and the target solution. If there is a change in the refractive index of the sensing medium, plasmon cannot be formed (Homola et al. 1999). This detection method detected whole cells of *E. coli* O157:H7, *Salmonella*, and *Listeria* at traceable concentrations. Also, small quantities of staphylococcal or botulinum toxins were detected (Zhang et al. 2014b).

14.7.4 Piezoelectric Biosensors

Piezoelectric biosensors are pathogen detection biosensors considered as mass-sensitive sensors, which can detect additional mass attached to the sensor and detect changes in the frequency of quartz crystal microbalance (QCM) (Lazcka et al. 2007). The sensor is generally coated with specific antibodies. When the bacteria bind with the antibodies, the mass on QCM increases the transducer surface. The probes were modified with protein-A antibody for detecting pathogen *Salmonella typhimurium* (Taitt et al. 2004).

14.7.5 Electrochemical DNA Biosensors

In electrochemical DNA biosensors, single-stranded nucleic acids or aptamers are used as receptors, and commonly used transducers are of gold electrodes (Nazari-Vanani et al. 2018). The basic principle of these biosensors is that the biological reaction between bioreceptor and analyte can produce electrons, which change the solution's electric current, potentiality, or other electrical properties. In this method, the bioreceptor is DNA, i.e., either naturally occurring DNA elements (nanosensors) or artificially synthesized aptamers (aptasensors) (Gaudin 2017). The probes immobilized on the electrode surface recognize and hybridize with target DNA by complementary base pairing.

Aptamers: Aptamers are single-stranded DNA/RNA molecules that bind to the target molecules as antigen-antibody molecules (Bini et al. 2008). In general, aptamers can target diverse molecules, as they can bind to a wide range of targets, including proteins, drugs, and organic compounds (Zhou et al. 2014). The main advantages of aptamers are that they can be amplified using PCR, very stable at extreme environmental conditions with a long lifetime and are economical for detection. When recognizing and interacting with target molecules, aptamers are folded into 3D form. The elongated primary structure of aptamer is normally unstable and interacts with each other and forms a tertiary structure. To avoid these problems, three different binding methods are being used in pathogen detection: (1) direct binding mechanism, the aptamer immobilized onto the electrode that binds to the pathogen directly and leads to conformational changes; (2) target-induced dissociation method, the aptamer is in a complementary state when no targets are available, and when the target binds to the aptamer, it releases from the self-complementary pairing and changes the electrochemical signal (Ge et al. 2018); and (3) dual aptamer detection mechanism, these are of sandwich-type aptasensors types, in which the first aptamer is immobilized as capture probe to bind with the target and the second aptamer acts as a signal probe. By using these methods, nanoporous glassy carbon electrodes to detect *Salmonella* DNA sequences (Amouzadeh Tabrizi and Shamsipur 2015), electrochemical DNA-based biosensors for detection of *Bacillus cereus* in milk and infant formula (Izadi et al. 2016), and label-free impedimetric biosensor to detect *S. typhimurium* in apple juice (Sheikhzadeh et al. 2016) were developed.

14.7.6 Impedimetric Biosensor

In an impedimetric biosensor, a small amplitude sinusoidal excitation in the system measures the changes in the electrical impedance of the medium. Based on the changes in conductance, capacitance and impedance analysis has to be done. In this biosensor, conductance and capacitance increase, whereas impedance decreases when applied to microbial metabolic processes (Ivnitski et al. 2000). These methods are more reliable diagnostic technologies in pathogen detection. The major application of this technique is to detect *Salmonella* among the various foodborne pathogens. In 1992, the Association of Analytical Communities International (AOAC) also accepted the impedance technique as the initial screening method for *Salmonella* in food (Bolton and Gibson 1995) and as a final action method for *Salmonella* in 1996.In addition, these methods are also useful for detecting Enterobacteriaceae, coliforms, and *L. monocytogenes*. Applying these methods can also analyze more samples in a single run. The main disadvantage of this technique is that the sensitivity of impedimetric sensors is less when compared to other sensors.

14.8 Microfluidic Diagnostic Technologies

Microfluidic methods are recognized as advanced platforms for detecting bloodstream pathogens. Pathogen concentration in the bloodstream is significantly less when compared with whole blood components. Moreover, many blood components have similar physical characteristicsto *E. coli*. These two issues made microfluidic methods of detecting and isolating pathogens from whole blood more difficult and challenging than detecting pathogens from other simple fluids. For rapid detection and isolation of pathogens, a variety of microfluidic methods were developed, which include acoustophoresis, dielectrophoresis, immunoaffinity-based methods, inertial fractionation, and adhesion-based separation methods, and all of these methods are considered as commonly applied approaches for on-chip microfluidic diagnosis (Cho et al. 2007; Hwang et al. 2008; Ai et al. 2013; Kang et al. 2014).

14.8.1 Antimicrobial Peptides as Diagnostic Tools

Antimicrobial peptides (AMP) are mainly used as diagnostic tools in patients suffering from infections by bearing biomedical devices like catheters, artificial heart valves, prosthetic joints, and other implants, as these patients develop a biomaterial-associated infection (BAI). Bacterial species such as *Staphylococcus aureus* and *Staphylococcus epidermidis* are the most causative organisms for BAI in patients (Zimmerli 2006). Bacterial biofilm formation and colonization around the tissue are considered to play a major role in the pathogenesis of BAI. As per the World Health Organization, bacteria belonging to ESKAPE panel (*Enterococcus faecium*, *Staphylococcus aureus*, *Klebsiella pneumoniae*, *Acinetobacter baumannii*, *Pseudomonas aeruginosa*, and *Enterobacter* spp.) are increasingly prevalent and resistant to antibiotics, thereby known as a specifically dangerous group of bacteria (Rice 2008). Antimicrobial peptides are one of the best methods to avoid the post-antibiotic era, especially in developing an antimicrobial drug with a different mode of action (World Health Organization 2015). Antimicrobial peptides have been used to design biosensors by the solid-phase peptide synthesis method. AMP binds to the receptor surface with the chemical groups such as spacers and anchors, and small molecule-sized peptides allow efficient immobilization on the sensor surface. The main advantage of these AMP biosensors is their stability, as they are more stable than enzymes and antibodies, even in harsh environments (Zasloff 2002) (Fig. 14.1).

The antimicrobial peptides magainin I is used in a fluorescence-based biosensor to detect *Escherichia coli* (*E. coli*) O157:H7 and *Salmonella typhimurium* (*S. typhimurium*). The binding of peptide on sensor surface allowed to bind 6.5×10^4 cells/mL for *E. coli* and *S. typhimurium*, respectively. Kulagina and colleagues (2006) also studied different antimicrobial peptides magainin I, cecropin A, parasin, polymyxin B, and polymyxin E with a different concentration on silanized glass slides against the same pathogens; they observed that the nonpathogenic *E.coli* did not interact with these peptides and concluded that AMP-based sensors could be

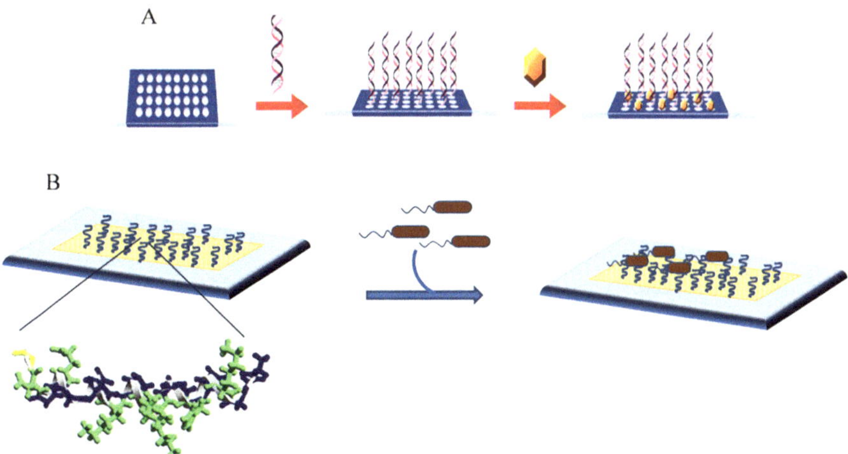

Fig. 14.1 Schematic representation of antimicrobial peptides' (AMPs') biosensor used for bacterial detection. (**a**) Design biosensors by the solid-phase peptide synthesis (AMP) covalently linked to the receptor surface with the spacers and anchors. (**b**) Binding of live bacterial cells to immobilized AMPs attched to biosensor chip, thus enabling the detection

used to differentiate bacterial species as well different strains in the same species. The selectivity and sensitivity of the AMPs can be improved by using nanomaterials as signal amplifiers (Cui et al. 2012; Hammond et al. 2016). Using carbon nanotubes and antimicrobial peptides to detect *Klebsiella pneumoniae*, *Enterococcus faecalis*, *Escherichia coli*, and *Bacillus subtilis*, they developed nanostructured biosensors based on carbon nanotubes and surface-immobilized clavanin A antimicrobial peptides. This method detected the bacterial concentration from 10^2 to 10^6 per ml and diagnosed gram-positive and gram-negative bacteria; clavanin A showed different affinities toward the pathogenic bacteria species (Andrade et al. 2015)..

14.8.2 Microbiome Engineering in Diseases Diagnostics

Most human diseases are aroused due to alterations in the gut microbiome population and metabolic alterations (Bober et al. 2018). Such diseases can be diagnosed by synthetically constructed bacteria engineered with genetically encoded sensors and genetic circuits. Bacteria naturally have evolved with several genetic markers, which can apparently be used as biosensors for diagnostic purposes against a wide range of diseases. This can be achieved by understanding the host's microbiome and biological process.

While genetically engineered bacteria are the potential to diagnose a variety of diseases associated with gastrointestinal tract, diabetes, cancer, and viral diseases (Lim and Song 2019), synthetic biology is widely used to engineer indigenous microbiota and probiotics for therapeutics to improve human health and various biomedical applications (Bober et al. 2018). Briefly, synthetic biology is a fast-growing discipline that evolved to develop programmed cellular behavior, which has enormous biomedical applications, including therapeutics. This can be achieved by using natural and synthetic biological means (Ullah et al. 2016). Previous studies on microbial communities focused on metagenomics and transcriptomics to identify genes expressed in microbes to find a correlation between microbes and health. Sequencing studies, however, limit our knowledge to understand individual microbes' relationship with their host fully. Recent developments in synthetic biology enable scientists to engineer bacteria with sensors, genetic circuits, and genes necessary for diagnostics and therapeutics. To achieve this facility, identification of host-adapted strains, stability, and performance of programmed microbe like effect of growth rate, detection threshold, circuit computation speed, and stability of genetic system of engendered organism must be considered. Engineering of smart probiotics with both capabilities of detection and treat diseases can be employed to reach such performance metrics (Landry and Tabor 2017). The details of the engineered bacteria in disease diagnosis and therapeutic are given in Table 14.2.

Microbial engineering is the technology that modifies microorganisms by manipulating the genes that are important in a trait. The developed recombinant technology and microbial application make a unique technology in microbial engineering. For example, a novel phenotype had developed by adopting a functional or

Table 14.2 Few examples of engeneered becteria for diagnosis and theraputics

Engeneered bacteria	Target recognization	Therapeutic compound	Disease targeted	Reference
E. coli BW25113	Fucose		*Citrobacter rodentium*	Pickard et al. (2014)
E. coli Nissle 1917	Tetrathionate		Colitis	Daeffler et al. (2017)
E. coli NGF-1	Tetrathionate		Colitis	Riglar et al. (2017)
E. coli Nissle 1917	Thiosulfate		Colitis	Daeffler et al. (2017)
Streptococcus thermophilus	Lactose		Diet sensor	Drouault et al. (2002)
E. coli NGF-1	ATC		Diet sensor	Kotula et al. (2014)
B. thetaiotaomicron	Arabinogalactan, IPTG, rhamnose		Diet sensor	Mimee et al. (2015)
Lactococcus lactis		Secreted hIL-10 (C) Secreted mIL-27 (C) Secreted hTFF1 (C) Secreted hIL-10; proinsulin (C)	Crohn's disease Colitis/IBD Oral mucositis Type 1 diabetes	Forkus et al. (2017), Hanson et al. (2014), Caluwaerts et al. (2010) and Takiishi et al. (2017)
E. coli E56b		Chimeric lipopolysaccharide	Cholera	Focareta et al. (2006)
L. lactis		Insulin	Diabetes	Ng and Sarkar (2011)
E. coli Nissle 1917		GLP-1	Diabetes	Duan et al. (2008)
E. coli MG1655		Bacteriocin	*P. aeruginosa*	Gupta et al. (2013)

phenotypic characteristic gene into a host gene for heterologous expression. The capability of synthesizing DNA molecules is a key technology for new rDNA-based applications in microbial research, such as synthetic biology, which has revolutionized microbial engineering, whereas molecular modeling technology anchors synthetic biology toward designing, fabricating, and installing de novo genetic devices and pathways in a plug-and-play fashion (Ellis et al. 2009). Microorganisms can grow quickly, potentially making them feasible for large and broad applications extended. For example, delivering drugs for cancer treatment; in colon cancer treatment, therapeutic products of bacteria hold them as emerging novel anticancer agents. In targeted cancer therapy, bacteria target cancer cells without affecting normal cells. Some other bacteria can multiply selectively in tumors and inhibit their growth (Yaghoubi et al. 2020).

14.9 Genetic Circuits and Encoded Sensors to Program Cellular Behavior

Genetic circuits are synthetically designed to control cellular behavior for diagnostic and therapeutic use. Genetic circuits consist of processors, sensors, and therapeutic actuators which act as switches, oscillators, logic gates, and biosensors to alter cellular behavior (Brophy and Voigt 2014). Component systems expressed in the bacteria are the naturally evolved sensors categorized into one-component (OCSs) and two-component systems (TCSs). Cytoplasmic transcription factors (OCSs) are allosterically altered by interacting with chemical or physical inputs, which are considered as large class of sensors on bacteria (Ulrich et al. 2005). This sensing capability can be used in gut microbiome-related applications. In an in vitro study, biologists have monitored communication between *E. coli* and *Pseudomonas aeruginosa* by altering the cellular communication of *E.coli* by using acyl-homoserine lactone quorum-sensing responding OCSs (Gupta et al. 2013). Various experiments have been conducted on dietary issues to sense lactose, fucose, rhamnose, and xylan by programing administered gut-adopted bacteria (Drouault et al. 2002; Bäckhed et al. 2004; Hamady et al. 2010; Mimee et al. 2016). However, it has limitations in applying biomedical gut-associated conditions like inflammations, and pH has become challenging. But a study was done to develop a gut inflammation EcN (*Escherichia coli* Nissle) biosensor to sense pro-inflammatory thiosulfate and tetrathionate. The programmed thiosulfate biosensor has successfully detected inflammation by expressing a florescent protein in inflammation-induced mouse GIT. Still, the tetrathionate sensor failed to detect the pro-inflammatory molecule (Daeffler et al. 2017).

14.10 Conclusions and Future Perspectives

A revolution in designing microbial diagnosis tools has occurred by using advanced technologies of the modern era to control disease-causing organisms. Though several technologies have erased the conventional culturing methods, these methods still have an important role in identifying and characterizing novel bacterial pathogens. PCR-based methods can be considered a boon to diagnose novel or outbreaks of viruses and bacteria. However, these PCR methods have limitations, especially since tests cannot be performed in required places and also the results cannot be obtained on time. The sensor-based and microfluidic methods can replace the molecular methods. Different sensor-based methods have been developed as per the requirement and targeted samples. The expanding genetic tools and knowledge in the field of diagnosis of the microbiome can create good opportunities for microbial engineering, diagnostic designs, and health applications.

References

Ai Y, Sanders CK, Marrone BL (2013) Separation of Escherichia coli bacteria from peripheral blood mononuclear cells using standing surface acoustic waves. Anal Chem 85:9126–9134. https://doi.org/10.1021/ac4017715

Albrecht V et al (2006) Easy and fast detection and genotyping of high-risk human papillomavirus by dedicated DNA microarrays. J Virol Methods 137:236–244. https://doi.org/10.1016/j.jviromet.2006.06.023

Amouzadeh Tabrizi M, Shamsipur M (2015) A label-free electrochemical DNA biosensor based on covalent immobilization of salmonella DNA sequences on the nanoporous glassy carbon electrode. Biosens Bioelectron 69:100–105. https://doi.org/10.1016/j.bios.2015.02.024

Andrade CA et al (2015) Nanostructured sensor based on carbon nanotubes and clavanin a for bacterial detection. Colloids Surf B Biointerfaces 135:833–839. https://doi.org/10.1016/j.colsurfb.2015.03.037

Arora P, Sindhu A, Dilbaghi N, Chaudhury A (2011) Biosensors as innovative tools for the detection of food borne pathogens. Biosens Bioelectron 28:1–12. https://doi.org/10.1016/j.bios.2011.06.002

Babamiri B, Salimi A, Hallaj R (2018) A molecularly imprinted electrochemiluminescence sensor for ultrasensitive HIV-1 gene detection using EuS nanocrystals as luminophore. Biosens Bioelectron 117:332–339. https://doi.org/10.1016/j.bios.2018.06.003

Bäckhed F et al (2004) The gut microbiota as an environmental factor that regulates fat storage. Proc Natl Acad Sci U S A 101:15718–15723. https://doi.org/10.1073/pnas.0407076101

Baldwin CD et al (2009) Usefulness of multilocus polymerase chain reaction followed by electrospray ionization mass spectrometry to identify a diverse panel of bacterial isolates. Diagn Microbiol Infect Dis 63:403–408. https://doi.org/10.1016/j.diagmicrobio.2008.12.012

Barreiros dos Santos M, Agusil JP, Prieto-Simón B, Sporer C, Teixeira V, Samitier J (2013) Highly sensitive detection of pathogen Escherichia coli O157:H7 by electrochemical impedance spectroscopy. Biosens Bioelectron 45:174–180. https://doi.org/10.1016/j.bios.2013.01.009

Beccati D et al (2005) SPR studies of carbohydrate-protein interactions: signal enhancement of low-molecular-mass analytes by organoplatinum(II)-labeling. Chembiochem 6:1196–1203. https://doi.org/10.1002/cbic.200400402

Bini A, Centi S, Tombelli S, Minunni M, Mascini M (2008) Development of an optical RNA-based aptasensor for C-reactive protein. Anal Bioanal Chem 390:1077–1086. https://doi.org/10.1007/s00216-007-1736-7

Bober JR, Beisel CL, Nair NU (2018) Synthetic biology approaches to engineer probiotics and members of the human microbiota for biomedical applications. Annu Rev Biomed Eng 20:277–300. https://doi.org/10.1146/annurev-bioeng-062117-121019

Bolton FJ, Gibson DM (1995) Automated electrical techniques in microbiological analysis. In: Patel PD (ed) Rapid analysis techniques in food microbiology. Springer, Boston, pp 131–169

Brophy JA, Voigt CA (2014) Principles of genetic circuit design. Nat Methods 11:508–520. https://doi.org/10.1038/nmeth.2926

Caglayan MO, Üstündağ Z (2020) Spectrophotometric ellipsometry based tat-protein RNA-aptasensor for HIV-1 diagnosis. Spectrochim Acta A Mol Biomol Spectrosc 227:117748. https://doi.org/10.1016/j.saa.2019.117748

Cai R, Yin F, Zhang Z, Tian Y, Zhou N (2019) Functional chimera aptamer and molecular beacon based fluorescent detection of Staphylococcus aureus with strand displacement-target recycling amplification. Anal Chim Acta 1075:128–136. https://doi.org/10.1016/j.aca.2019.05.014

Caluwaerts S et al (2010) AG013, a mouth rinse formulation of Lactococcus lactis secreting human trefoil factor 1, provides a safe and efficacious therapeutic tool for treating oral mucositis. Oral Oncol 46:564–570. https://doi.org/10.1016/j.oraloncology.2010.04.008

Chen Z-Z, Cai L, Chen M-Y, Lin Y, Pang D-W, Tang H-W (2015) Indirect immunofluorescence detection of E. coli O157:H7 with fluorescent silica nanoparticles. Biosens Bioelectron 66:95–102. https://doi.org/10.1016/j.bios.2014.11.007

Chiu CY, Miller SA (2019) Clinical metagenomics. Nat Rev Genet 20:341–355. https://doi.org/10.1038/s41576-019-0113-7

Cho YK, Lee JG, Park JM, Lee BS, Lee Y, Ko C (2007) One-step pathogen specific DNA extraction from whole blood on a centrifugal microfluidic device. Lab Chip 7:565–573. https://doi.org/10.1039/b616115d

Cho EJ, Do JH, Kim YS, Bae S, Ahn WS (2011) Evaluation of a liquid bead array system for high-risk human papillomavirus detection and genotyping in comparison with hybrid capture II, DNA chip and sequencing methods. J Med Microbiol 60:162–171. https://doi.org/10.1099/jmm.0.021642-0

Compton J (1991) Nucleic acid sequence-based amplification. Nature 350:91–92. https://doi.org/10.1038/350091a0

Coyte KZ, Schluter J, Foster KR (2015) The ecology of the microbiome: networks, competition, and stability. Science 350:663–666. https://doi.org/10.1126/science.aad2602

Cui Y, Kim SN, Naik RR, McAlpine MC (2012) Biomimetic peptide nanosensors. Acc Chem Res 45:696–704. https://doi.org/10.1021/ar2002057

D'Amico L, Ajami NJ, Adachi JA, Gascoyne PR, Petrosino JF (2017) Isolation and concentration of bacteria from blood using microfluidic membraneless dialysis and dielectrophoresis. Lab Chip 17:1340–1348. https://doi.org/10.1039/c6lc01277a

Daeffler KN, Galley JD, Sheth RU, Ortiz-Velez LC, Bibb CO, Shroyer NF (2017) Engineering bacterial thiosulfate and tetrathionate sensors for detecting gut inflammation. Mol Syst Biol 13:923. https://doi.org/10.15252/msb.20167416

de Miranda JL, Oliveira MDL, Oliveira IS, Frias IAM, Franco OL, Andrade CAS (2017) A simple nanostructured biosensor based on clavanin a antimicrobial peptide for gram-negative bacteria detection. Biochem Eng J 124:108–114. https://doi.org/10.1016/j.bej.2017.04.013

Drancourt M, Bollet C, Carlioz A, Martelin R, Gayral JP, Raoult D (2000) 16S ribosomal DNA sequence analysis of a large collection of environmental and clinical unidentifiable bacterial isolates. J Clin Microbiol 38:3623–3630

Drouault S, Anba J, Corthier G (2002) <em>Streptococcus thermophilus</em> is able to produce a β-Galactosidase active during its transit in the digestive tract of germ-free mice. Appl Environ Microbiol 68:938–941. https://doi.org/10.1128/aem.68.2.938-941.2002

Duan F, Curtis KL, March JC (2008) Secretion of Insulinotropic proteins by commensal bacteria: rewiring the gut to treat diabetes. Appl Environ Microbiol 74:7437–7438. https://doi.org/10.1128/aem.01019-08

Ellis T, Wang X, Collins JJ (2009) Diversity-based, model-guided construction of synthetic gene networks with predicted functions. Nat Biotechnol 27:465–471. https://doi.org/10.1038/nbt.1536

Erb EM et al (2000) Characterization of the surfaces generated by liposome binding to the modified dextran matrix of a surface plasmon resonance sensor chip. Anal Biochem 280:29–35. https://doi.org/10.1006/abio.1999.4469

Etayash H, Jiang K, Thundat T, Kaur K (2014) Impedimetric detection of pathogenic gram-positive bacteria using an antimicrobial peptide from class IIa bacteriocins. Anal Chem 86:1693–1700. https://doi.org/10.1021/ac4034938

Fahy E, Kwoh DY, Gingeras TR (1991) Self-sustained sequence replication (3SR): an isothermal transcription-based amplification system alternative to PCR. PCR Methods Appl 1:25–33. https://doi.org/10.1101/gr.1.1.25

Faria HAM, Zucolotto V (2019) Label-free electrochemical DNA biosensor for zika virus identification. Biosens Bioelectron 131:149–155. https://doi.org/10.1016/j.bios.2019.02.018

Focareta A, Paton JC, Morona R, Cook J, Paton AW (2006) A recombinant probiotic for treatment and prevention of cholera. Gastroenterology 130:1688–1695. https://doi.org/10.1053/j.gastro.2006.02.005

Fong CC, Lai WP, Leung YC, Lo SC, Wong MS, Yang M (2002) Study of substrate-enzyme interaction between immobilized pyridoxamine and recombinant porcine pyridoxal kinase using surface plasmon resonance biosensor. Biochim Biophys Acta 1596:95–107. https://doi.org/10.1016/s0167-4838(02)00208-x

Forkus B, Ritter S, Vlysidis M, Geldart K, Kaznessis YN (2017) Antimicrobial probiotics reduce salmonella enterica in Turkey gastrointestinal tracts. Sci Rep 7:40695. https://doi.org/10.1038/srep40695

Gaudin V (2017) Advances in biosensor development for the screening of antibiotic residues in food products of animal origin – a comprehensive review. Biosens Bioelectron 90:363–377. https://doi.org/10.1016/j.bios.2016.12.005

Ge C et al (2018) Target-induced aptamer displacement on gold nanoparticles and rolling circle amplification for ultrasensitive live salmonella typhimurium electrochemical biosensing. J Electroanal Chem 826:174–180. https://doi.org/10.1016/j.jelechem.2018.07.002

Grahn N, Olofsson M, Ellnebo-Svedlund K, Monstein HJ, Jonasson J (2003) Identification of mixed bacterial DNA contamination in broad-range PCR amplification of 16S rDNA V1 and V3 variable regions by pyrosequencing of cloned amplicons. FEMS Microbiol Lett 219:87–91. https://doi.org/10.1016/s0378-1097(02)01190-4

Gupta S, Bram EE, Weiss R (2013) Genetically programmable pathogen sense and destroy. ACS Synth Biol 2:715–723. https://doi.org/10.1021/sb4000417

Hall TA et al (2009) Rapid molecular genotyping and clonal complex assignment of Staphylococcus aureus isolates by PCR coupled to electrospray ionization-mass spectrometry. J Clin Microbiol 47:1733–1741. https://doi.org/10.1128/jcm.02175-08

Hamady ZZR et al (2010) Xylan-regulated delivery of human keratinocyte growth factor-2 to the inflamed colon by the human anaerobic commensal bacterium <em>bacteroides ovatus</em>. Gut 59:461–469. https://doi.org/10.1136/gut.2008.176131

Hammond JL, Formisano N, Estrela P, Carrara S, Tkac J (2016) Electrochemical biosensors and nanobiosensors. Essays Biochem 60:69–80. https://doi.org/10.1042/ebc20150008

Hanson ML et al (2014) Oral delivery of IL-27 recombinant bacteria attenuates immune colitis in mice. Gastroenterology 146:210–221.e213. https://doi.org/10.1053/j.gastro.2013.09.060

Homola J, Yee SS, Gauglitz G (1999) Surface plasmon resonance sensors: review. Sensors Actuators B Chem 54:3–15. https://doi.org/10.1016/S0925-4005(98)00321-9

Hoyos-Nogués M et al (2016) Impedimetric antimicrobial peptide-based sensor for the early detection of periodontopathogenic bacteria. Biosens Bioelectron 86:377–385. https://doi.org/10.1016/j.bios.2016.06.066

Hwang KY et al (2008) Bacterial DNA sample preparation from whole blood using surface-modified Si pillar arrays. Anal Chem 80:7786–7791. https://doi.org/10.1021/ac8012048

Ivnitski D, Abdel-Hamid I, Atanasov P, Wilkins E, Stricker S (2000) Application of electrochemical biosensors for detection of food pathogenic bacteria. Electroanalysis 12:317–325

Izadi Z, Sheikh-Zeinoddin M, Ensafi AA, Soleimanian-Zad S (2016) Fabrication of an electrochemical DNA-based biosensor for Bacillus cereus detection in milk and infant formula. Biosens Bioelectron 80:582–589. https://doi.org/10.1016/j.bios.2016.02.032

Jin M et al (2012) A rapid subtractive immunization method to prepare discriminatory monoclonal antibodies for food E. coli O157:H7 contamination. PLoS One 7:e31352. https://doi.org/10.1371/journal.pone.0031352

Kaleta EJ et al (2011) Use of PCR coupled with electrospray ionization mass spectrometry for rapid identification of bacterial and yeast bloodstream pathogens from blood culture bottles. J Clin Microbiol 49:345–353. https://doi.org/10.1128/jcm.00936-10

Kang DK et al (2014) Rapid detection of single bacteria in unprocessed blood using integrated comprehensive droplet digital detection. Nat Commun 5:5427. https://doi.org/10.1038/ncomms6427

Kordas A, Papadakis G, Milioni D, Champ J, Descroix S, Gizeli E (2016) Rapid salmonella detection using an acoustic wave device combined with the RCA isothermal DNA amplification method. Sens BioSensing Res 11:121–127. https://doi.org/10.1016/j.sbsr.2016.10.010

Kotula JW et al (2014) Programmable bacteria detect and record an environmental signal in the mammalian gut. Proc Natl Acad Sci 111:4838–4843. https://doi.org/10.1073/pnas.1321321111

Kulagina NV, Shaffer KM, Anderson GP, Ligler FS, Taitt CR (2006) Antimicrobial peptide-based array for Escherichia coli and salmonella screening. Anal Chim Acta 575:9–15. https://doi.org/10.1016/j.aca.2006.05.082

Kulagina NV, Shaffer KM, Ligler FS, Taitt CR (2007) Antimicrobial peptides as new recognition molecules for screening challenging species. Sens Actuators B Chem 121:150–157. https://doi.org/10.1016/j.snb.2006.09.044

Landry BP, Tabor JJ (2017) Engineering diagnostic and therapeutic gut bacteria. Microbiol Spectr 5:5. https://doi.org/10.1128/microbiolspec.BAD-0020-2017

Lavania S et al (2018) Aptamer-based tb antigen tests for the rapid diagnosis of pulmonary tuberculosis: potential utility in screening for tuberculosis. ACS Infect Dis 4:1718–1726. https://doi.org/10.1021/acsinfecdis.8b00201

Lazcka O, Campo FJD, Muñoz FX (2007) Pathogen detection: a perspective of traditional methods and biosensors. Biosens Bioelectron 22:1205–1217. https://doi.org/10.1016/j.bios.2006.06.036

Leustean AM et al (2018) Implications of the intestinal microbiota in diagnosing the progression of diabetes and the presence of cardiovascular complications. J Diabetes Res 2018:5205126. https://doi.org/10.1155/2018/5205126

Li J, Convertino M (2019) Optimal microbiome networks: macroecology and criticality. Entropy 21:506. https://doi.org/10.3390/e21050506

Li Y et al (2014) Impedance based detection of pathogenic E. coli O157:H7 using a ferrocene-antimicrobial peptide modified biosensor. Biosens Bioelectron 58:193–199. https://doi.org/10.1016/j.bios.2014.02.045

Li L, Li Q, Liao Z, Sun Y, Cheng Q, Song Y (2018) Magnetism-resolved separation and fluorescence quantification for near-simultaneous detection of multiple pathogens. Anal Chem 90:9621–9628. https://doi.org/10.1021/acs.analchem.8b02572

Lim D, Song M (2019) Development of bacteria as diagnostics and therapeutics by genetic engineering. J Microbiol 57:637–643. https://doi.org/10.1007/s12275-019-9105-8

Lloyd-Price J, Abu-Ali G, Huttenhower C (2016) The healthy human microbiome. Genome Med 8:51. https://doi.org/10.1186/s13073-016-0307-y

Loewe RP, Nelson PJ (2011) Microarray bioinformatics. Methods Mol Biol 671:295–320. https://doi.org/10.1007/978-1-59745-551-0_18

World Health Organization (2015) Global action plan on antimicrobial resistance. In: In. World Health Organization, Geneva

Ma X, Jiang Y, Jia F, Yu Y, Chen J, Wang Z (2014) An aptamer-based electrochemical biosensor for the detection of salmonella. J Microbiol Methods 98:94–98. https://doi.org/10.1016/j.mimet.2014.01.003

Madeira A, Vikeved E, Nilsson A, Sjögren B, Andrén PE, Svenningsson P (2011) Identification of protein-protein interactions by surface plasmon resonance followed by mass spectrometry. Curr Protoc Protein Sci Chapter 19(Unit19):21. https://doi.org/10.1002/0471140864.ps1921s65

Majka J, Speck C (2007) Analysis of protein-DNA interactions using surface plasmon resonance. Adv Biochem Eng Biotechnol 104:13–36

Mannoor MS, Zhang S, Link AJ, McAlpine MC (2010) Electrical detection of pathogenic bacteria via immobilized antimicrobial peptides. Proc Natl Acad Sci U S A 107:19207–19212. https://doi.org/10.1073/pnas.1008768107

Martin CR, Osadchiy V, Kalani A, Mayer EA (2018) The brain-gut-microbiome axis. Cell Mol Gastroenterol Hepatol 6:133–148. https://doi.org/10.1016/j.jcmgh.2018.04.003

Miller MB, Tang YW (2009) Basic concepts of microarrays and potential applications in clinical microbiology. Clin Microbiol Rev 22:611–633. https://doi.org/10.1128/cmr.00019-09

Mimee M, Tucker Alex C, Voigt Christopher A, Lu Timothy K (2015) Programming a human commensal bacterium, bacteroides thetaiotaomicron, to sense and respond to stimuli in the murine gut microbiota. Cell Systems 1:62–71. https://doi.org/10.1016/j.cels.2015.06.001

Mimee M, Tucker Alex C, Voigt Christopher A, Lu Timothy K (2016) Programming a human commensal bacterium, bacteroides thetaiotaomicron, to sense and respond to stimuli in the murine gut microbiota. Cell Systems 2:214. https://doi.org/10.1016/j.cels.2016.03.007

Miyoshi H et al (2006) Binding analyses for the interaction between plant virus genome-linked protein (VPg) and plant translational initiation factors. Biochimie 88:329–340. https://doi.org/10.1016/j.biochi.2005.09.002

Nazari-Vanani R, Sattarahmady N, Yadegari H, Heli H (2018) A novel and ultrasensitive electrochemical DNA biosensor based on an ice crystals-like gold nanostructure for the detection of enterococcus faecalis gene sequence. Colloids Surf B Biointerfaces 166:245–253. https://doi.org/10.1016/j.colsurfb.2018.03.025

Ng DTW, Sarkar CA (2011) Nisin-inducible secretion of a biologically active single-chain insulin analog by Lactococcus lactis NZ9000. Biotechnol Bioeng 108:1987–1996. https://doi.org/10.1002/bit.23130

Ngwa GA, Schop R, Weir S, León-Velarde CG, Odumeru JA (2013) Detection and enumeration of E. coli O157:H7 in water samples by culture and molecular methods. J Microbiol Methods 92:164–172. https://doi.org/10.1016/j.mimet.2012.11.018

Pabbaraju K, Tokaryk KL, Wong S, Fox JD (2008) Comparison of the luminex xTAG respiratory viral panel with in-house nucleic acid amplification tests for diagnosis of respiratory virus infections. J Clin Microbiol 46:3056–3062. https://doi.org/10.1128/jcm.00878-08

Pickard JM et al (2014) Rapid fucosylation of intestinal epithelium sustains host–commensal symbiosis in sickness. Nature 514:638–641. https://doi.org/10.1038/nature13823

Prieto M, Colin P, Fernández-Escámez P, Alvarez-Ordóñez A (2015) Epidemiology, detection, and control of foodborne microbial pathogens. Bio Med Res Int 2015:617417–617417. https://doi.org/10.1155/2015/617417

Procop GW (2007) Molecular diagnostics for the detection and characterization of microbial pathogens. Clin Infect Dis 45(Suppl 2):S99–s111. https://doi.org/10.1086/519259

Rice LB (2008) Federal funding for the study of antimicrobial resistance in nosocomial pathogens: no ESKAPE. J Infect Dis 197:1079–1081. https://doi.org/10.1086/533452

Riglar DT et al (2017) Engineered bacteria can function in the mammalian gut long-term as live diagnostics of inflammation. Nat Biotechnol 35:653–658. https://doi.org/10.1038/nbt.3879

Saad SM, Abdullah J (2019) A fluorescence quenching based gene assay for Escherichia coli O157:H7 using graphene quantum dots and gold nanoparticles. Mikrochim Acta 186:804. https://doi.org/10.1007/s00604-019-3913-8

Sabounchi P, Morales AM, Ponce P, Lee LP, Simmons BA, Davalos RV (2008) Sample concentration and impedance detection on a microfluidic polymer chip. Biomed Microdevices 10:661–670. https://doi.org/10.1007/s10544-008-9177-4

Salamon Z, Cowell S, Varga E, Yamamura HI, Hruby VJ, Tollin G (2000) Plasmon resonance studies of agonist/antagonist binding to the human delta-opioid receptor: new structural insights into receptor-ligand interactions. Biophys J 79:2463–2474. https://doi.org/10.1016/s0006-3495(00)76489-7

Savas S, Altintas Z (2019) Graphene quantum dots as nanozymes for electrochemical sensing of Yersinia enterocolitica in milk and human serum. Materials 12:2189. https://doi.org/10.3390/ma12132189

Savory N et al (2014) Selection of DNA aptamers against uropathogenic Escherichia coli NSM59 by quantitative PCR controlled cell-SELEX. J Microbiol Methods 104:94–100. https://doi.org/10.1016/j.mimet.2014.06.016

Sheikhzadeh E, Chamsaz M, Turner APF, Jager EWH, Beni V (2016) Label-free impedimetric biosensor for salmonella typhimurium detection based on poly [pyrrole-co-3-carboxyl-pyrrole] copolymer supported aptamer. Biosens Bioelectron 80:194–200. https://doi.org/10.1016/j.bios.2016.01.057

Sibley CD, Peirano G, Church DL (2012) Molecular methods for pathogen and microbial community detection and characterization: current and potential application in diagnostic microbiology. Infect Genet Evol 12:505–521. https://doi.org/10.1016/j.meegid.2012.01.011

Singh A, Poshtiban S, Evoy S (2013) Recent advances in bacteriophage based biosensors for foodborne pathogen detection. Sensors 13:1763–1786. https://doi.org/10.3390/s130201763

Suehiro J, Hamada R, Noutomi D, Shutou M, Hara M (2003) Selective detection of viable bacteria using dielectrophoretic impedance measurement method. J Electrost 57:157–168. https://doi.org/10.1016/S0304-3886(02)00124-9

Suh SH, Choi SJ, Dwivedi HP, Moore MD, Escudero-Abarca BI, Jaykus LA (2018) Use of DNA aptamer for sandwich type detection of listeria monocytogenes. Anal Biochem 557:27–33. https://doi.org/10.1016/j.ab.2018.04.009

Sypabekova M, Jolly P, Estrela P, Kanayeva D (2019) Electrochemical aptasensor using optimized surface chemistry for the detection of mycobacterium tuberculosis secreted protein MPT64 in human serum. Biosens Bioelectron 123:141–151. https://doi.org/10.1016/j.bios.2018.07.053

Taitt CR, Shubin YS, Angel R, Ligler FS (2004) Detection of salmonella enterica serovar typhimurium by using a rapid, array-based immunosensor. Appl Environ Microbiol 70:152–158. https://doi.org/10.1128/aem.70.1.152-158.2004

Takiishi T et al (2017) Reversal of diabetes in NOD mice by clinical-grade proinsulin and IL-10-secreting Lactococcus lactis in combination with low-dose anti-CD3 depends on the induction of Foxp 3-positive T cells. Diabetes 66:448–459. https://doi.org/10.2337/db15-1625

Teengam P, Siangproh W, Tuantranont A, Vilaivan T, Chailapakul O, Henry CS (2018) Electrochemical impedance-based DNA sensor using pyrrolidinyl peptide nucleic acids for tuberculosis detection. Anal Chim Acta 1044:102–109. https://doi.org/10.1016/j.aca.2018.07.045

Trausch JJ, Shank-Retzlaff M, Verch T (2017) Development and characterization of an HPV Type-16 specific modified DNA Aptamer for the improvement of potency assays. Anal Chem 89:3554–3561. https://doi.org/10.1021/acs.analchem.6b04852

Ullah MW, Khattak WA, Ul-Islam M, Khan S, Park JK (2016) Metabolic engineering of synthetic cell-free systems: strategies and applications. Biochem Eng J 105:391–405. https://doi.org/10.1016/j.bej.2015.10.023

Ulrich LE, Koonin EV, Zhulin IB (2005) One-component systems dominate signal transduction in prokaryotes. Trends Microbiol 13:52–56. https://doi.org/10.1016/j.tim.2004.12.006

Vandecasteele SJ, Frans J, Van Ranst M (2002) Contamination management of broad-range ribosomal DNA PCR: where is the evidence? J Clin Microbiol 40:3885–3886. https://doi.org/10.1128/jcm.40.10.3885-3886.2002

Walker GT, Fraiser MS, Schram JL, Little MC, Nadeau JG, Malinowski DP (1992) Strand displacement amplification--an isothermal, in vitro DNA amplification technique. Nucleic Acids Res 20:1691–1696. https://doi.org/10.1093/nar/20.7.1691

Wang Z, Yang Q, Zhang Y, Meng Z, Ma X, Zhang W (2018) Saltatory rolling circle amplification (SRCA): a novel nucleic acid isothermal amplification technique applied for rapid detection of Shigella Spp. in vegetable salad. Food Anal Methods 11:504–513. https://doi.org/10.1007/s12161-017-1021-0

Wu R et al (2018) Efficient capture, rapid killing and ultrasensitive detection of bacteria by a nano-decorated multi-functional electrode sensor. Biosens Bioelectron 101:52–59. https://doi.org/10.1016/j.bios.2017.10.003

Wu Q, Zhang Y, Yang Q, Yuan N, Zhang W (2019) Review of electrochemical DNA biosensors for detecting food borne pathogens. Sensors 19:4916. https://doi.org/10.3390/s19224916

Xi Z, Gong Q, Wang C, Zheng B (2018) Highly sensitive chemiluminescent aptasensor for detecting HBV infection based on rapid magnetic separation and double-functionalized gold nanoparticles. Sci Rep 8:9444. https://doi.org/10.1038/s41598-018-27792-5

Yager P et al (2006) Microfluidic diagnostic technologies for global public health. Nature 442:412–418. https://doi.org/10.1038/nature05064

Yaghoubi A, Khazaei M, Avan A, Hasanian SM, Soleimanpour S (2020) The bacterial instrument as a promising therapy for colon cancer. Int J Color Dis 35:595–606. https://doi.org/10.1007/s00384-020-03535-9

Yoon T, Moon HS, Song JW, Hyun KA, Jung HI (2019) Automatically controlled microfluidic system for continuous separation of rare bacteria from blood. Cytometry A 95:1135–1144. https://doi.org/10.1002/cyto.a.23909

Zasloff M (2002) Antimicrobial peptides of multicellular organisms. Nature 415:389–395. https://doi.org/10.1038/415389a

Zhang H et al (2014a) Investigation of biological cell-protein interactions using SPR sensor through laser scanning confocal imaging-surface plasmon resonance system. Spectrochim Acta A Mol Biomol Spectrosc 121:381–386. https://doi.org/10.1016/j.saa.2013.10.100

Zhang X et al (2014b) Development of a simultaneous detection method for foodborne pathogens using surface plasmon resonance biosensors. Food Sci Technol Res 20:317–325. https://doi.org/10.3136/fstr.20.317

Zhou L, Wang J, Li D, Li Y (2014) An electrochemical aptasensor based on gold nanoparticles dotted graphene modified glassy carbon electrode for label-free detection of bisphenol a in milk samples. Food Chem 162:34–40. https://doi.org/10.1016/j.foodchem.2014.04.058

Zhou Y et al (2018) Gut microbiota offers universal biomarkers across ethnicity in inflammatory bowel disease diagnosis and infliximab response prediction. mSystems 3:e00188–e00117. https://doi.org/10.1128/mSystems.00188-17

Zhou P et al (2020) A pneumonia outbreak associated with a new coronavirus of probable bat origin. Nature 579:270–273. https://doi.org/10.1038/s41586-020-2012-7

Zimmerli W (2006) Prosthetic-joint-associated infections. Best Pract Res Clin Rheumatol 20:1045–1063. https://doi.org/10.1016/j.berh.2006.08.003

Microbiome Therapeutics: Emerging Concepts and Challenges in Translational Microbial Research

15

Mani Jayaprakashvel, Swarnakala Thamada, Kuraganti Gunaswetha, and Veera Bramhachari Pallaval

Abstract

The microbiome is an umbrella terminology that refers to an ecological community of symbiotic and pathogenic microorganisms. The microbiome is crucial in ecology, environment, human health, and disease. The human microbiome is one of the largest microbial communities of bacteria. The human microbiome has extensive functions such as the development of immunity, defense against pathogens, host nutrition, synthesis of vitamins, and fat storage, making it an essential body organ without which we would not function correctly. Hence, attempts have been to leverage the associations between the microbiome and the human host. Microbiome therapeutics implies the modulation of microbiomes with microbes, microbiomes, molecules, and synthetics. Besides, treatment of some internal diseases or ailments through microbiomes is also a factor in microbiome therapeutics. The microbiome therapeutic approach is used to diagnose diseases, alter microbial community ecology, and enhance the production of inhibitory or therapeutic proteins. The agents or principles used in microbiome therapeutics include genetically engineered probiotics, engineered consortia, chemicals, peptides, bacteriophages, bacteriocins, small-molecule antibiotics, etc. Recent advances in the field of synthetic biology and our knowledge and understanding of the host-associated microbiome have enhanced the scope and applications of microbiome therapeutics. However, there have been many challenges to be

M. Jayaprakashvel · Swarnakala Thamada
Department of Marine Biotechnology, AMET Deemed to be University, Kanathur, Tamil Nadu, India

K. Gunaswetha
Department of Microbiology, Kakatiya University, Warangal, Telangana, India

V. B. Pallaval (✉)
Department of Biosciences & Biotechnology, Krishna University, Machilipatnam, Andhra Pradesh, India

P. Veera Bramhachari (ed.), *Human Microbiome in Health, Disease, and Therapy*, https://doi.org/10.1007/978-981-99-5114-7_15

addressed. Many research studies on microbiome therapeutics were done on animal models. Being very distinctive to every individual, the suitability of the research answers to the personalized usage of microbiome therapeutics is a major challenge. A comprehensive assay model needs to be developed. Genetic knowledge of the facts and principles behind the host-microbiome association is essential for the implementation of microbiome therapeutics. In the personalized therapeutic era, the development of suitable biosensors and genetic circuits is essentially needed for the development of completely autonomous cellular therapies. A translational approach to developing basic scientific research into technologically intense applied products is the need of the hour. In this context, this chapter would provide an insight into the basic science behind microbiome therapeutics, the challenges faced in the field, and the scope and diverse applications of microbiome therapeutics.

Keywords

Microbiome · Therapeutics · Microbiota · Synthetic biology · Human

15.1 Introduction

The microbiome evolves from the first moments of life and gradually diversifies along with the physiological development of the host individual (Sommer et al. 2017). The microbiome is an umbrella terminology that refers to an ecological community of a variety of symbiotic and pathogenic microorganisms (Tipton et al. 2019). Microbiome composition is influenced by several environmental factors, including breastfeeding, mode of delivery, gestational age, diet, and antibiotic use (Dong and Gupta 2019). The microbiome is crucial in ecology, environment, human health, and disease. The human microbiome is the collection of microorganisms in and on the human body, such as bacteria, viruses, archaea, and single-celled eukaryotes (Dekaboruah et al. 2020). The human microbiome has extensive functions such as the improvement of immunity, defense against pathogens, host nutrition, synthesis of vitamins, and fat storage, making it an essential body organ without which we would not function correctly (Amon and Sanderson 2017). However, an imbalance in commensal microbiota has been associated with a wide range of medical conditions such as diabetes (Giongo et al. 2011), irritable bowel syndrome (Saulnier et al. 2011), and different types of cancer (Schwabe and Jobin 2013). Hence, attempts have been made to leverage the associations between the microbiome and the human host.

The human gut microbiome is a huge supporter of human digestion and prosperity, comprising of trillions of microorganisms colonized in the human gut (Milani et al. 2017). The microbiota in the gastrointestinal tract can play out an assortment of capacities for the human host, including physiological, nutritional, and immunological capacities independent of the host's inborn assets. As a result, the gut microbiome is viewed as a human organ with special capacities and intricacies. Because of the trouble of culturing many of these gut microbial species in lab conditions, gut microbiome research has been restricted before (Brüssow 2020). The advancement of next-generation sequencing-based metagenomics has empowered us to acquire a superior comprehension of the gut microbiome's diversity, structure, and roles in human well-being and illness. Such genome-driven high-throughput strategies, then again, offer minimal unthinking bits of knowledge into how gut microbiota communicate with one another and with the host and what these connections mean for the host metabolome.

Microbiome therapeutics implies the modulation of microbiomes with either microbes, microbiomes, molecules, or synthesis (Li et al. 2017). Besides, treatment of some internal diseases or ailments through microbiomes is also a factor in microbiome therapeutics. The microbiome therapeutic approach is used to diagnose diseases, alter microbial community ecology, and enhance the production of inhibitory therapeutic proteins (Mimee et al. 2016). The agents or principles used in microbiome therapeutics include genetically engineering probiotics, engineered consortia, chemicals, peptides, bacteriophages, bacteriocins and small-molecule antibiotics, etc. In this context, this chapter would provide insight into the basic science behind microbiome therapeutics and the diverse applications of microbiome therapeutics.

15.2 Microbiome

The microorganisms living on or inside another organism are called a "microbiome" (Riiser 2015). They can be sorted as symbiotic or pathogenic depending on how they communicate with one another and their host. From birth to death, the microbiome calibrates itself inside a steady host to maintain homeostatic equilibrium with the immune system (Rooks and Garrett 2016). The adaptive and innate immune systems, just as external factors such as medication, diet, and toxin exposure, impact the human microbiome's advancement after birth. In people, the microbiome represents 90% of the cells in a 10:1 proportion (Ley et al. 2006). As indicated by ongoing research, the quantity of microorganisms in the body is comparable to the number of human cells (Sender et al. 2016). Most of these microorganisms are gut inhibitors. The microbiome adds numerous qualities to the human genome, possibly growing it by 200 overlaps (Maurice et al. 2013). As an outcome, the human microbiome's organization could be fundamental regarding health and sickness.

15.3 Human Microbiome

Notwithstanding how people have almost indistinguishable hereditary makeup, the slight contrasts in DNA bring about colossal phenotypic variety across the populace. The human microbiome's metagenome, or absolute DNA substance of organisms occupying our bodies, is substantially more factor, with only 33% of its constituent genes present in most healthy people. The full genome of all microbes in and on the human body is called the human microbiome, which normally refers to the population of all microorganisms in this habitat (Rajpoot et al. 2018). The human microbiome consists of 10–100 trillion symbiotic microbial cells in which the trillions of microbes would struggle to break down essential nutrients (O'Neal 2017). At the point when we liken the human gene list to our microbiota, we find that our human hereditary variety could not hope to compare. There are 3.3 million qualities in the gut alone, which houses most of our microbiota, contrasted with 22,000 qualities in the human genome (Munro 2016).

15.4 Gut Microbiome

The human gastrointestinal tract is a mind-boggling biological system of billions of organisms. Changes in supplement supply, pH, oxygen fixation, and bile salts impact the overall abundance of bacteria, archaea, protists, and parasites from all domains of life (Carey and Assadi-Porter 2017). Bacteria are the most very concentrated of these microorganisms, representing most of the DNA arrangements and biomass. This bacterial gathering is likewise fundamental for mammalian gut physiology, as it helps metabolic capacities, ensures against microbes, and balances the immune system. Likewise, the bacterial community's composition and plenitude changes are much connected to infections like colorectal cancer, inflammatory bowel disease, obesity, and neurological issues. The powers that structure the design of these bacterial species' design are still ineffectively comprehended, which has frustrated the creation of microbiome-based therapeutics.

The human gut microbiome hugely supports human digestion and prosperity, with trillions of microorganisms colonizing the human gut (Milani et al. 2017). Given the significance of the gut microbiome in human well-being, a superior comprehension of the transient elements of intestinal microbial species, just like the host, is required (Lamont et al. 2018). The microbiota in the gastrointestinal tract can play out an assortment of capacities for the human host, including physiological, dietary, and immunological capacities that are discrete from the host's intrinsic assets (Ewald and Sumner 2018). Accordingly, the gut microbiome is viewed as a human organ with exceptional capacities and intricacies. Because of the trouble of refining large numbers of these gut microbial species in lab conditions, gut microbiome research has been restricted previously. The advancement of next-generation sequencing-based metagenomics has empowered us to better comprehend the gut microbiome's structure, variety, and parts in human well-being and illness. Such genome-driven high-throughput strategies, then again, offer minimal robotic experiences into how gut microbiota communicate with each other and with the host and how these connections identify with the host metabolic machinery.

15.5 Microbiome and Human Health

The gut microbiome is fundamental for human well-being (Fig. 15.1), as it helps the host collect and store energy through different metabolic functions (Clemente et al. 2012). The gut microbiota was characterized by two gatherings of good health: commensal pathobionts and commensal symbionts. Commensal microbes have for quite some time been perceived as valuable to have physiology by providing essential supplements and securing against opportunistic pathogen colonization (Kamada et al. 2013). Ecological factors, for example, way of life and a Westernized diet, have generally affected the gut microbiota (Voreades et al. 2014). Shockingly, there is mounting proof that natural components like obesity and diet are connected to improving colorectal cancer (CRC). Since the reason for irregular CRC is obscure, a person's gut microbiota can address dietary propensities that advance or ensure against the disease (Bultman 2017).

The bacterial synthesis of the gut assists with safeguarding the mucosal and systemic immunity homeostasis of its hosts, forestalling invulnerable triggers that may cause physiological hindrance. Dysbiosis is an illness set apart by adjusting the gut

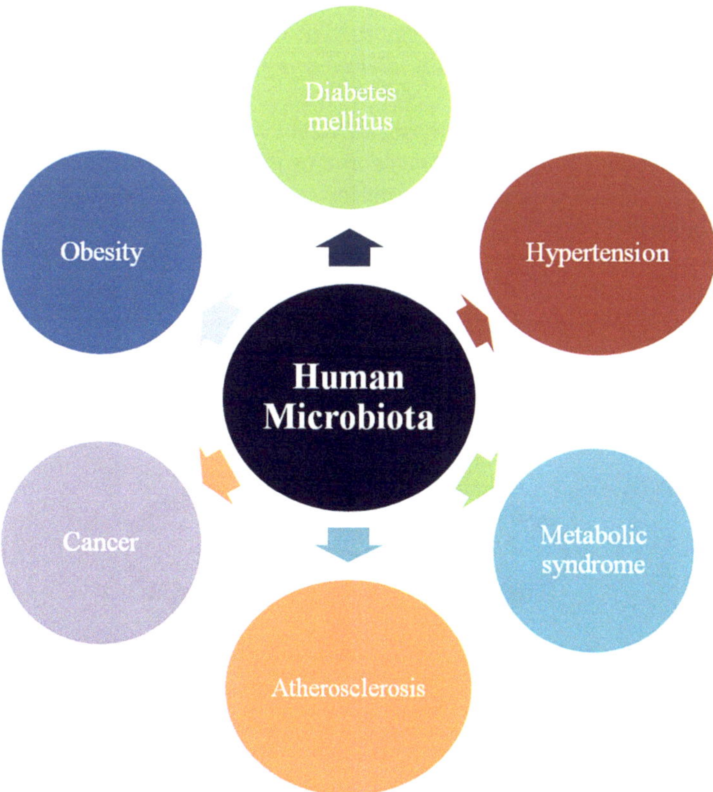

Fig. 15.1 Microbiome and human health

commensal microbiota toward opportunistic pathogenic microorganisms (Barman et al. 2008). As indicated by an ongoing examination, dysbiosis influences an assortment of physiological capacities and irritates the colon, which raises the danger of CRC (Nistal et al. 2015). Numerous new examinations have uncovered that the gut microbiome is significant in oncogenesis, where their relationship with the immune system can either keep the host healthy or trigger tumor progression (Gagliani et al. 2014).

Alongside acknowledging the gut microbiome's role in health and infection, progresses in next-generation sequencing have brought about various forward leaps in phylogenetic, taxonomic, and functional profiling of the gut microbiome. A metagenomic way to deal with gut microbiome profiling has the additional advantage of giving answers on metagenome populace characterization and physiological impacts on the human host.

15.6 Microbiome Therapeutics

In various gastrointestinal and non-gastrointestinal diseases, the microbiome has become a more appealing focus for future therapeutics (Fig. 15.2). Exogenous administration of live microbes within the gut is the aim of current microbiome-based therapeutics. These slants are referred to as probiotics, and they have grown in popularity over the last decade. Nonetheless, determining cause-effect relationships and designing microbiome-based therapies to achieve predictable effects on the microbial community and host health is a major challenge in microbiome study. Due to less evidence to support the efficacy of probiotics, an alternative approach termed prebiotics was evolved. Prebiotics are compounds that are consumed to induce the growth or activity of beneficial microbes (Böger et al. 2019). The most communal example is in the gastrointestinal tract, where prebiotics can alter the composition of organisms in the gut microbiome. Further studies are warranted to

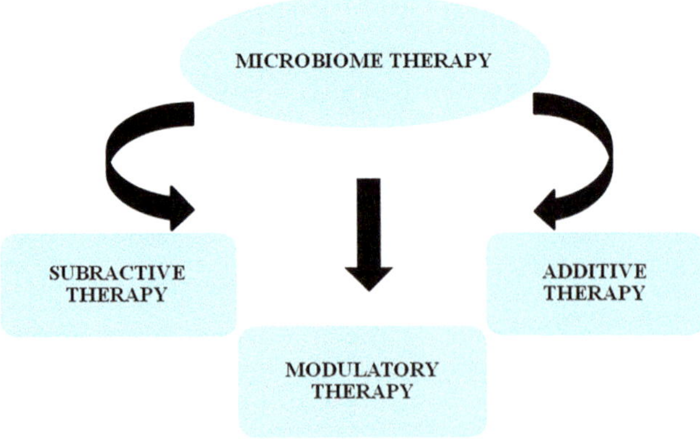

Fig. 15.2 Microbiome therapy

complete the characterization of the effect of prebiotics on different bacterial species (Holscher 2017).

15.7 Efforts to Harness and Engineer the Microbiota

The microbiome's importance to human health and disease has been established through several decades of study. Recent attempts to capitalize on this relationship to develop microbiome-based therapies have been made. There are three major paradigms for these treatments: modulatory, subtractive, and additive therapies. The host-microbiota is, however, replaced with individual strains or a mixture of natural and engineered microorganisms in additive care. To treat the disease, subtractive therapies lead to the specific elimination of harmful microbiome members. Nonetheless, modulatory treatments require the utilization of nonliving synthetic substances, for example, prebiotics, to modify the construction or activity of the endogenous microbiome. The original microbiome treatments, probiotics, and prebiotics have been widely investigated elsewhere. With recombinant probiotics, particular antimicrobials, and microbial consortia, we will zero in with the microbiome treatment. Up to this point, restorative bacterial consortia were mostly comprised of regular strains, though genetically engineered bacteria have generally been utilized as monotherapies. Genetically engineered networks, then again, can be intended to join the variety and vigor of microbial consortia with the additional adequacy and controllability of synthetic gene circuits. We'll begin with ongoing progressions in microbiome treatment and proceed to the incredible difficulties of transforming the microbiota's capacity into natural treatments.

15.8 Genetically Engineered Probiotics

The rationale of probiotic therapies is that naturally occurring human-associated microbes provide myriad of health benefits. Many diseases have been scientifically proven to be remedied, such as oral ingestion of *Lactobacillus* spp., *E. coli*, and *Bifidobacterium* spp. The use of cells as drug delivery vehicles may allow for in situ biotherapeutic production, addressing issues like bioavailability and drug inactivation that can occur with oral administration. Protein therapy synthesis may depend upon identifying and incorporating particular disease-related environmental indications. This conditional, on-demand release of drugs is an especially enticing advantage of cell-based therapies, which could require for new pharmacological paradigms. If the therapeutic organism can stably colonize the host, the engineered microbe may be able to dynamically correct disease-induced disturbances and restore homeostasis in the host. The idea of completely autonomous "smart" cell-based therapy is still far off. The ongoing examination has shown that cell-based bacterial treatments are powerful in forestalling infection, treating metabolic disorders, and reducing inflammation.

Natural colonization resistance offered by native members of normal flora has been improved, thanks to cellular engineering. Ozdemir et al. (2018) looked into probiotic *E. coli* for prophylaxis; Nissle 1917 was created to reduce *Vibrio cholerae* virulence in infant mouse models. Extracellular quorum sensing molecules, which modulate density-dependent gene expression, play a role in *V. cholerae* virulence. At the point when bacterial numbers are low, *V. cholerae* communicates the virulence factors expected to infect the host; when numbers are high, the virulence factors are quelled, permitting the bacteria to get away. *E. coli* has been intended to meddle with quorum sensors, forestalling disease. The utilization of therapeutic cells brought about an expansion in endurance just as a diminishing in bacterial weight and cholera toxin articulation. In a rhesus macaque model, hereditarily altered *Lactobacillus jensenii* was used to forestall the chimeric simian/human immunodeficiency infection (SHIV) transmission. Bacteria are genetically engineered to produce the antiviral cyanovirin-N. Despite many problems, prophylactic macaque therapy reduced both the incidence of SHIV and the peak viral load. As a result, engineered bacteria may be used to treat both bacterial and viral infections. IBD has gotten a ton of consideration as a promising possibility for cell-based treatments because of the role of gut microbiota in infection and the absence of long-term, cost-effective treatments.

Preliminary studies looked at using *Lactococcus lactis* to secrete recombinant interleukin-10 (IL-10), a powerful anti-inflammatory cytokine that is deficient in IBD patients. Breyner et al. (2019) used chemical and genetically induced mouse models of colitis to show that recombinant microorganisms may be used to reduce pathology and suppress pro-inflammatory cytokine secretion. The recombinant *L. lactis* therapy was also well tolerated in a small Crohn's disease population in Phase I clinical trials, though efficacy was modest. IL-10-secreting *L. lactis* was further modified to develop either autoantigenic proinsulin or glutamic acid decarboxylase-65 to treat autoimmune diabetes. When utilized for anti-CD3 treatment, both recombinant species had the option to induce tolerance, increment the number of administrative T cells, and opposite hyperglycemia in mice. Microbial improvement of mitigating cytokines, including changing development factor-β1 and against tumor necrosis factor α-nanobodies, just as the tissue repair factor keratinocyte development factor-2, has appeared to protect mice from colitis in IBD models. Furthermore, the protease inhibitor elafin produced by lactic acid bacteria has been shown to restore proteolytic homeostasis and protect against inflammation in mouse colitis models.

Oral mucositis, a condition described by ulcerative lesions and a successive symptom of chemotherapy, was additionally treated with recombinant bacteria. In hamster models, effective utilization of *L. lactis* designed to emit factor-1 was powerful in treating oral mucositis. Early clinical preliminary discoveries demonstrate that the drug is much endured and might be viable in lessening occurrence. Recombinant cell therapies can be useful therapeutic agents for treating inflammation, according to this report. In treating metabolic disorders, including obesity and diabetes, recombinant microbes have been successfully incorporated into the host microbiota. Probiotics gave standard taking care of *E. coli* changed to integrate anorexigenic lipid precursors decreased weight, adiposity, and food utilization in mice with high-fat eating regimens. Such defensive impacts were kept up for weeks after bacterial treatment ended. *Lactobacillus gasseri* has additionally been utilized as a delivery medium for GLP-1, a protein that makes intestinal epithelial cells convert into insulin-producing cells. In a rodent model, the organization of the designed probiotic expanded the number of cells producing intestinal insulin while additionally bringing down hyperglycemia.

15.9 Microbiome Engineered Consortia

Designing the entire microbial populace is another strategy to treating intermittent *C. difficile* infection that has had much achievement in the facility. A fecal microbiota transplant, which includes injecting stool from healthy donors into unhealthy patients, has a triumph pace of more than 90% in settling intermittent diseases and is more than twice as compelling as an antimicrobial treatment alone. Regardless of their clinical achievement, fecal microbiota transplants present security worries inspired by a paranoid fear of pathogens or opportunists being presented, which could worsen the sickness. A regulatory structure and strict donor screening guidelines were established, but deciphering the minimum subset of therapeutic microbes was a priority for mitigating safety issues and treatment reliability. Except for the occasional *C. difficile* infections, many agree that fecal microbiota transplants are promising in treating IBD and early studies have shown moderate success. Due to the more complex nature of the disease and a higher frequency of adverse effects, more research is required to determine if stool transplants or infusions of proven microbial communities are feasible treatment option for IBD. Identifying and tailoring microbial communities that can tackle the dynamics of human disease and human-associated microbiota diversity will continue to be an ongoing challenge in the production of microbiota-based therapies. In mouse models, the reconstitution of microbiota has proven effective in altering urea's metabolic activity across the population. Systemic ammonia accumulation correlates with neurotoxicity and encephalopathy in hepatic deficient patients. In a model of hepatic injury, the reclassified microbiota expanded endurance and shielded against cognitive deficits brought about by hyperammonemia. This examination exhibits the attainability of reasonably chiseling a host-related microbial culture to secure against metabolic illnesses.

The construction and capacity of the gut microbiome have critical ramifications for human health. The gut epithelium's respectability, energy equilibrium, and host immune reactions are influenced by the huge network of intestinal microbial metabolites (Fessler et al. 2019). While certain genera are known to prevail in many adults' microbiomes, the variety of microorganisms colonizing the human digestive tract is an exceptional factor, especially at the species level. A dysbiosis of the gut microbiota, or the breakdown of homeostasis among destructive and defensive intestinal bacteria, can be connected to and even reason for those infections (Verdugo-Meza et al. 2020). These alterations were related, among others to inflammatory bowel disease (IBD), diabetes, asthma, obesity, and allergy (Gholizadeh et al. 2019).

Efforts were made to recover toxic microbiomes using probiotics. Probiotics are live microorganisms that provide beneficial health effects when ingested in adequate quantities (Kleerebezem et al. 2019). They've appeared to improve manifestations of pouchitis, bacterial diarrhea, peevish inside disorder, *Helicobacter pylo*ri disease, *Clostridium difficile* contamination, and antimicrobial-related diarrhea (Cameron et al. 2017). However, they also colonize the host only transiently and are not maintained in the long run (Zhang et al. 2016). In addition, current probiotics are not intended to treat a particular condition but provide general health benefits. This issue raises the potential to use genetic engineering to create more functional probiotics capable of generating substances important for treating specific conditions. With enhanced awareness of the gut microbiome and the role of different keystone microbes in our health, coupled with the creation of modern synthetic biology tools, probiotic microorganisms were engineered to diagnose and treat inflammation of the intestines. These microorganisms were designed to detect in situ inflammation-related biomarkers with sensitivity and precision. Besides, live biotherapeutics were made with various capacities going from constitutive therapeutic specialist articulation to more complex sensing components. This audit aims to give a state-of-the-art outline of late advancements in the finding and treatment of inflammatory bowel diseases utilizing live biotherapeutics.

15.10 Microbiome-Targeted Drug Delivery

Microorganisms with distinct, targeted therapeutic roles are the objective of another field of microbial drug development. Our capacity to design works all the more accurately on account of the approach of synthetic biology strategies, just as our capacity to comprehend the mechanistic effect of microorganisms on human health because of microbiome study, has powered this arranged activity (Mimee et al. 2016). Drug companies normally plan their medications for systemic absorption to accomplish the ideal outcomes. Drug companies forming treatments for conveyance into the intestinal microbiome, then again, commonly mean to restrict systemic absorption and keep restorative impacts inside the digestion tracts' lumen. Various contemplations are required for these two entirely opposed strategies for successful medication convenience (Bristol and Hubert 2019). Creating therapeutics that is dynamic against the intestinal microbiota requires managing them to the substance and lining of the digestion tracts in an operational structure. Managing prescriptions to the intestines accompanies its assortment of troubles. Most of the materials in the digestive tract are catabolized into basic structure impedes, that can also be retained and utilized by the body (Finlay and Finlay 2019).

15.10.1 Synthetic Microbes as Optimistic Drug Delivery System

Synthetic cell therapy is a region with a great deal of guarantee for treating human diseases later on. Designed bacterial strains equipped for diagnosing illness, creating and distributing therapeutics, and checking their numbers to meet regulation and security concerns will make up next-generation therapies (Claesen and Fischbach 2015). The advantages of in vivo synthesis and delivery through cell therapy are numerous, for example, by reducing the requisite dose of the therapeutic agent by many orders of magnitude to achieve a comparable therapeutic effect or by reducing undesired side effects, both at the site of delivery and elsewhere throughout the body (Senapati et al. 2018). Multiple therapeutic agents can be generated by the same cell simultaneously as a combination therapy, and certain diseases of the gastrointestinal tract could be treated by oral administration of a synthetic bacterium that can traverse to the target site, engraft, and start delivering a medication (Claesen and Fischbach 2015). Notwithstanding the numerous benefits, manufactured cell treatment raises issues encompassing well-being, regulation, and the general assessment of utilizing hereditarily altered life forms in medication. Flow instances of utilized modules incorporate recognizing small molecules or cell markers connected to a particular human disease (Verstraelen et al. 2016). A serious level of molecular particularity, affectability, and dose-dependency are alluring attributes of such a module. When detection systems for the ideal molecules or conditions are accessible, they can be utilized to assemble diagnostic sensing modules (Chellappan et al. 2019). The synthesis and dispersion of the active compound is an introductory module in a synthetic therapeutic system. Heterologous articulation of a therapeutic protein, a gene cluster encoding the biosynthesis of a small molecule, or knockdown of eukaryotic gene expression by bacterially delivered small RNAs is a regular strategy for production (Trosset and Carbonell 2015).

15.11 Conclusions and Future Perspectives

Microbial species' tremendous effects on human health are opening up new diagnostic and therapeutic avenues for disease treatment. Conversely, current therapeutic methods for modulating microbiomes in the clinic are still fairly crude. Improved designing ways to deal with permit the alteration of an expansive scope of bacterial hosts, just as the advancement of disease-relevant sensors that can drive contingent heterologous therapeutic yield can aid in implementing microbiome therapeutics in the real world. A solid comprehension of the fundamental molecular basis for human infections and the environment of sound bacterial populaces is basic for the headway of manufactured bacterial "physicians." Not only can these bits of knowledge assist with the advancement of explicit therapeutic systems, but engineered science can assist with these key investigations. The determination of a powerful suspension that engrafts steadily and discusses profitably with the occupant populace in novel body specialties would profit from a point-by-point comprehension of the microbial environment of the human body and the collaboration of microbes with the immune system.

Acknowledgments The authors gratefully acknowledge AMET Deemed to be University, Tamil Nadu, India, and
Krishna University, Machilipatnam, for the support extended.

Conflict of Interest The authors declare that they have no competing interests.

References

Amon P, Sanderson I (2017) What is the microbiome? Arch Dis Child Educ Pract 102(5):257–260
Barman M, Unold D, Shifley K, Amir E, Hung K, Bos N, Salzman N (2008) *Enteric salmonellosis* disrupts the microbial ecology of the murine gastrointestinal tract. Infect Immun 76(3):907–915
Böger M, van Leeuwen SS, Lammerts van Bueren A, Dijkhuizen L (2019) Structural identity of galactooligosaccharide molecules selectively utilized by single cultures of probiotic bacterial strains. J Agric Food Chem 67(50):13969–13977
Breyner NM, Vilas Boas PB, Fernandes G, de Carvalho RD, Rochat T, Michel ML et al (2019) Oral delivery of pancreatitis-associated protein by *Lactococcus lactis* displays protective effects in dinitro-benzenesulfonic-acid-induced colitis model and is able to modulate the composition of the microbiota. Environ Microbiol 21(11):4020–4031
Bristol, A., & Hubert, S (2019). Formulation development of microbiome therapeutics localized in the intestines.
Brüssow H (2020) Problems with the concept of gut microbiota dysbiosis. Microb Biotechnol 13(2):423–434
Bultman SJ (2017) Interplay between diet, gut microbiota, epigenetic events, and colorectal cancer. Mol Nutr Food Res 61(1):1500902
Cameron D, Hock QS, Musal Kadim NM, Ryoo E, Sandhu B, Yamashiro Y et al (2017) Probiotics for gastrointestinal disorders: proposed recommendations for children of the Asia-Pacific region. World J Gastroenterol 23(45):7952
Carey HV, Assadi-Porter FM (2017) The hibernator microbiome: host-bacterial interactions in an extreme nutritional symbiosis. Annu Rev Nutr 37:477–500
Chellappan DK, Ning QLS, Min SKS, Bin SY, Chern PJ, Shi TP et al (2019) Interactions between microbiome and lungs: paving new paths for microbiome based bio-engineered drug delivery systems in chronic respiratory diseases. Chem Biol Interact 310:108732
Claesen J, Fischbach MA (2015) Synthetic microbes as drug delivery systems. ACS Synth Biol 4(4):358–364
Clemente JC, Ursell LK, Parfrey LW, Knight R (2012) The impact of the gut microbiota on human health: an integrative view. Cell 148(6):1258–1270
Dekaboruah E, Suryavanshi MV, Chettri D, Verma AK (2020) Human microbiome: an academic update on human body site specific surveillance and its possible role. Arch Microbiol 202:1–21
Dong TS, Gupta A (2019) Influence of early life, diet, and the environment on the microbiome. Clin Gastroenterol Hepatol 17(2):231–242
Ewald DR, Sumner SC (2018) Human microbiota, blood group antigens, and disease. Wiley Interdiscip Rev Syst Biol Med 10(3):e1413
Fessler J, Matson V, Gajewski TF (2019) Exploring the emerging role of the microbiome in cancer immunotherapy. J Immunother Cancer 7(1):108
Finlay BB, Finlay JM (2019) The whole-body microbiome: how to harness microbes—inside and out—for lifelong health. The Experiment
Gagliani N, Hu B, Huber S, Elinav E, Flavell RA (2014) The fire within: microbes inflame tumors. Cell 157(4):776–783

Gholizadeh P, Mahallei M, Pormohammad A, Varshochi M, Ganbarov K, Zeinalzadeh E et al (2019) Microbial balance in the intestinal microbiota and its association with diabetes, obesity and allergic disease. Microb Pathog 127:48–55
Giongo A, Gano KA, Crabb DB, Mukherjee N, Novelo LL, Casella G et al (2011) Toward defining the autoimmune microbiome for type 1 diabetes. ISME J 5:82–91
Holscher HD (2017) Dietary fiber and prebiotics and the gastrointestinal microbiota. Gut Microbes 8(2):172–184
Kamada N, Chen GY, Inohara N, Núñez G (2013) Control of pathogens and pathobionts by the gut microbiota. Nat Immunol 14(7):685–690
Kleerebezem M, Binda S, Bron PA, Gross G, Hill C, van Hylckama Vlieg JE et al (2019) Understanding mode of action can drive the translational pipeline towards more reliable health benefits for probiotics. Curr Opin Biotechnol 56:55–60
Lamont RJ, Koo H, Hajishengallis G (2018) The oral microbiota: dynamic communities and host interactions. Nat Rev Microbiol 16(12):745–759
Ley RE, Peterson DA, Gordon JI (2006) Ecological and evolutionary forces shaping microbial diversity in the human intestine. Cell 124(4):837–848
Li X, Watanabe K, Kimura I (2017) Gut microbiota dysbiosis drives and implies novel therapeutic strategies for diabetes mellitus and related metabolic diseases. Front Immunol 8:1882
Maurice CF, Haiser HJ, Turnbaugh PJ (2013) Xenobiotics shape the physiology and gene expression of the active human gut microbiome. Cell 152(1–2):39–50
Milani C, Duranti S, Bottacini F, Casey E, Turroni F, Mahony J et al (2017) The first microbial colonizers of the human gut: composition, activities, and health implications of the infant gut microbiota. Microbiol Mol Biol Rev 81(4):e00036–e00017
Mimee M, Citorik RJ, Lu TK (2016) Microbiome therapeutics—advances and challenges. Adv Drug Deliv Rev 105:44–54
Munro N (2016) Gut microbiota: its role in diabetes and obesity. Diabetes Prim Care 18(4):1–6
Nistal E, Fernández-Fernández N, Vivas S, Olcoz JL (2015) Factors determining colorectal cancer: the role of the intestinal microbiota. Front Oncol 5:220
O'Neal, L. (2017). The human gastrointestinal microbiome: insights into the microbial community structure and antibiotic resistome
Ozdemir T, Fedorec AJ, Danino T, Barnes CP (2018) Synthetic biology and engineered live biotherapeutics: toward increasing system complexity. Cell Syst 7(1):5–16
Rajpoot M, Sharma AK, Sharma A, Gupta GK (2018) Understanding the microbiome: emerging biomarkers for exploiting the microbiota for personalized medicine against cancer. Semin Cancer Biol 52:1–8
Riiser A (2015) The human microbiome, asthma, and allergy. Allergy, Asthma Clin Immunol 11(1):35
Rooks MG, Garrett WS (2016) Gut microbiota, metabolites and host immunity. Nat Rev Immunol 16(6):341–352
Saulnier DM, Riehle K, Mistretta TA, Diaz MA, Mandal D, Raza S et al (2011) Gastrointestinal microbiome signatures of pediatric patients with irritable bowel syndrome. Gastroenterology 141:1782–1791
Schwabe RF, Jobin C (2013) The microbiome and cancer. Nat Rev Cancer 13:800–812
Senapati S, Mahanta AK, Kumar S, Maiti P (2018) Controlled drug delivery vehicles for cancer treatment and their performance. Signal Transduct Target Ther 3(1):1–19
Sender R, Fuchs S, Milo R (2016) Revised estimates for the number of human and bacteria cells in the body. PLoS Biol 14(8):e1002533
Sommer F, Anderson JM, Bharti R, Raes J, Rosenstiel P (2017) The resilience of the intestinal microbiota influences health and disease. Nat Rev Microbiol 15(10):630–638
Tipton L, Darcy JL, Hynson NA (2019) A developing symbiosis: enabling cross-talk between ecologists and microbiome scientists. Front Microbiol 10:292
Trosset JY, Carbonell P (2015) Synthetic biology for pharmaceutical drug discovery. Drug Des Devel Ther 9:6285

Verdugo-Meza A, Ye J, Dadlani H, Ghosh S, Gibson DL (2020) Connecting the dots between inflammatory bowel disease and metabolic syndrome: a focus on gut-derived metabolites. Nutrients 12(5):1434

Verstraelen H, Vervaet C, Remon JP (2016) Rationale and safety assessment of a novel intravaginal drug-delivery system with sustained DL-lactic acid release, intended for long-term protection of the vaginal microbiome. PLoS One 11(4):e0153441

Voreades N, Kozil A, Weir TL (2014) Diet and the development of the human intestinal microbiome. Front Microbiol 5:494

Zhang C, Derrien M, Levenez F, Brazeilles R, Ballal SA, Kim J et al (2016) Ecological robustness of the gut microbiota in response to ingestion of transient food-borne microbes. ISME J 10(9):2235–2245

Insights on the New-Generation Technologies and Role of Bioinformatics Tools to Understand Microbiome Research and the Microbial World

16

Satyanagalakshmi Karri, Manohar Babu Vadela, and Vijay A. K. B. Gundi

Abstract

Over the decades, omics-based research intensely changed our understanding of microorganisms and their importance in animal and human life. Different strategies are exploited to understand microbes and their mechanisms in various environments. The microbiome research enhanced our knowledge from understanding the microbes to surprising associations with Parkinson's disease and depression. It is also understood that the microbiome can change health status by influencing the life bodies regarding several diseases such as allergies, cancer, cardiometabolic disorders, and obesity. In such cases, nutrients, metabolites, and microorganisms play an important role. Moreover, microbial characteristics can change rapidly with environmental conditions, including temperature and air. The progress of microbiome research depends on designing and standardizing methodologies and protocols, modifying existing procedures, or adapting novel technologies and models. Thus, it is necessary to develop standard protocols for microbiome research, including ideal protocols for sampling microbiomes and their data analysis. To develop standard protocols for metagenomics, Human Microbiome Project Consortium (NIH) established quality-controlled high-throughput metagenomic data for scientific communities. The questions raised toward microbiome standard protocol are developing an ideal protocol for collecting a microbiome sample for analysis and proper tools for data analysis. In this chapter, we described the new-generation technologies and capabilities of bioinformatic tools to understand the microbiome and microbial world.

S. Karri · M. B. Vadela · V. A. K. B. Gundi (✉)
Department of Biotechnology, Vikrama Simhapuri University, Nellore, Andhra Pradesh, India

P. Veera Bramhachari (ed.), *Human Microbiome in Health, Disease, and Therapy*, https://doi.org/10.1007/978-981-99-5114-7_16

Keywords
Microbiome · Protocol · Next-generation sequencing · Standards · Benchwork challenges

16.1 Introduction

Depending on the type of study and specimen collections, the microbial research process leads to a variability between labs and results in bias microbial profile, sometimes complete loss of the required important information. For example, in a publication from an American gut project (McDonald et al. 2018), the unwanted bacteria grew that flourished due to the protocol followed to collect the fecal sample, transportation, and compromises with the quality of the microbiome profile estimation. So, immediate sample preservation is important because it plays a key role in the results. The sample collection should be static to avoid temperature fluctuations and freeze-thawing issues from collecting samples for DNA extraction.

DNA extraction is also the most malign source of variability from the microbial population in a sample. In metagenomic analysis, DNA extraction is the most significant factor to consider, such as the size of the microorganisms, cellular structure, and methods used for lysis (Costea et al. 2017). For example, few DNA extraction methods favor gram-negative bacteria more than gram-positive bacteria due to their cell wall so these species will be underestimated in the resulting analysis. Other microbiota, such as yeast, are also difficult to lysis. DNA amplification is also an important step in creating a genomic library. The amplification of DNA also sometimes produces false results. For example, metagenomic studies can be conducted based on the small ribosomal subunit called 16S rDNA genome-based sequencing. In the region selected for the amplification and sequencing for complete microbial population diversity, commonly used primers may amplify bacterial species but may lose the data of other microbiomes. The bioinformatic tools used to classify bacteria based on sequencing, 11 tools interpreting shotgun metagenomic data, concluded with different conclusions (McIntyre et al. 2017).

The rising interest in microbial research has impacted experimental practice; microbial research has been flagged with improved molecular and analytical tools. The progression in the research area of gut microbiome depends on the computational and other developed techniques. Researchers weigh the strengths and weaknesses of their methods. So, “normalize our methods to the science,” says Cano. Vast data produced in microbial studies has to be analyzed using bioinformatic tools; statistical tools help to evaluate microbial association. After characterization and functional modeling of the experiments, the in vitro and in vivo experiments should be established to consider the cause and effect of microbial communities in pathogenicity. An important step was established to regulate the marketing of microbial products by adopting the standard operating procedures (SOPs) and also important in producing reliable, enabling comparisons between studies.

16.2 Applications of DNA Sequencing Technologies

Microbial identification and genotyping of microbes are carried out by marker gene amplification and shotgun metagenomics.

In the early days, microbial characterization was performed by conventional culture methods; this studying of genotyping was difficult. For the last 25 years, molecular sequencing has been the primary method for identifying and taxonomically classifying microbes and commonly used marker genes16S ribosomal RNA genome for bacterial sequencing and internal transcribed spacer (ITS) region for fungal identification.

The highly conserved 16S rRNA gene (~1542 bp) consists of nine variable regions (V1–V9). Most 16S rRNA-based genotyping protocols use V5–V6, V3–V4, or V4 hypervariable regions for bacterial taxonomic classification (Woo et al. 2008). Identification of rare species and phenotypic characterization using these marker genes can identify >90% of cases; 65–83% of these sequences are identified at the species level (Drancourt et al. 2000). Woo PC et al. studied that amplification of a first 527-bp fragment of the 16S rRNA genes using conventional sequencing method and sequencing also biochemical profile was generated. Among 37 clinically significant bacterial strains, 37 aerobic gram-positive, gram-negative, and *Mycobacterium* species were characterized. MicroSeq 500 16S rRNA-based bacterial sequencing method also identified 30 (81.1%) of them properly, and five (13.5%) isolates were misidentified at the genus level. The misidentification of these isolates was due to a lack of the 16S rRNA gene sequences of these bacteria in the database of the MicroSeq 500 16S rDNA-based bacterial identification system. This study concluded the MicroSeq 500 16S rDNA-based method for bacterial identification, but MicroSeq 500 16S rDNA-based databases must be expanded to cover the rare bacterial species (Woo et al. 2003).

16.3 Protocols for Microbial Analysis

16.3.1 Sample Collection Methods

The first step in microbial analysis is establishing the proper pipelines for sample collection. Samples, environmental samples such as water, soil, or animals, and normal controls are collected from the patients. Proper handling of the samples, like transportation, temperature, and transport time, can promote microbes' growth. The sample collection allows the transfer of contaminated microbes from the environment to the laboratory, is an important step to prevent a fundamental awareness of sample collection from various specimens.

16.3.2 The Basic Protocol for Environmental Sample Collection

The standard method for a sample collection from the environment contains the collection of samples, which are kept with ice packs or dry ice and then transported to the laboratory. Recently, a biopolymer called acacia gum was replaced with the methods mentioned above; it uses as a preservative agent in collected samples without damage during transport. For example, a sample is filtered, and the filter with microbes is immobilized in a biopolymer solution and transferred to the laboratory in small ziplock bags. This method is significantly used in the environmental sampling process. In this method, the transport weightage can be reduced, and loss or damage of the microbial population can be controlled; thereby, chance of false-positive/negative results can be controlled. In an experiment, a water sample with *E. coli* transported with biopolymer increased the efficiency by 260% and viability by four times (Krumnow et al. 2009). The samples can be recovered by adding water, and microbes can be alive for the 6 months at room temperature and up to 16 months at refrigerated temperatures in the immobilized biopolymers (Krumnow et al. 2009; Sorokulova et al. 2012). It mainly involves immobilization and polymerization; it provides structural integrity to the microbial cells and protects them from mechanical stress during transportation. It stabilizes keeping in a dormant state by decreasing the microbial cells' metabolic rate, water retention, and molecular stability. It protects from structural changes and releases mechanical stress.

The air samples are collected by filtering the large volumes of air using pumps; the filtered samples should be positioned at the bottom of the petri dish, and add liquid polymer. If the water is of interest, the sample is collected using a dipper, added to a volumetric flask, and filtered through a syringe filter; the filter should be kept in the liquid polymer. Soil samples were inoculated using sterile water and followed the procedure as a liquid sample (Sorokulova et al. 2015).

16.3.3 The Basic Protocol for Clinical Sample Collection

Clinical samples are collected from patients; it is also important to collect normal samples from healthy humans as the control. Clinical samples are collected from different body parts of the patients for diagnostic purposes. However, oral, fecal, and virgin clinical samples are predominant in microbial research for comparative, probiotic development and engineering the indigenous organism to improve health.

Oral sample collection. Oral samples can be collected in two ways*: one is the collection of saliva in sterile collection tubes, and the other* is the collection of swabs by swabbing inside the cheeks.

16.4 Materials

Collection tubes: Many sterile plastic tubes are commercially available, but researchers use 50 mL conical tubes in some laboratories.

Swabs: Any sterile cotton can be used, and today, suitable collection swabs are available and can be obtained from scientific companies like Thermo Fisher Scientific.

16.4.1 Directions to Collect the Saliva

1. Wait 30–40 min before eating, drinking, or chewing anything.
2. Take the sterile plastic or conical tube and open it.
3. Carefully split into tubes until the required quantity of liquid saliva is reached (5 mL is enough).
4. Close the tube with its original lid and store it at −20 or −80° until the sample is processed.

16.4.2 Directions to Collect Swabs

1. Carefully insert the cotton swabs into the mouth and rub smoothly on the inner surface of the cheeks to collect microbes attached to the cheeks.
2. Store the swabs at −20 or −80° until further processing.

16.5 Basic Protocol on Media Preparation for Bacteria Culturing

For the required media (minimal media/selective media), prepare in a flask and heat till all the media components dissolve. Transfer into a separate bottle and autoclave at 15 lb./in2 for 15 min. Antibiotic and mineral solutions should be added after reaching a temperature < 50 °C, mixed thoroughly, and plated into petri plates. The streak plate method has to be carried out for single colony isolation. Characterization of isolated colonies is carried out using biochemical methods. The conventional culture methods have not been detailed here; the molecular techniques used for genotyping are explained in this book chapter.

16.6 Molecular Techniques for Analysis of Microbes

16.6.1 Applications of Molecular Techniques in Microbial Analysis

Due to the developed technologies, microarray, next-generation sequencing, etc. the complex microbial communities that cause diseases in a human were studied vastly. The questions raised to study mainly a group of microbes in any region, such as organ specific, habitat, or any specific environment, are further studied to identify genes and their products in a habitat. These investigations enable the creation of the metagenomic field; it helps bypass the conventional culture analysis of microbiota to characterize the metabolic and functional activity of the microbial community in an ecosystem (Riesenfeld et al. 2004). Metagenomics is the analysis of genetic material extracted directly from environmental samples (Tringe and Rubin 2005). These developed technologies improved our knowledge of the diversity and function of microbes and developed applications in human health and agriculture. In human health, research focuses on the impact of microbial communities on the disease; also, understanding the imbalance in the microbial communities and individual species is related to different disease states; altered gut microbes leads to obesity (Maruvada et al. 2017), diabetes (Jamshidi et al. 2019), cardiovascular disease (Ahmadmehrabi and Tang 2017), cancer (Scott et al. 2019), and other neurodegenerative conditions (Shen and Ji 2019). In agriculture, the soil- and plant-associated bacteria and its benefits in the growth and resilience of crops have been studied, further potentially improving the crop yield.

Protocol for amplification and sequencing involves the following steps: (1) DNA extraction or isolation from the bacterial cell culture. (2) quantification of the DNA and amplification, (3) sequencing and taxonomical classification, and (4) functional or metabolic or rare phenotypic character gene identification using computational methods.

16.6.2 Basic Protocol on DNA Isolation

Marmur developed a protocol for DNA extraction from bacteria (Marmur 1961). It was an invaluable contribution to microbiology and has been modified by researchers for their specific needs (Amaro et al. 2008; Ogg and Patel 2009; Adelskov and Patel 2016). We present a modified DNA extraction method from Wright et al. (2017).

Reagents

1. 10% sodium dodecyl sulfate (SDS): It solubilizes cell membrane lipids.
2. RNase A (100 U/mL) dissolves in 50 mM Tris-Cl (pH 8.0), 10 mM EDTA: Degrades single-stranded RNA.
3. Achromopeptidase (50 kU/mL): Lysis of gram-positive bacterial cell walls (if the target bacteria is gram-negative only).
4. Lysozyme (24,000 kU/mL): Lysis of enzyme with bacteriolytic activity against gram-negative bacterial cell walls (if the target bacteria is gram-positive only).
5. Proteinase K (20 mg/mL): Digestion of proteins.
6. Phenol-chloroform-isoamyl alcohol (PCI) solution (25:24:1): Separation of DNA from other cellular components.
7. Ethanol (100%): Precipitates DNA from solution.
8. Tris-EDTA (TE) buffer [10 mM Tris (pH 8.0), 1 mM EDTA]: Used to store purified DNA.

Procedure

1. Take 10 mL of late log phase culture into a falcon tube and centrifuge at 7500 rpm for 10 min. Discard the supernatant and wash the pellet with phosphate-bufferred saline by repeating the same step.

2. Resuspend the pellet with 4 μL RNase A, 8 μL lysozyme, and 5 μL achromopeptidase; mix gently and incubate for 60 min at 37 °C.
3. After incubation, add 30 μL 10% SDS and 3 μL proteinase K, mix properly, and incubate at 50 °C for 60 min.
4. Add 525 μL of PCI (phenol-chloroform-isoamyl) solution and mix properly for 10 min by gentle inversion. After mixing, centrifuge the mixture at 12,000 rpm for 15 min (phenol is corrosive and may cause skin damage. Proper care should be taken while working with phenol).
5. Separate the upper layer into a sterile microcentrifuge without disturbing the layers.
6. Mix thoroughly with an equal volume of chilled ethanol (100%). Now, centrifuge at 12,000 rpm for 20 min.
7. Discard the supernatant and dry the pellet at room temperature or in a dry bath at 50 °C.
8. Resuspend the pellet in 50 μL TE (Tris-EDTA) buffer, allow the pellet to dissolve, and store it in the refrigerator. For long-time storage, keep it in −20 °C refrigerator. Confirm the quality of the bacterial DNA by running 5 μL of product on a 1.5% agarose gel.

Different kit methods are also developed to eliminate the laborious work and time taking protocols. Kit methods are easy to use and can be completed in hours, and good yield quality can be observed in these methods. Kit methods include Qiagen kit, Promega kit, Sigma-Aldrich, Zymo Research, etc. All the necessary reagents are provided along with the kit.

16.7 Amplification of Desired Gene Product Using PCR

Once the DNA quantity is checked, PCR is used to amplify the bacterial DNA. Generally, 16S rRNA gene should be selected to design primers in bacterial identification. The V4 to V9 regions are used most frequently for amplification. Different kit methods are available to amplify the desired gene to create an amplicon library. The PCR product was confirmed by running a sample in an agarose gel and visualized by UV illumination—the purified PCR product sequencing for identifying bacterial species.

16.8 Materials Required

PCR primers (10 μM/rxn): Primers order at the concentration of 50 nmol scale with desalting purification. The stock solution should be diluted to 10 μM for use in PCR reactions.

Taq DNA polymerase (5 units/μL): A thermostable enzyme derived from the *Thermus aquaticus* bacteria that amplify DNA fragments in the PCR reaction and can withstand up to 95 °C without losing activity. PCR reaction will be carried out using kit methods. Here, we are introduced to Sigma. This protocol was adapted from the Sigma protocols.

16.9 How to Perform PCR?

PCR setup consists of four steps: (1) Add all the reagents or ready-to-use mixture along with the template. (2) Mix and centrifuge thoroughly. (3) Amplify per thermo cycler. (4) Evaluate amplified DNA by agarose gel electrophoresis.

16.10 Procedure

Add the reagents to an appropriately sized tube. If different reactions are needed, a mix without the template should be prepared and aliquoted into reaction tubes. In the end, the template should be added to the appropriate tubes (Table 16.1).

After preparing the reaction mixture, mix gently by vortex and briefly centrifuge to collect all components to the bottom of the tube, and then amplify. The amplification parameters will vary depending on the amplification primers and the thermal cycle used. It is important to optimize the amplification conditions. The PCR conditions for the 16S rRNA amplification conditions are as follows; [Initial denaturation at 94 °C/1 min (denaturation at 94 °C/30s, annealing at 50 °C/1 min, extension at 65 °C/1 min), 65 °C/3 min hold at 4 °C.

Load the PCR reaction on a 1.0% agarose/Tris-borate-EDTA gel and electrophorese. The PCR product is approximately 380 base pair predicted product size. Visualize by UV illumination and photograph. The targeted DNA band will be excised and purified using commercially available kits.

16.11 Sequencing of the Amplified Product

The order of polynucleotide chains has the information for the hereditary and life. Hence, the ability to infer such sequence is dominant and essential in biological research.

Table 16.1 PCR reaction mixture components and final concentration

Component	Final concentration
Water	X volume
10x PCR buffer (P2192 or P2317)	1x
Deoxynucleotide mix	200 μM (1 μL)
Forward primer (typically 15–30 bases in length)	0.1–0.5 μM
Reverse primer (typically 15–30 bases in length)	0.1–0.5 μM
Taq DNA polymerase	0.05 units/μL
Template DNA (typically 10 ng)	200 pg/μL
25 mM MgC12 (use only with buffer P2317)	0.1–0.5 mM

[A] knowledge of sequences could contribute much to our understanding of living matter. Frederick Sanger (Frederick 1980).

Watson and Crick discovered the 3D structure of DNA in 1953 by working from Rosalind Franklin and Maurice Wilkins' crystallographic data (Watson and Crick 1953; Zallen 2003). Later, the strategies developed to analyze the sequence of protein chains did not work out with the nucleic acid chain due to longer chain than protein. In Maxam and Gilbert's technique, instead of depending on DNA polymerase to generate fragments, radiolabeled DNA is treated with chemicals that break the chain at specific bases. This was the *first-generation DNA sequencing* method (Maxam and Gilbert 1977). In 1977, a major invention that forever altered DNA sequencing technology's progress was developed by Sanger's "chain termination" or dideoxy technique (Sanger et al. 1977). The next developed sequencing method relies on the luminescent method, which has two enzymes. ATP sulfurylase is used to convert pyrophosphate into ATP. This product is used as a substrate for luciferase; hence, the amount of pyrophosphate produced is proportional to the light emitted (Nyrén and Lundin 1985). The Sanger's and pyrosequencing methods require DNA polymerase to produce the visible output. These two methods are sequence-by-synthesis (SBS) techniques and observed in real time; it is established by PålNyrén and colleagues (Nyrén 1987; Ronaghi et al. 1996; Ronaghi et al. 1998).

Later, pyrosequencing was licensed to 454 Life Sciences, which evolved as a next-generation sequencing (NGS), also called the second-generation sequencing method. The recently developed advanced sequencing technologies of single molecule sequencing in real time, are undeniably different from previous sequencing technology platforms and have been defined as third-generation sequencing methods. This technology was first developed in the lab of Stephen Quake (Braslavsky et al. 2003). The Roche/454 FLX, the Illumina/Solexa Genome Analyzer, and the Applied Biosystems SOLiD™ Systems are the three commonly used platforms in next-generation sequencing. Recently, the HelicosHeliscopeTM and Pacific Biosciences SMRT instruments have been commercialized for sequencing. Though different NGS platforms have been established, the working protocol follows three steps: sample preparation, (2) sequencing, and (3) data analysis. The platform and its specifications and sequencing methods were reviewed by Goodwin et al. (2016). The comparison of technologies and workflow application of technologies is given in Table 16.2 (Kumar et al. 2019) (Fig 16.1).

Roche/454 FLX Workflow Adopted from https://allseq.com/

1. The first step of the DNA is breakup into more fragments of approximately 200–600 base pairs.
2. A short stretch of DNA called an adaptor attaches to the DNA fragments. Incubate the fragments with sodium hydroxide to make a single strand of the adaptor and DNA fragment.
3. The DNA fragments are washed across the flow cell, the complementary binds to primer stay on the surface of the flow cell, and the unbonded will be washed out.

Table 16.2 The comparison of technologies and workflow application of technologies

NGS technology	Instrumentation	Maximum reads per run	Maximum output range (gigabases)	Sequencing run time (hours)	Major applications
Illumina	Iseq 100	4–8 M	0.14–1.2	9–17.5	Small genome sequencing, targeted gene sequencing, long-range PCR, amplicon sequencing
	MiniSeq	25–50 B	1.65–7.5	7–24	Small genome sequencing, targeted gene sequencing, targeted gene expression profiling, 16S metagenomic sequencing
	NextSeq 550	0.4–0.8 B	25–120	11–29	Small genome sequencing, targeted gene sequencing, mRNA-Seq gene expression profiling, miRNA and small RNA analysis
	HiSeq 3000	2.5–5 B	105–750	24–84	Exome sequencing, whole transcriptome sequencing
	HiSeq 4000	5–10 B	210–1500	24–84	Exome sequencing, whole transcriptome sequencing
	HiSeqX	6 B	1600–1800	72	Whole-genome sequencing
	NovaSeq 6000	2.6–8.2 B	134–1250	13–36	Whole-genome sequencing, exome sequencing, whole transcriptome sequencing, methylation sequencing
		16–20 B	1600–3000	25–44	
Ion torrent	Ion 510 Chip	2–3 M	0.3–1	2.5–4	Targeted, exome, transcriptome, small genome
	Ion 520 Chip	3–6 M	0.5–2	2.5–4	
	Ion 530 Chip	9 20 M	3–8	2.5–4	
	Ion 540 Chip	60–80 M	10–15	2.5–4	
	Ion 550 Chip	100–130 M	20–25	2.5–4	
Nanopore	MinION Mk 1B	0.5 M	50	Up to 72	Whole-genome sequencing, targeted sequencing, gene expression, and RNA sequencing, epigenetic interrogation
	GridION X5	2.5 M	250	Up to 72	
	PromethION (Beta)	375 M	15,000	Up to 64	

NGS technology	Instrumentation	Maximum reads per run	Maximum output range (gigabases)	Sequencing run time (hours)	Major applications
PacBio	Sequel	0.5 M	20	10–20	Whole-genome sequencing, RNA sequencing, targeted sequencing, complex population sequencing, epigenetics
BGI	MGISEQ-200	300 M	60	48	Small genome sequencing, targeted sequencing
	BGISEQ-500	1300 M	520	45–213	Whole-genome sequencing, whole-exome sequencing, targeted sequencing, RNA sequencing
	MGISEQ-2000	1800 M	1080	48	Whole-genome sequencing, whole-exome sequencing, transcriptome sequencing
	MGISEQ-T7	–	6000	24	Whole-genome sequencing, deep exome sequencing, transcriptome sequencing, targeted panel sequencing

B billion, *M* million

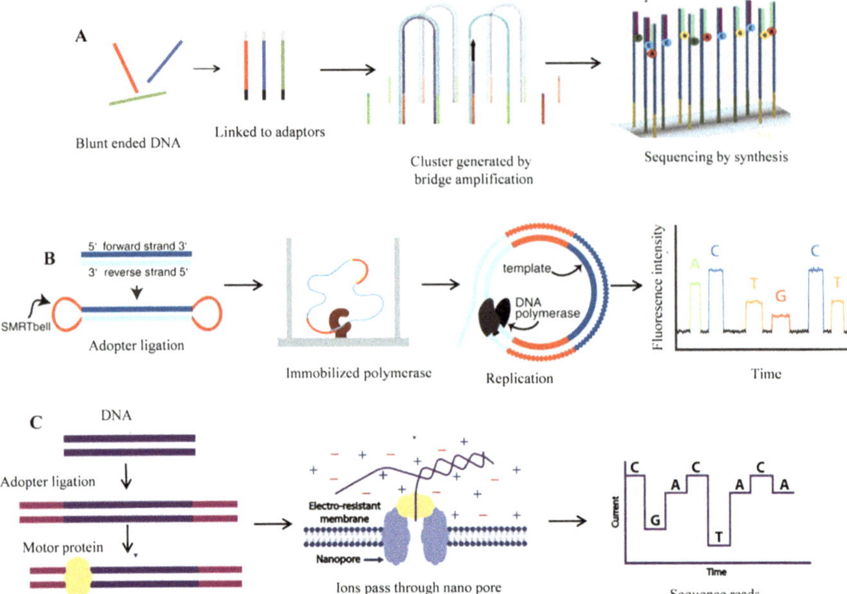

Fig. 16.1 Short-read and long-read sequencing technologies. (**a**) Illumina sequencing; (**b**) PacBio sequencing involves a circular consensus sequencing (CCS) SMRTbell technique. (**c**) Nanopore sequencing

4. DNA replicates to form a small cluster with identical sequences. These clusters emit a signal that a camera can detect.
5. Unlabeled nucleotides and DNA polymerase are added to lengthen and join the strands of DNA attached to the flow cell. This establishes bridges of dsDNA between primers.
6. The double-stranded DNA is then broken down into single-stranded DNA using heat, leaving several million dense clusters of identical DNA sequences.
7. Labeled terminators (terminators are a version of nucleotide base—A, C, G, or T—that stop DNA synthesis) are added to the flow cell.
8. The DNA polymerase binds to the primer and adds the first fluorescently labeled terminator to the new DNA strand. Once a base has been added, no more bases can be added to the DNA strand until the terminator base is cut.
9. Lasers are passed over the flow cell to activate the fluorescent label on the nucleotide base. This fluorescence is detected by a camera and recorded on a computer. Each terminator base (A, C, G, and T) gives off a different color.
10. The fluorescently labeled terminator group is then removed from the first base, and the next fluorescently labeled terminator base can be added. And so, the process continues until millions of clusters have been sequenced.
11. The DNA sequence is analyzed base-by-base during Illumina sequencing, making it a highly accurate method. The sequence generated can then be aligned to a reference sequence; this looks for matches or changes in the sequenced DNA (Fig. 16.2).

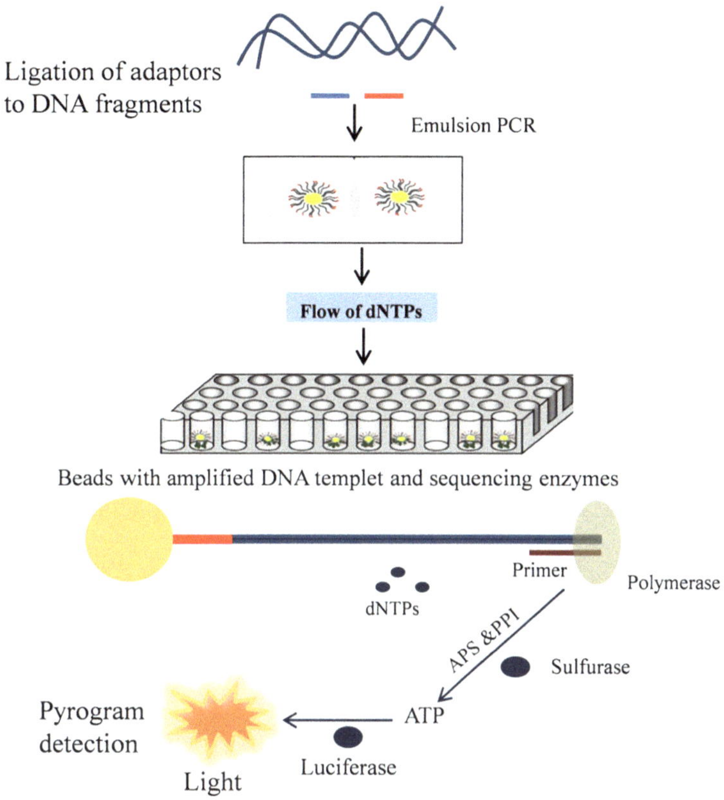

Fig. 16.2 Overview of Roche/454 FLX pyrosequencing platform workflow

16.12 Microbiome Data Analysis

The data generated analysis by QIIME (Quantitative Insights Into Microbial Ecology) has many useful and important bioinformatic tools needed for microbiome analysis. The data generated is also compatible with other program tools (Navas-Molina et al. 2013). The data analysis should be supported by different tools listed here. QIIME wrapper QWRAP guides users to use publicly available tools like FASTQC for quality checking and FASTX quality filtering, followed by QIIME tools to perform microbiome analysis. QWRAP used the bash, Perl, python, and R languages to work on Illumina fastq files (GitHub at https://github.com/QWRAP/QWRAP). QWRAP is designed for de-multiplexed datasets.

The user should have ideas on software on a Linux operating system. These can be installed in Linux or QIIME virtual box (see http://qiime.org/install/virtual_box.html):

(a) QIIME: There are several ways to install QIIME (http://qiime.org/install/index.html).

(b) USEARCH (http://www.drive5.com/usearch/). After downloading the 32bit Linux binary, rename the binary file to usearch61.
(c) FASTQC (http://www.bioinformatics.babraham.ac.uk/projects/fastqc/).
(d) FASTX (http://hannonlab.cshl.edu/fastx_toolkit/).
(e) R (http://www.r-project.org/).

Protocol

1. First, always create a new directory for storing files created during the analysis.
2. Transfer all the raw data files to a fastq format folder. One fastq file indicates one sample here. The QWRAP program only works on the files in a given folder with fastq/fastq.gz format.
3. We run the quality check on the data in the files; it is performed in two steps.
4. First, all the sequences are trimmed to a defined sequence length as programmed by the user.
5. The final check generates the quality report for the filtered data reads, followed by further checks with a QScore>20 over at least 80% of the bases retained. The generated input files compatible with QIIME files present in the dictionary will be included in the analysis.
6. The program runs to analyze the data and generate several files. It generates seqs.fna;mapping.txt; sample_order.txt; script.sh.
7. The generated results can be summarized into an HTML report. This format is stable and easy to transfer to another user. The HTML report is generated by running the program (`report_microbiome.sh.`)
8. Then, it creates a html file named " microbiome_report.html" in the Analysis directory (Fig 16.3).

16.13 Standards Followed in the Microbiome Research

Microbial research has tremendously increased after next-generation sequencing technology development. DNA amplicon sequencing is a key stem for microbial characterization and has complications too. It is clear and well said that DNA extraction results from microbiome analysis (Sinha et al. 2017). Due to the different methods used to process the samples, the same and reliable results do not overlap for the same samples (Angelakis et al. 2016), and interpretation will be difficult. A major problem faced is the lack of control in the sample processing. It is a very important and good practice to perform control in all the steps of experimental methods. The significance of microbial studies identified might have been due to a lack of knowledge and inaccessibility of positive controls (Hornung et al. 2019). There are several factors that must be considered in microbial research.

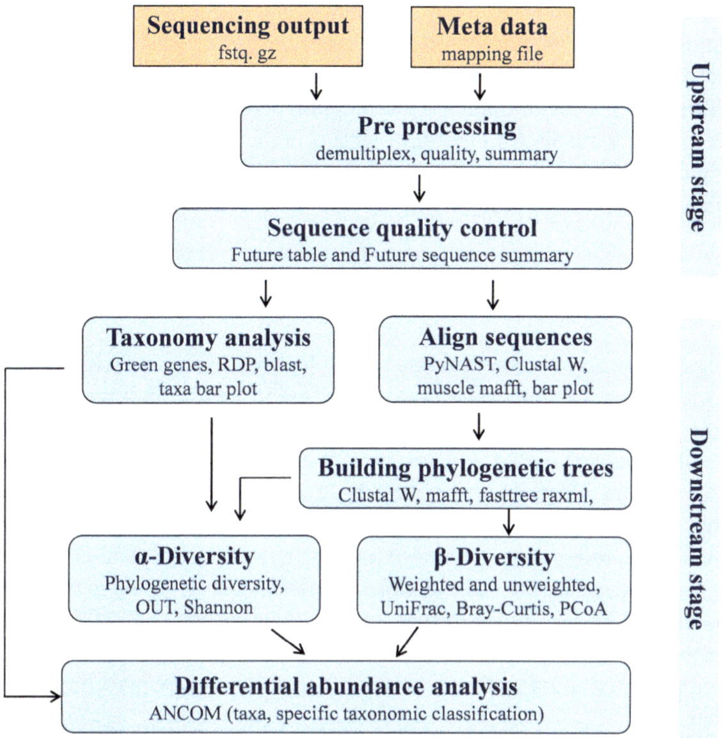

Sequencing output
fstq. gz
Meta data
mapping file
Pre processing
demultiplex, quality, summary
Sequence quality control
Future table and Future sequence summary
Taxonomy analysis
Green genes, RDP, blast, taxa bar plot
Align sequences
PyNAST, Clustal W, muscle mafft, bar plot
Building phylogenetic trees
Clustal W, mafft, fasttree raxml,
α-Diversity
Phylogenetic diversity, OUT, Shannon
β-Diversity
Weighted and unweighted, UniFrac, Bray-Curtis, PCoA
Differential abundance analysis
ANCOM (taxa, specific taxonomic classification)
Upstream stage
Downstream stage

Fig. 16.3 QIIME (Quantitative Insights Into Microbial Ecology) workflow overview. PCoA, principal coordinates analysis; OTU, operational taxonomic unit; ANCOM, analysis of composition of microbiomes

16.14 Positive Controls Should Be Considered in Microbial Research

The selection of positive organisms is important in an experimental pool. We should know what microbes are present in the sample before using a positive control. Most of the microorganisms are commercially available to select as a positive control. If commercially not available, custom-designed positive controls could be needed. Interdependency between the DNA extraction kit and positive control is important (Hornung et al. 2019). The kit methods will be standardized using the positive control developed by the manufacturer. They can extract the DNA from the positive control manufacturer's use, but the sample results can't be guaranteed while using it for the other microbial communities. Many factors influencing the DNA extraction metabolites produced by microbial communities may interact with the extraction process (Angelakis et al. 2016). Amplification of the microbiome with bias and errors will be observed if an amplification protocol is adjusted to an average GC content. Microbes containing high and low GC content may not be amplified. It results if amplification biases exist (Benjamini and Speed 2012). Positive controls can identify these problems, which helps distinguish amplification bias and errors from DNA isolation protocols (Costea et al. 2017). Computational analysis challenges influence the communities' classification and taxonomic assignment. Using multiple microorganisms with the same strains in the sample of different richness in a positive control can help to get accurate binning results during the parameter like GC content as a specific parameter in the binning process (Sangwan et al. 2016).

16.15 Negative Controls Considered in Microbiome Analysis

Also, negative controls in microbiome research play a key role. It should be considered at which step it is important to consider and in what way it has to be used—a few of the points discussed here. Sampling is an important step; a negative sample should always be considered. For instance, if a cohort sampling is obtained from a patent in a given location, there may be a chance of contamination from the researcher who is collecting a sample and the equipment used for sampling. Additionally, the researchers must also handle negative swabs, and DNA extraction must be carried out with the same samples. If no DNA is observed in the control, the experiment shall be carried for further analysis (Vandeputte et al. 2017). Index hopping is another problem in data analysis; if samples are multiplexed during the same run, the non-ligated adopters from one sample will vary and be inaccurately assigned to another.

In some cases, the negative controls might have the data as of the sample profile. In these cases, it is difficult to decide the true contamination and index hopping, and carrying a control sample from the first step of the experiment is useless (Edmonds and Williams 2017). The figure below indicates the possible chances of contamination and bias resulting in stages while working in microbiome research and analysis (Fig. 16.4).

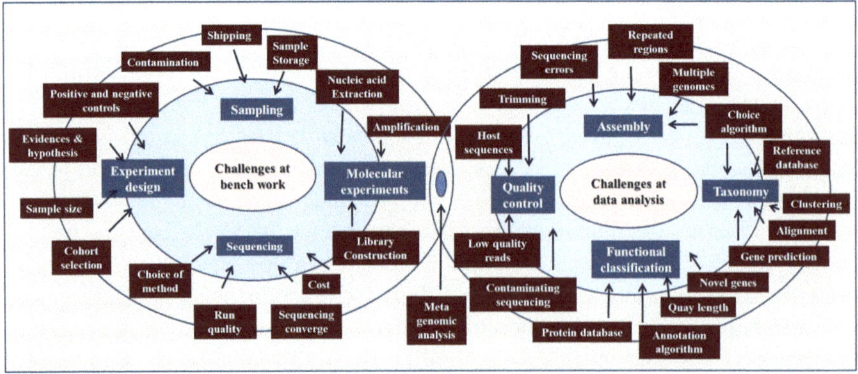

Fig. 16.4 Schematic overview of experimental and data analysis challenges associated with microbiome research

Solving all the associated problems is difficult, but negative control is always important (Zhong et al. 2018). New methods developed to reduce contamination are also tried, but the standardization of new method in laboratories is not easy (Minich et al. 2018). A negative control is crucial for data analysis, and number reads are also considered. While extracting DNA from a sample, we should ensure that the desired microbes are abundant enough while using positive control. In sequencing, the positive control should not introduce any sequences or errors. Good scientific practice should be applied throughout the experiment. Trials with newly developed methods in research and publishing papers are important; it will help to interpret other data. Microbiome research has become an emerging and important area of research in human health and benefit; developed and modified methods without a proper control sample and standards generated biased results.

References

Adelskov J, Patel BKC (2016) A molecular phylogenetic framework for bacillus subtilis using genome sequences and its application to bacillus subtilis subspecies stecoris strain D7XPN1, an isolate from a commercial food-waste degrading bioreactor. 3Biotech 6(1):96–96

Ahmadmehrabi S, Tang WHW (2017) Gut microbiome and its role in cardiovascular diseases. Curr Opin Cardiol 32(6):761–766

Amaro A, Duarte E, Amado A, Ferronha H, Botelho A (2008) Comparison of three DNA extraction methods for mycobacterium bovis, mycobacterium tuberculosis and mycobacterium avium subsp. avium. Lett Appl Microbiol 47(1):8–11

Angelakis E, Bachar D, Henrissat B, Armougom F, Audoly G, Lagier J-C, Robert C, Raoult D (2016) Glycans affect DNA extraction and induce substantial differences in gut metagenomic studies. Sci Rep 6:26276–26276

Benjamini Y, Speed TP (2012) Summarizing and correcting the GC content bias in high-throughput sequencing. Nucleic Acids Res 40(10):e72–e72

Braslavsky I, Hebert B, Kartalov E, Quake SR (2003) Sequence information can be obtained from single DNA molecules. Proc Natl Acad Sci U S A 100(7):3960–3964

Costea PI, Zeller G, Sunagawa S, Pelletier E, Alberti A, Levenez F, Tramontano M, Driessen M, Hercog R, Jung F-E, Kultima JR, Hayward MR, Coelho LP, Allen-Vercoe E, Bertrand L, Blaut M, Brown JRM, Carton T, Cools-Portier S, Daigneault M, Derrien M, Druesne A, de Vos WM, Finlay BB, Flint HJ, Guarner F, Hattori M, Heilig H, Luna RA, van Hylckama Vlieg J, Junick J, Klymiuk I, Langella P, Le Chatelier E, Mai V, Manichanh C, Martin JC, Mery C, Morita H, O'Toole PW, Orvain C, Patil KR, Penders J, Persson S, Pons N, Popova M, Salonen A, Saulnier D, Scott KP, Singh B, Slezak K, Veiga P, Versalovic J, Zhao L, Zoetendal EG, Ehrlich SD, Dore J, Bork P (2017) Towards standards for human fecal sample processing in metagenomic studies. Nat Biotechnol 35(11):1069–1076

Drancourt M, Bollet C, Carlioz A, Martelin R, Gayral J-P, Raoult D (2000) 16S ribosomal DNA sequence analysis of a large collection of environmental and clinical unidentifiable bacterial isolates. J Clin Microbiol 38(10):3623–3630

Edmonds KP, Williams LL (2017) The role of the negative control in microbiome analyses. FASEB J 31:940.3

Frederick, S. (1980). "Frederick Sanger—biographical." http://www.nobelprize.org/nobel_prizes/chemistry/laureates/1980/sanger-bio.html

Goodwin S, McPherson JD, McCombie WR (2016) Coming of age: ten years of next-generation sequencing technologies. Nat Rev Genet 17(6):333–351

Hornung BVH, Zwittink RD, Kuijper EJ (2019) Issues and current standards of controls in microbiome research. FEMS Microbiol Ecol 95(5):fiz045

Jamshidi P, Hasanzadeh S, Tahvildari A, Farsi Y, Arbabi M, Mota JF, Sechi LA, Nasiri MJ (2019) Is there any association between gut microbiota and type 1 diabetes? A systematic review. Gut Pathogens 11:49–49

Krumnow AA, Sorokulova IB, Olsen E, Globa L, Barbaree JM, Vodyanoy VJ (2009) Preservation of bacteria in natural polymers. J Microbiol Methods 78(2):189–194

Kumar KR, Cowley MJ, Davis RL (2019) Next-generation sequencing and emerging technologies. Semin Thromb Hemost 45(07):661–673

Marmur J (1961) A procedure for the isolation of deoxyribonucleic acid from micro-organisms. J Mol Biol 3(2):208-IN201

Maruvada P, Leone V, Kaplan LM, Chang EB (2017) The human microbiome and obesity: moving beyond associations. Cell Host Microbe 22(5):589–599

Maxam AM, Gilbert W (1977) A new method for sequencing DNA. Proc Natl Acad Sci U S A 74(2):560–564

McDonald D, Hyde E, Debelius JW, Morton JT, Gonzalez A, Ackermann G, Aksenov AA, Behsaz B, Brennan C, Chen Y, DeRight Goldasich L, Dorrestein PC, Dunn RR, Fahimipour AK, Gaffney J, Gilbert JA, Gogul G, Green JL, Hugenholtz P, Humphrey G, Huttenhower C, Jackson MA, Janssen S, Jeste DV, Jiang L, Kelley ST, Knights D, Kosciolek T, Ladau J, Leach J, Marotz C, Meleshko D, Melnik AV, Metcalf JL, Mohimani H, Montassier E, Navas-Molina J, Nguyen TT, Peddada S, Pevzner P, Pollard KS, Rahnavard G, Robbins-Pianka A, Sangwan N, Shorenstein J, Smarr L, Song SJ, Spector T, Swafford AD, Thackray VG, Thompson LR, Tripathi A, Vázquez-Baeza Y, Vrbanac A, Wischmeyer P, Wolfe E, Zhu Q, American Gut C, Knight R (2018) American gut: an open platform for citizen science microbiome research. mSystems 3(3):e00031–e00018

McIntyre ABR, Ounit R, Afshinnekoo E, Prill RJ, Hénaff E, Alexander N, Minot SS, Danko D, Foox J, Ahsanuddin S, Tighe S, Hasan NA, Subramanian P, Moffat K, Levy S, Lonardi S, Greenfield N, Colwell RR, Rosen GL, Mason CE (2017) Comprehensive benchmarking and ensemble approaches for metagenomic classifiers. Genome Biol 18(1):182–182

Minich JJ, Zhu Q, Janssen S, Hendrickson R, Amir A, Vetter R, Hyde J, Doty MM, Stillwell K, Benardini J, Kim JH, Allen EE, Venkateswaran K, Knight R (2018) KatharoSeq enables high-throughput microbiome analysis from low-biomass samples. mSystems 3(3):e00218–e00217

Navas-Molina JA, Peralta-Sánchez JM, González A, McMurdie PJ, Vázquez-Baeza Y, Xu Z, Ursell LK, Lauber C, Zhou H, Song SJ, Huntley J, Ackermann GL, Berg-Lyons D, Holmes S, Caporaso JG, Knight R (2013) Advancing our understanding of the human microbiome using QIIME. Methods Enzymol 531:371–444

Nyrén P (1987) Enzymatic method for continuous monitoring of DNA polymerase activity. Anal Biochem 167(2):235–238

Nyrén P, Lundin A (1985) Enzymatic method for continuous monitoring of inorganic pyrophosphate synthesis. Anal Biochem 151(2):504–509

Ogg CD, Patel BK (2009) Caloramator australicus sp. nov., a thermophilic, anaerobic bacterium from the great Artesian Basin of Australia. Int J Syst Evol Microbiol 59(Pt 1):95–101

Riesenfeld CS, Schloss PD, Handelsman J (2004) Metagenomics: genomic analysis of microbial communities. Annu Rev Genet 38:525–552

Ronaghi M, Karamohamed S, Pettersson B, Uhlén M, Nyrén P (1996) Real-time DNA sequencing using detection of pyrophosphate release. Anal Biochem 242(1):84–89

Ronaghi M, Uhlén M, Nyrén P (1998) A sequencing method based on real-time pyrophosphate. Science 281(5375):363, 365

Sanger F, Nicklen S, Coulson AR (1977) DNA sequencing with chain-terminating inhibitors. Proc Natl Acad Sci U S A 74(12):5463–5467

Sangwan N, Xia F, Gilbert JA (2016) Recovering complete and draft population genomes from metagenome datasets. Microbiome 4:8–8

Scott AJ, Alexander JL, Merrifield CA, Cunningham D, Jobin C, Brown R, Alverdy J, O'Keefe SJ, Gaskins HR, Teare J, Yu J, Hughes DJ, Verstraelen H, Burton J, O'Toole PW, Rosenberg DW,

Marchesi JR, Kinross JM (2019) International cancer microbiome consortium consensus statement on the role of the human microbiome in carcinogenesis. Gut 68(9):1624–1632
Shen L, Ji HF (2019) Associations between gut microbiota and Alzheimer's disease: current evidences and future therapeutic and diagnostic perspectives. J Alzheimers Dis 68(1):25–31
Sinha R, Abu-Ali G, Vogtmann E, Fodor AA, Ren B, Amir A, Schwager E, Crabtree J, Ma S, Microbiome Quality Control Project, Abnet CC, Knight R, White O, Huttenhower C (2017) Assessment of variation in microbial community amplicon sequencing by the microbiome quality control (MBQC) project consortium. Nat Biotechnol 35(11):1077–1086
Sorokulova I, Watt J, Olsen E, Globa L, Moore T, Barbaree J, Vodyanoy V (2012) Natural biopolymer for preservation of microorganisms during sampling and storage. J Microbiol Methods 88(1):140–146
Sorokulova I, Olsen E, Vodyanoy V (2015) Biopolymers for sample collection, protection, and preservation. Appl Microbiol Biotechnol 99(13):5397–5406
Tringe SG, Rubin EM (2005) Metagenomics: DNA sequencing of environmental samples. Nat Rev Genet 6(11):805–814
Vandeputte D, Tito RY, Vanleeuwen R, Falony G, Raes J (2017) Practical considerations for large-scale gut microbiome studies. FEMS Microbiol Rev 41(Supp_1):S154–S167
Watson JD, Crick FH (1953) Molecular structure of nucleic acids; a structure for deoxyribose nucleic acid. Nature 171(4356):737–738
Woo PCY, Ng KHL, Lau SKP, Yip K-t, Fung AMY, Leung K-w, Tam DMW, Que T-l, Yuen K-y (2003) Usefulness of the microseq 500 16S ribosomal DNA-based bacterial identification system for identification of clinically significant bacterial isolates with ambiguous biochemical profiles. J Clin Microbiol 41(5):1996–2001
Woo PCY, Lau SKP, Teng JLL, Tse H, Yuen KY (2008) Then and now: use of 16S rDNA gene sequencing for bacterial identification and discovery of novel bacteria in clinical microbiology laboratories. Clin Microbiol Infect 14(10):908–934
Wright MH, Adelskov J, Greene AC (2017) Bacterial DNA extraction using individual enzymes and phenol/chloroform separation. J Microbiol Biol Educ 18(2):18.12.48
Zallen DT (2003) Despite Franklin's work, Wilkins earned his Nobel. Nature 425(6953):15
Zhong Z-P, Solonenko NE, Gazitúa MC, Kenny DV, Mosley-Thompson E, Rich VI, Van Etten JL, Thompson LG, Sullivan MB (2018) Clean low-biomass procedures and their application to ancient ice core microorganisms. Front Microbiol 9:1094–1094